Modern Sensor Systems: Design and Applications

Modern Sensor Systems: Design and Applications

Edited by Bob Tucker

CLANRYE INTERNATIONAL
www.clanryeinternational.com

Clanrye International,
750 Third Avenue, 9th Floor,
New York, NY 10017, USA

ISBN: 978-1-63240-656-9

Cataloging-in-Publication Data

Modern sensor systems : design and applications / edited by Bob Tucker.
p. cm.
Includes bibliographical references and index.
ISBN 978-1-63240-656-9
1. Detectors. 2. Engineering instruments. I. Tucker, Bob.
TK7871.674 .M63 2018
681.2--dc23

For information on all Clanrye International publications visit our website at www.clanryeinternational.com

CLANRYE INTERNATIONAL

Printed in China.

Contents

Preface

This book has been an outcome of determined endeavour from a group of educationists in the field. The primary objective was to involve a broad spectrum of professionals from diverse cultural background involved in the field for developing new researches. The book not only targets students but also scholars pursuing higher research for further enhancement of the theoretical and practical applications of the subject.

The objects used to detect changes in the environment are known as sensors. They transfer the observed data to an electronic device. In modern times, the use of sensors has increased and it is used in objects such as airplanes, cars, robots, etc. This book contains some path-breaking studies in the field of modern sensor systems. The various studies that are constantly contributing towards advancing technologies and evolution of this field are examined in detail. This book, with its detailed analyses and data, will prove immensely beneficial to professionals and students involved in this area at various levels.

It was an honour to edit such a profound book and also a challenging task to compile and examine all the relevant data for accuracy and originality. I wish to acknowledge the efforts of the contributors for submitting such brilliant and diverse chapters in the field and for endlessly working for the completion of the book. Last, but not the least; I thank my family for being a constant source of support in all my research endeavours.

Editor

1

Instantaneous Real-Time Kinematic Decimeter-Level Positioning with BeiDou Triple-Frequency Signals over Medium Baselines

Xiyang He [1,2], Xiaohong Zhang [1,2,*], Long Tang [1,2,*] and Wanke Liu [1,2]

Academic Editor: Anton de Ruiter

1 School of Geodesy and Geomatics, Wuhan University, 129 Luoyu Road, Wuhan 430079, China; hexiyang704814@126.com (X.H.); wkliu@sgg.whu.edu.cn (W.L.)
2 Collaborative Innovation Center for Geospatial Technology, 129 Luoyu Road, Wuhan 430079, China
* Correspondences: xhzhang@sgg.whu.edu.cn (X.Z.); ltang@whu.edu.cn (L.T.);

Abstract: Many applications, such as marine navigation, land vehicles location, *etc.*, require real time precise positioning under medium or long baseline conditions. In this contribution, we develop a model of real-time kinematic decimeter-level positioning with BeiDou Navigation Satellite System (BDS) triple-frequency signals over medium distances. The ambiguities of two extra-wide-lane (EWL) combinations are fixed first, and then a wide lane (WL) combination is reformed based on the two EWL combinations for positioning. Theoretical analysis and empirical analysis is given of the ambiguity fixing rate and the positioning accuracy of the presented method. The results indicate that the ambiguity fixing rate can be up to more than 98% when using BDS medium baseline observations, which is much higher than that of dual-frequency Hatch-Melbourne-Wübbena (HMW) method. As for positioning accuracy, decimeter level accuracy can be achieved with this method, which is comparable to that of carrier-smoothed code differential positioning method. Signal interruption simulation experiment indicates that the proposed method can realize fast high-precision positioning whereas the carrier-smoothed code differential positioning method needs several hundreds of seconds for obtaining high precision results. We can conclude that a relatively high accuracy and high fixing rate can be achieved for triple-frequency WL method with single-epoch observations, displaying significant advantage comparing to traditional carrier-smoothed code differential positioning method.

Keywords: BeiDou; triple frequency; carrier-smoothed code (CSC); extra-wide-lane; wide-lane; ambiguity resolution; differential positioning

---

1. Introduction

Over the past decades, different GNSS techniques have been developed by researchers, which have made remarkable contributions to scientific applications (e.g., geodesy, remote sensing, space and fundamental physics) [1,2] and engineering services (e.g., surveying, navigation, and timing) [3,4]. With regard to the navigation and positioning techniques of GNSS, they can be divided into two main categories: absolute positioning and relative positioning. Absolute positioning mainly includes standard point positioning (SPP) and precise point positioning (PPP). Relative positioning mainly includes differential DGNSS (DGNSS) and real-time kinematic (RTK) positioning techniques. Different advantages as well as disadvantages are shown for these techniques.

SPP, as the simplest positioning technique, can realize real-time positioning using a single receiver with broadcast ephemeris. Its applications are mainly in navigation. The main disadvantage of SPP is that the accuracy is limited to several to tens meters [5]. Using the PPP technique, positioning with a single receiver can also be implemented, but providing higher centimeter level accuracy [6,7]. However, precise ephemeris and clock products must be provided for PPP. In addition, a convergence time of about thirty minutes is required to reach high precision results [8].

The real-time kinematic (RTK) positioning technique, which was first investigated by Kleusberg [9], Cannon [10] and Schwarz *et al.* [11], implements positioning using two or more receivers. The rover position is determined by forming double-differenced observations with the reference station. Using the RTK technique, one can realize fast precise centimeter level positioning [12]. However, the applications of RTK are limited to short distances (e.g., <10 km). For medium and long distances (e.g., >10 km), a certain time duration (usually 10–40 min for GPS) is needed first for fixing the ambiguities due to the existence of residual atmospheric errors [13].

Using multiple reference stations with inter-receiver distances of several tens of kilometers, the network-RTK [14] and PPP-RTK [15] techniques can be implemented, which are extensions of RTK and PPP, respectively. In this way, the atmospheric errors in a rover station can be modeled or eliminated rapidly, thus improving the positioning performance. For network-RTK, the distance can be extended to several tens of kilometers [14]. For PPP-RTK, high-precision results can be obtained rapidly, e.g., within several seconds [15,16]. The main disadvantage of these two techniques is that they are unavailable in some areas, like marine areas, due to the difficulty of establishing reference stations. PPP-RTK using global-scale networks were also proposed by some researchers [17]. However, at least five to ten minutes are needed for ambiguity fixing for this technique.

Differential GNSS (DGNSS) positioning is a positioning technique where the user's position is determined more accurately by applying differential corrections, derived from a fixed ground-based reference station [18]. DGNSS was invented in the mid-1980s, and was called DGPS at that time. When implementing DGNSS positioning, both original code and carrier-smoothed code (CSC) measurements can be used. The positioning accuracy of original code-based DGNSS is usually at meter level [18]. If one requires decimeter level positioning, CSC measurements usually are needed. The reason is that the impact of noise and multipath code errors can be reduced by applying carrier-smoothing techniques [19,20]. Apparently but unfortunately, a certain time duration (usually several minutes) is also needed for carrier smoothing the code measurements to achieve a relatively high accuracy.

As can be deduced, all these GNSS-based positioning techniques operate under a set of constraints. These constraints include baseline length, attainable accuracy, time-to-solution, instrumentation, operational modes, and so on [21]. Some applications, such as marine navigation, aircraft landing guidance and land vehicles location, *etc.*, require obtaining the user positions in a rather short time with certain positioning accuracy (decimeter-level) and certain coverage (e.g., <100 km). In some areas like marine areas, it is also hard to establish multiple-reference stations in the surrounding areas. For such applications, CSC-based DGNSS is currently the most suitable method, since it can obtain the user positions in a relatively short time with decimeter level accuracy using just a single or a few reference stations. However, this technique still suffers some time cost for code carrier smoothing, which has been described above.

Currently, with the two new and emerging BeiDou and Galileo constellations as well as the ongoing modernization of GPS and GLONASS, more multi-frequency signals will become available for most users [22]. As to the current BDS constellation, all BDS satellites can transmit triple-frequency signals to the users. With triple-frequency signals, more useful combinations can be formed, which can improve the performance of ambiguity resolution (AR). Significant research efforts have been made in the past decade to make use of the multi-frequency signals for improving AR (ambiguity resolution). Harris [23] and Forssell *et al.* [24] proposed the Three Carrier Ambiguity Resolution (TCAR) method. Jung [25] proposed the Cascade Integer Resolution (CIR) method. Feng [26] presented the triple-frequency AR method using geometry-based model. Feng and Li [27] studied the positioning accuracy of wide-lane (WL) signals using GPS measurements. In these studies, only theoretical analysis or simulation test performed. With regard to BDS, some studies [28,29] with real observations are implemented. However, these studies concentrated on the positioning for short baselines or fixing the ambiguities of narrow-lane over medium and long distances, which still requires long observation time.

In this contribution, we aim to develop a new real time differential GNSS positioning technique by using WL measurements obtained from BDS triple-frequency signals. This technique can realize instantaneous real-time kinematic positioning over medium distances with decimeter level precision. The fixing rate and positioning accuracy are investigated in detail. The performance is compared with those of dual-frequency WL case and CSC differential positioning case.

2. Wide-Lane Based Differential Positioning

In this section, the models of differential GNSS positioning using WL measurements are introduced. The case with regard to medium distances is focused on here.

2.1. Ambiguity Resolution Models

In medium and long baseline (>10 km) cases, residual atmospheric delays will affect the ambiguity resolution, therefore, the atmospheric errors have to be taken into account. There are two methods to deal with the residual atmospheric errors. The first one is to form combinations that can eliminate atmospheric errors or can reduce the impact of atmospheric errors to an extent that can be ignored, e.g., the Hatch-Melbourne-Wübbena (HMW) combination [19,30,31]. The second one is to estimate the residual atmospheric delays as unknown parameters. Here, we focus on the method of forming combinations.

With triple-frequency or multi-frequency signals, more useful combinations can be formed [26,32,33], such as EWL combinations. General triple-frequency phase combinations can be defined as [26]:

$$\Delta\nabla\phi_{(i,j,k)} = \frac{i \cdot f_1 \cdot \Delta\nabla\phi_1 + j \cdot f_2 \cdot \Delta\nabla\phi_2 + k \cdot f_3 \cdot \Delta\nabla\phi_3}{i \cdot f_1 + j \cdot f_2 + k \cdot f_3} \tag{1}$$

where the combination coefficients i, j, k are integers, and f_1, f_2, f_3, are three signal frequencies corresponding to B1, B3 and B2 of BDS or L1, L2 and L5 of GPS, respectively. The DD (double-differenced) phase measurements of three original DD observations in meters are $\Delta\phi_1$, $\Delta\phi_2$, $\Delta\phi_3$, corresponding to, f_1, f_2 and f_3 According to [26], the phase linear combinations can be classified as four categories: extra-wide-lane (EWL, $\lambda \geqslant 2.93$ m), wide-lane (WL, $0.75 \leqslant \lambda < 2.93$ m), medium-lane (ML, 0.19 m $\leqslant \lambda < 0.75$ m) and narrow-lane (NL, $\lambda < 0.19$ m) combinations.

Taking atmospheric errors into account, double-differenced phase and code combinations can be expressed as:

$$\begin{aligned} \Delta\nabla\phi_{(i,j,k)} &= \Delta\nabla\rho - \beta_{(i,j,k)}\Delta\nabla I + \Delta\nabla T + \lambda_{(i,j,k)}N_{(i,j,k)} + \varepsilon_{\Delta\nabla\phi} \\ \Delta\nabla P_{(l,m,n)} &= \Delta\nabla\rho + \beta_{(l,m,n)}\Delta\nabla I + \Delta\nabla T + \varepsilon_{\Delta\nabla P} \end{aligned} \tag{2}$$

where β is the ionospheric scale factor, the combination coefficients l, m, n are integers, ρ is the geometrical distance, I is the ionospheric delay on the first frequency, T is the tropospheric delay, N is the ambiguity of phase combination, λ is the wavelength of phase combination and ε is the residual error.

Based on the former studies [26,33], several optimal and useful EWL combinations can be formed, for example, (0, 1, −1) and (1, −5, 4) of BDS. Considering the effects of the phase noise standard deviation and atmospheric errors, total noise level can be defined [26], which is the ratio of the sum of errors and the wavelength. According to [26], the total noise level of these EWL combinations is lower than 0.3 cycles in medium baselines case. Therefore, when implementing AR using these combinations, it is safe to ignore the atmospheric errors.

Theoretically, at least two independent EWL combinations can be formed using triple-frequency signals. The ambiguities of these two independent EWL combinations are determined first. A WL combination then can be derived by combining any of two independent EWL combinations. Therefore, we can fix the ambiguities of two EWL combinations firstly, and then form a WL signals from these two ambiguity-recovered EWL combinations for positioning. Theoretically, implementing wide-lane ambiguity resolution using this method is easier than that in dual-frequency case, because such EWL combinations cannot be found in dual-frequency case.

We will introduce the procedures of this method in detail in the following paragraphs. Two independent EWL combinations (EWL-I and EWL-II) have to be selected first, as listed in Table 1. The characteristics of these selected combinations are also shown. As can be seen, the wavelength of the EWL combinations is up to several meters. Code combinations also have to be formed to constrain the phase combinations. Table 2 shows the combinations we selected.

Table 1. Characteristics of the selected phase combinations.

Constellation	Combinations		Wavelength (m)	Ionospheric Scale Factor (m)	Noise Scale Factor (m)
BDS	EWL-I	$(0, 1, -1)$	4.8842	−1.5915	28.5287
	EWL-II	$(1, -5, 4)$	6.3707	0.6521	172.6135
GPS	EWL-I	$(0, 1, -1)$	5.8610	−1.7186	33.24
	EWL-II	$(1, -6, 5)$	3.2561	−0.0744	103.80

Table 2. Characteristics of the selected code combinations.

Constellation	Combinations	Ionospheric Scale Factor (m)	Noise Scale Factor (m)
BDS	$(0, 1, 1)$	1.5915	28.5287
	$(1, 0, 0)$	1.0000	1.0
GPS	$(0, 1, 1)$	1.7186	33.24
	$(1, 0, 0)$	1.0000	1.0

The first step is to resolve the ambiguity of the first EWL combination $\Delta\nabla\phi_{EWL-I}$. Both geometry-free and geometry based models can be used. The geometry-free model for EWL-I combination is as follows:

$$\breve{N}_{EWL-I} = \left[\frac{\Delta\nabla P_{(0,1,1)} - \Delta\nabla\phi_{EWI-I}}{\lambda_{EWL-I}}\right]_{round-off} \tag{3}$$

where P indicates the code combinations. λ indicates the wavelength $\breve{N}$ indicates the fixed ambiguities. The geometry-based model for EWL-I combination can be written as:

$$\begin{bmatrix} \Delta\nabla P_{(0,1,1)} \\ \Delta\nabla\phi_{EWL-I} \end{bmatrix} = \begin{bmatrix} A & 0 \\ A & E\cdot\lambda_{EWL-I} \end{bmatrix}\begin{bmatrix} X \\ N_{EWL-I} \end{bmatrix} + \begin{bmatrix} \varepsilon_{\Delta\nabla P_{(0,1,1)}} \\ \varepsilon_{\Delta\nabla\phi_{EWL-I}} \end{bmatrix} \tag{4}$$

where E is the identity matrix. N indicates the ambiguities. X indicates the baseline estimates, such as position. A indicates the design matrix. ε is the residual errors. As the ionospheric scale factor of $\Delta\nabla P_{(0,11)}$ is equal to that of $\Delta\nabla\varphi_{EWL-I}$, the ionspheric error can be completely eliminated. The tropospheric error is also eliminated. When using geometry-free model, rounding method is always used. When using geometry-based model, the Integer Least Squares (ILS) method is always used, which was investigated by many researchers [34–36].

The second step is to resolve the ambiguity of the second EWL combination $\Delta\nabla\varphi_{EWL-\Pi}$ according to the following equation. Also, both geometry-free and geometry based models can be used. The geometry-free model is as follows:

$$\breve{N}_{EWL-II} = \left[\frac{\Delta\nabla P_{(1,0,0)} - \Delta\nabla\phi_{EWL-II}}{\lambda_{EWL-II}} \right]_{round-off} \tag{5}$$

The geometry-based model can be written as:

$$\begin{bmatrix} \Delta\nabla P_{(1,0,0)} \\ \Delta\nabla\phi_{EWL-II} \end{bmatrix} = \begin{bmatrix} A & 0 \\ A & E\cdot\lambda_{EWL-II} \end{bmatrix} \begin{bmatrix} X \\ N_{EWL-II} \end{bmatrix} + \begin{bmatrix} \varepsilon_{\Delta\nabla P_{(1,0,0)}} \\ \varepsilon_{\Delta\nabla\phi_{EWL-II}} \end{bmatrix} \tag{6}$$

As the ionospheric factor of $\Delta\nabla P_{(0,11)}$ is not equal to that of $\Delta\nabla\varphi_{EWL-\Pi}$, the ionspheric error cannot be completely eliminated. However, the tropospheric error is eliminated.

Based on the above two independent resolved EWL ambiguities, the ambiguity of WL combination $\Delta\nabla\varphi_{(1,0,-1)}$ or $\Delta\nabla\varphi_{(1,-1,0)}$ can be computed according to the linear relationship, e.g., $\breve{N}_{(1,0,-1)} = 5\cdot\breve{N}_{(0,1,-1)} + \breve{N}_{(1,-5,4)}$. Then, we can derive the WL observables which can be considered as precise pseudorange observables:

$$\Delta\nabla\tilde{\Phi}_{WL} = \Delta\nabla\phi_{WL} + \lambda_{WL}\breve{N}_{WL} \tag{7}$$

The success rate of WL AR is dependent on the success rate of the two EWL combinations. For dual-frequency signals, HMW combinations can be formed, which eliminate both geometric terms (including clock errors, tropospheric delay, *etc.*) and ionospheric delays (first-order). In the double-difference case, only wide-lane ambiguities, multipath and measurement noise remain. Therefore, the ambiguities of WL measurement can be computed conveniently. Since the wavelength of the WL measurement is longer than 80 cm, we can fix its ambiguity much easier than that of original measurements [17,37].

The formula of DD HMW combination can be expressed as:

$$\Delta\nabla L_{HMW} = \left[\frac{(f_1\phi_1 - f_2\phi_2)}{f_1 - f_2} - \frac{(f_1P_1 + f_2P_2)}{f_1 + f_2} \right] / \lambda_{WL12} = N_{WL12} + \varepsilon_{HMW} \tag{8}$$

where f_1 and f_2 are the frequencies of the signals; ε_{HMW} denotes the residual errors and observational noise of HMW combination, which is in units of wide-lane cycles. λ_{WL12} is the wavelength of wide-lane measurements, which can be computed as follows:

$$\lambda_{WL12} = \frac{c}{(f_1 - f_2)} \tag{9}$$

where c denotes the speed of light. N_{WL12} is the ambiguity of wide-lane measurements, which is defined as:

$$N_{WL12} = N_1 - N_2 \tag{10}$$

where N_1 and N_2 are the ambiguities of original measurements.

Using geometry-free model, WL ambiguities can be obtained:

$$\breve{N}_{WL} = \left[\Delta\nabla\overline{L}_{HMW}\right]_{round-off} \tag{11}$$

where $\Delta\nabla\overline{L}_{HMW}$ indicates the averaged value of DD HMW combination. According to former studies [17,37], ten to thirty minutes are usually needed for averaging.

2.2. Positioning Model

After the ambiguities of wide-lane observations are fixed, positioning with wide-lane observations can be implemented. The model is as follows:

$$\Delta\nabla\widetilde{\Phi}_{WL} = AX + \varepsilon_{\Delta\nabla\phi_{WL}} \tag{12}$$

The estimated parameters are only user positions. Since the atmospheric errors are not parameterized or eliminated, this model still suffers from the impact of atmospheric errors. The positioning accuracy are analyzed in detail in the section entitled "Theoretical analysis".

3. Code-based Differential Positioning

The positioning accuracy of the proposed method will be compared with the code-based differential positioning, therefore, the code-based DGNSS model will be briefly introduced here. Code-based DGNSS uses code measurements for positioning. Based on the similarity of the atmospheric errors of base and rover stations, the corrections produced by base station are used for the positioning of rover station.

In base station, corrections at time t_0 are computed as follows:

$$PRC^S(t_0) = \widetilde{P} - \rho = dt^s - dt^r + T + I + \varepsilon_{\widetilde{P}} \tag{13}$$

where $\widetilde{P}$ represents the original code or carrier-smoothed code; ρ represents the geometrical distance between satellite and receiver; dt^s represents the satellite clock errors; dt^r represents the receiver clock errors; T represents the tropospheric delays; I represents the ionospheric delays; $\varepsilon_{\widetilde{P}}$ represents the measurement noise.

The positioning model of the rover station is as follows:

$$v = \widetilde{P} - (dt^s - dt^r + T + I) + \varepsilon_{\widetilde{P}} = P - PRC^S(t) + \varepsilon_{\widetilde{P}} \tag{14}$$

where t indicates the time when a user received the correction for positioning. $PRC^S(t)$ is computed according to the following equation:

$$PRC^S(t) = PRC^S(t_0) + RRC^S(t_0)(t - t_0) \tag{15}$$

When $\widetilde{P}$ represents the original code, the model is original code based differential positioning method. When P represents the carrier-smoothed code, the model become carrier-smoothed code based differential positioning method. Applying least square method, rover position can be obtained.

For code carrier-smoothing, the general formula is as follows:

$$\overline{P}_{(n)} = \phi_{(n)} + \frac{1}{n}\sum_{i=1}^{n}\left(P_{(i)} - \phi_{(i)}\right) \tag{16}$$

where n represents the number of epochs. i indicates at ith epoch. P and ϕ can be original measurements or combinations. Its recursive form can be written as:

$$\begin{cases} \overline{P}_{(n)} = P_{(1)} & \text{for } n = 1 \\ \overline{P}_{(n)} = \frac{1}{n}P_{(n)} + \frac{n-1}{n}\left(\overline{P}_{(n-1)} + \phi_{(n)} - \phi_{(n-1)}\right) & \text{otherwise} \end{cases} \tag{17}$$

There are several computing methods for code carrier-smoothing. Hatch [19], Goad [20] and McGraw [38] investigated the carrier-smoothing algorithm. McGraw [38] proposed an ionospheric divergence-free carrier-smoothing algorithm, P and ϕ adopt the following form:

$$\begin{cases} P = P_1 \\ \phi = \dfrac{f_1^2 + f_2^2}{f_1^2 - f_2^2}\phi_1 - \dfrac{2f_2^2}{f_1^2 - f_2^2}\phi_2 \end{cases} \tag{18}$$

Because the ionospheric scale factor of the carrier phase combination part is equal to 1, and its sign is same as that of code measurement, this algorithm will not be affected by the ionospheric delay divergence.

4. Theoretical Analysis

For carrier phase measurements, high positioning accuracy can be achieved only when its ambiguities are fixed. Therefore, the fixing rate of the ambiguities decides the accuracy and reliability of the positioning results in large extent. In this section, the fixing rate of WL AR using dual-frequency signals and triple-frequency signals are analyzed theoretically. The wide-lane observational errors are larger than those of narrow-lane or original carrier phase measurements, thus the positioning accuracy is lower than that of narrow-lane measurements. It is necessary to figure out what positioning accuracy the wide-lane signals can achieve over medium distances. Therefore, the positioning accuracy is also evaluated theoretically, and compared with that of carrier-smoothed code differential positioning in this section.

For convenience, we define:

(a) DFWL-DGNSS: WL differential positioning with dual-frequency signals;
(b) TFWL-DGNSS: WL differential positioning with triple-frequency signals;
(c) ORC-DGNSS: original code based differential positioning;
(d) CSC-DGNSS: carrier-smoothed-code based differential positioning.

4.1. Theoretical Fixing Rate

According to [39], when using rounding method, the theoretical fixing rate can be expressed as:

$$P_{succ} = \int_{-\infty}^{\frac{0.5-\mu_{\hat{N}}}{\sigma_{\hat{N}}}} \frac{1}{\sqrt{2\pi}} e^{-z^2/2} dz - \int_{-\infty}^{\frac{-0.5-\mu_{\hat{N}}}{\sigma_{\hat{N}}}} \frac{1}{\sqrt{2\pi}} e^{-z^2/2} dz \tag{19}$$

where $\sigma_{\hat{N}}$ is the standard deviation of the errors of the float ambiguities, $\mu_{\hat{N}}$ is the bias of the float ambiguities. We use this formula to assess the theoretical fixing rate of TFWL-DGNSS and DFWL-DGNSS. Since Equation (19) corresponds to a rounding method, some difference will be caused to the fixing rate when TFWL-DGNSS uses ILS method. However, the fixing rate of ILS method used in geometry-based model will be higher or at least equivalent to that of rounding method used in geometry-free model [40]. In addition, DFWL-DGNSS uses the rounding method. Therefore, it is still reasonable for us to evaluate the fixing rate using Equation (19).

From the analysis of Section 2.1, we can learn that the float ambiguities of the EWL-I combination are not affected by the atmospheric errors, whereas the float ambiguities of the EWL-II combination are affected by the residual atmospheric errors.

Assuming that the standard errors of undifferenced phase and code measurements are $\sigma_{P_u} = 0.3$ m and $\sigma_{\phi_u} = 0.003$ m, respectively and taking the double-differenced (DD) case into account, the standard errors of the float ambiguities for BDS WL (computed based on the HMW combination), EWL-I and EWL-II combinations will be $\sigma_{\hat{N}_{WL}} = 0.565$ cycles, $\sigma_{\hat{N}_{EWL-I}} = 0.087$ cycles and $\sigma_{\hat{N}_{EWL-II}} = 0.188$ cycles, respectively. The values are computed based on Equations (3), (5) and (8) according to the error propagation law. Since the float ambiguities of EWL-II are affected by the atmospheric errors, biases might exist in the float ambiguities. Assuming that residual ionospheric delay is 0.40 m, the caused biases for EWL-II will not exceed 0.11 cycles. We roughly assume that the residual atmospheric errors cause a bias of 0.11 cycles to the EWL-II float ambiguities. Therefore, the forms of the distribution for the float ambiguities of WL, EWL-I and EWL-II are $\varepsilon_{\hat{N}_{WL}} \sim N(0, 0.565)$, $\varepsilon_{\hat{N}_{EWL-I}} \sim N(0, 0.087)$ and $\varepsilon_{\hat{N}_{EWL-II}} \sim N(0.11, 0.188)$, respectively. In a similar way, the distributions of the float ambiguities for GPS WL, EWL-I and EWL-II combinations can be obtained.

Based on Equation (19), we computed the theoretical fixing rate of WL and two EWL combinations, as shown in Table 3. As can be seen, the fixing rates of the BDS EWL-I and EWL-II combinations are higher than 98.0%, whereas the fixing rate of the BDS WL combination is lower than 65%. The values indicate that the ambiguities of the selected two EWL combinations can be resolved with a rather high fixing rate.

Table 3. Theoretical fixing rate of WL and EWL combinations.

Combinations	Theoretical Fixing Rate	
	BDS	**GPS**
WL	62.38%	62.65%
EWL-I	~100.00%	~100.00%
EWL-II	98.04%	91.67%

4.2. Theoretical Accuracy

We first analyze the accuracy of the CSC measurements. Equation (16) can be rewritten as:

$$\overline{P}_{(n)} = \frac{n-1}{n}\phi_{(n)} - \frac{1}{n}\sum_{i=1}^{n-1}\phi_{(i)} + \frac{1}{n}\sum_{i=1}^{n}P_{(i)} \tag{20}$$

Assuming that the accuracy of DD code measurement is σ_P, the accuracy of DD carrier phase measurement is σ_{ϕ_c}, then the accuracy of the carrier-smoothed code measurements can be computed as follows:

$$\sigma_{\overline{P}} = \sqrt{\frac{n-1}{n}\sigma_{\phi_c}^2 + \frac{1}{n}\sigma_P^2} \tag{21}$$

where:

$$\sigma_{\phi_c} = \sqrt{\left(\frac{f_1^2+f_2^2}{f_1^2-f_2^2}\right)^2 + \left(\frac{2f_2^2}{f_1^2-f_2^2}\right)^2}\,\sigma_\phi = 6.24\sigma_\phi \tag{22}$$

where σ_ϕ indicates the accuracy of original DD carrier phase measurement. Then, we have:

$$\sigma_{\overline{P}} = \sqrt{\frac{n-1}{n}\left(6.24\sigma_\phi\right)^2 + \frac{1}{n}\sigma_P^2} \tag{23}$$

For wide lane measurements, only carrier phase measurements have to be taking into account, the accuracy can be computed as follows:

$$\sigma_{\phi_{WL}} = \sqrt{\left(\frac{f_1}{f_1-f_2}\right)^2\sigma_{\phi_1}^2 + \left(\frac{f_2}{f_1-f_2}\right)^2\sigma_{\phi_2}^2} = \sqrt{\left(\frac{f_1}{f_1-f_2}\right)^2 + \left(\frac{f_2}{f_1-f_2}\right)^2}\,\sigma_\phi = 6.87\sigma_\phi \tag{24}$$

Assuming that $\sigma_{P_u} = 0.3$ m and $\sigma_{\phi_u} = 0.003$ m and taking the double-differenced (DD) case into account, we calculated the standard deviation of CSC measurements and WL measurements *versus* the number of epochs, as shown in Figure 1. From the above two equations and Figure 1, we can learn that the accuracy of the WL measurement is comparable to that of the CSC measurement when a sufficient number of epochs are used. However, the accuracy of the CSC measurement is lower than that of the WL measurement for the first several hundred epochs. In other words, certain time smoothing is needed for CSC measurements to reaching a relatively high accuracy. In addition,

CSC measurements can be regarded as a phase measurement with ambiguity without fixing to integer, but roughly calculated by the following expression:

$$\frac{1}{n}\sum_{i=1}^{n}\left(P_{(i)} - \phi_{(i)}\right) \tag{25}$$

Therefore, the accuracy of CSC measurement also depends on whether the hardware delays of the signals are relatively stable and how long the multipath errors and observational noise can be eliminated or largely reduced. According to [41], the stability of BeiDou between-receiver differential code biases shows a dependence on satellite type, which has impacts on the carrier-smoothing performance.

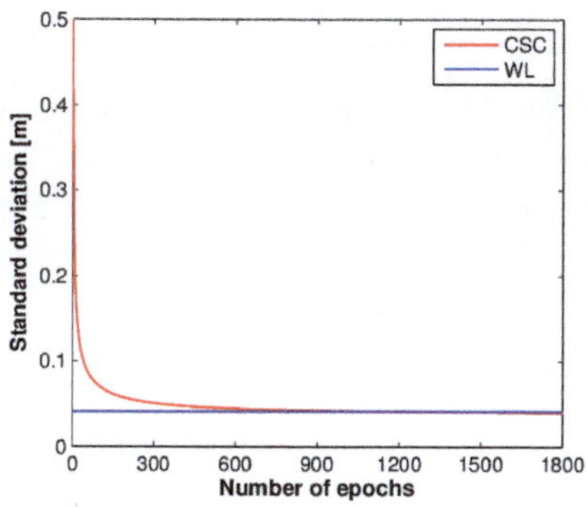

Figure 1. Standard deviation of CSC measurements and WL measurements *versus* the number of epochs.

In the medium baselines case, atmospheric errors also have an impact on positioning, because the atmosphere delays are not considered when implementing positioning for both algorithms. For tropospheric delays, the impacts is identical for both methods. For ionospheric delays, however, the impacts are different, because they dependent on frequency. The ionospheric scale factors of WL and CSC measurements are as follows:

$$\begin{cases} \beta_{\phi WL} = \dfrac{f_1^2}{f_1 - f_2}\left(\dfrac{1}{f_1} - \dfrac{1}{f_2}\right) = -1.293 \\ \beta_{\overline{P}} = 1 \end{cases} \tag{26}$$

As can be noticed, the ionospheric scale factor of WL measurements is a little higher than that of CSC measurements. Taking atmospheric errors into account, the user range error (URE) of WL and CSC measurements can be computed as follows:

$$URE_{WL} = \sqrt{\sigma_{\phi_{WL}}^2 + \sigma_{trop}^2 + \beta_{WL}^2 \cdot \sigma_{iono}^2} \tag{27}$$

$$URE_{\overline{P}} = \sqrt{\sigma_{\overline{P}}^2 + \sigma_{trop}^2 + \sigma_{iono}^2} \tag{28}$$

Table 4 gives two sets of error budgets to roughly represent the short and medium distance cases. The errors refer to original measurements on the first frequency. Based on these error budgets, URE_{WL} and $URE_{\overline{P}}$ for both short baselines and medium baselines are computed, as shown in Table 5. For $URE_{\overline{P}}$, we assume that the measurements have been implemented enough time carrier smoothing. The results indicate that the positioning accuracy of both methods will be mainly affected by the atmospheric errors in medium baselines. Therefore, the positioning accuracy will be approximately in the decimeter level.

Table 4. DD phase error budgets for short and medium distance [26].

Error Factors	Noise	Troposphere	Ionosphere
Short distance (0–10 km)	≈0.6 cm	<1 cm	<10 cm
Medium distance (10–100 km)	≈0.6 cm	<5 cm	<40 cm

Table 5. URE of CSC and WL measurements.

Case	$URE_{\overline{P}}$	URE_{WL}
Short distance (0–10 km)	0.107 m	0.136 m
Medium distance (10–100 km)	0.405 m	0.520 m

5. Data Collection and Processing Strategies

Observations of one short baseline and six medium baselines collected in static mode are used for our experiments. The baseline information is presented in Table 6. CUTA-CUT2 is located at the Curtin University of Technology (CUT, Perth, Australia). The data were collected on July 20 (DOY 201), 2015. The stations of the six medium baselines are located in Wuhan (China). The data were collected on February 22 (DOY 053), 2014. All of the data were obtained using Trimble NetR9 receivers, which can receive both GPS and BDS signals. The sampling rate of is 30 s for CUTA-CUT2 and 1 s for the six medium baselines.

Table 6. Baseline information.

Baseline	Distance
CUTA-CUT2	8.4 m
WHCD-WHHP	47.2 km
WHHN-WHHP	68.8 km
WHEZ-WHCD	83.9 km
WHEZ-WHHN	77.0 km
WHEZ-WHHP	74.5 km
WHHN-WHCD	28.7 km

For the DFWL-DGNSS method, observations on B1 and B2 frequencies are used for the test. For ORC-DGNSS method, we use the code measurement on B1 frequency for positioning. As to CSC-DGNSS, code measurements on one frequency and carrier phase measurement on two frequencies have to be used for carrier-smoothing. For GPS, we use code measurement on L1 and carrier phase measurements on L1 and L2. For BDS, we use code measurement on B1 and carrier phase measurements on B1 and B2. Details of the data processing strategy are summarized in Table 7.

Table 7. Processing strategy.

Options	Processing Strategy
Ephemeris	BeiDou broadcast ephemeris
Elevation cutoff angle	15°
Observation weighting	Elevation-dependent weightingRatio between standard deviation of phase and standard deviation of code: 1:100

6. Empirical Analysis

In this section, the fixing rate and positioning accuracy are evaluated empirically with observational data.

6.1. Empirical Fixing Rate

Using observations of the six medium baselines presented in Table 6, the biases of the float ambiguities for the WL, EWL-I and EW-II combinations relative to their corresponding true integer ambiguities are computed. The float ambiguities of the WL combination are computed based on the HMW combination. The float ambiguities of the EWL-I and EWL-II combinations are computed based on Equations (3) and (5), respectively.

Figure 2 shows the biases of the WL float ambiguities of WHCD-WHHP in DOY-053 of 2014 for different types of satellites as examples. It can be noticed that the WL float ambiguities exhibit large errors with respect to the WL integer ambiguities. The maximum error is close to three cycles. Obvious systematic errors can be observed from the WL float ambiguities of the BDS GEO satellites (C01-C02), which are mainly caused by the systematic multipath errors of GEO satellites [42,43].

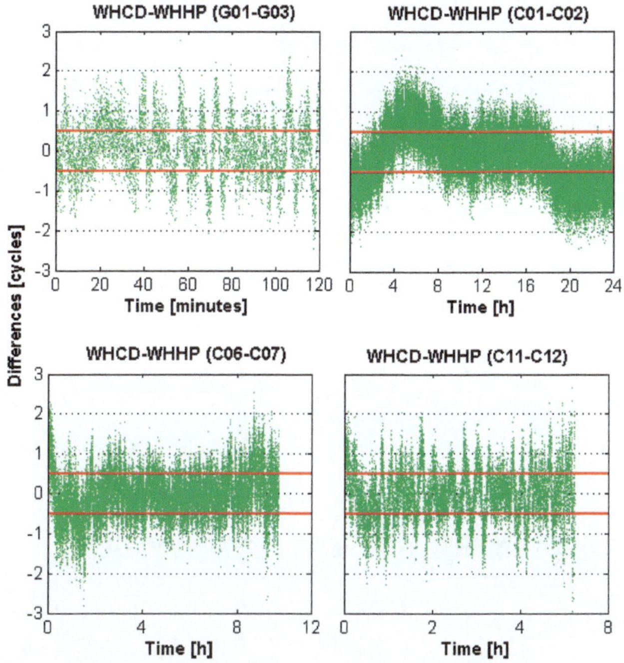

Figure 2. Dual-frequency WL float ambiguities of the GPS and BDS satellites for baseline WHCD-WHHP. The red line indicates the threshold for rounding. Green dots indicate the values of WL float ambiguities.

Figures 3 and 4 present the biases of the EWL-I and EWL-II float ambiguities of WHCD-WHHP in DOY-053 of 2014 for different types of satellites. It can be noticed that the errors of the EWL-I and EWL-II float ambiguities with respect to their integer ambiguities is smaller than 0.5 cycles most of the time, which is much smaller than those of the WL float ambiguities. The results of the EWL-II combination are a little lower than that of the EWL-I combination, which is mainly because the atmospheric errors are not fully eliminated for the EWL-II combination model. Furthermore, the fixing rate of GPS results is a little lower than that of BDS. This may be caused by the larger multipath errors of GPS code measurements. Fortunately, the fixing rate is extremely high for both the EWL-I and EWL-II combinations.

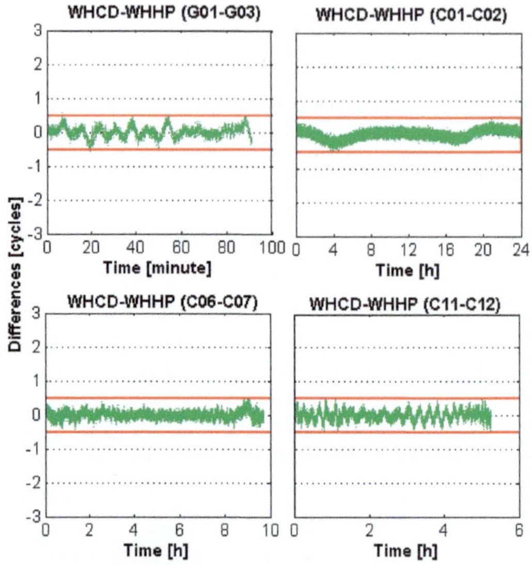

Figure 3. EWL-I time series of the GPS and BDS satellites for baseline WHCD-WHHP. The red line indicates the threshold of rounding. Green dots indicate the values of EWL-I combination.

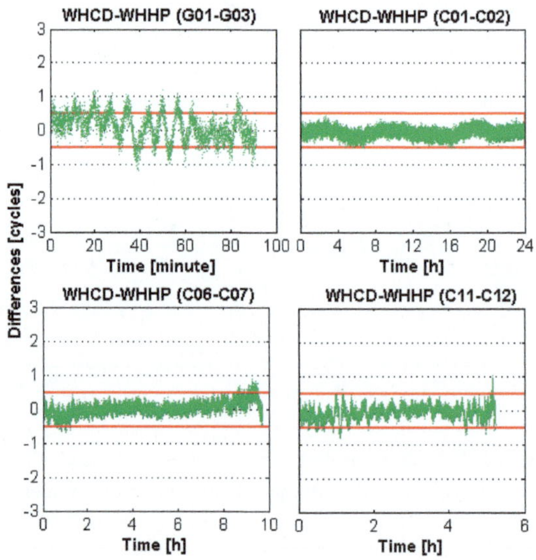

Figure 4. EWL-II time series of the GPS and BDS satellites for baseline WHCD-WHHP. The red line indicates the threshold of rounding. Green dots indicate the values of EWL-II combination.

We also computed the AR fixing rate of WL, EWL-I and EWL-II using observations of the six medium baselines, as shown in Table 8. The fixing rate of the WL combination is computed based on the HMW combination. The fixing rates of the EWL-I and EWL-II combinations are computed based on Equations (3) and (5), respectively. As can be noticed from this table, the fixing rate of both BDS EWL-I AR and EWL-II AR is close to 100%, though the fixing rate of EWL-II AR is a little lower. The fixing rate of WL AR using the HMW combination, however, is lower than 65%. Specifically, the fixing rate of GPS WL AR is even lower than 60%. These results indicated that we can obtain WL measurements using single-epoch triple-frequency signals with extremely high success rate, which is impossible in the dual-frequency case.

Table 8. Fixing rate of GPS/BDS AR. The fixing rate of the WL combination is computed based on the HMW combination. The fixing rates of theb EWL-I and EWL-II combinations are computed based on Equations (3) and (5), respectively.

Combinations	GPS MEO	BDS GEO	BDS IGSO	BDS MEO
WL	59.30%	62.45%	65.11%	63.31%
EWL-I	99.90%	100.00%	99.98%	99.99%
EWL-II	82.18%	99.98%	98.57%	99.35%

Multi-epoch smoothing or averaging algorithms also can be used to obtain the right solution. In this case, the smoothing time is very important. We compared the time needed for smoothing for different combinations. The time needed for achieving 100% fixing rate for dual-frequency signals and triple-frequency signals are computed. Figure 5 presents the averaged double-differenced HMW series of baseline WHCD-WHHP *versus* e time as an example. As can be noticed, several minutes to several tens of minutes are needed to obtain the correct wide-lane integer ambiguities for the BDS IGSO and MEO satellites. The time needed for GEO satellites is even longer. The minimum time for reaching 100% fixing rate is listed in Table 9. As can be noticed from this table, the needed time for obtaining the right solution for GPS dual-frequency signals and BDS dual-frequency signals are tens of minutes and several hours, respectively, which are much longer than that of GPS triple-frequency signals and BDS triple-frequency signals. These results indicate that implementing WL AR in a short time span using dual-frequency signals cannot be achieved. This is rather severe for BDS GEO satellites, since its multipath impact is large [42,43]. However, when triple-frequency signals are provided, the impact of noise and multipath of code measurements can be largely reduced by forming EWL combinations. WL measurements can thus be obtained with high fixing rates by linear derivation, which can be used for positioning.

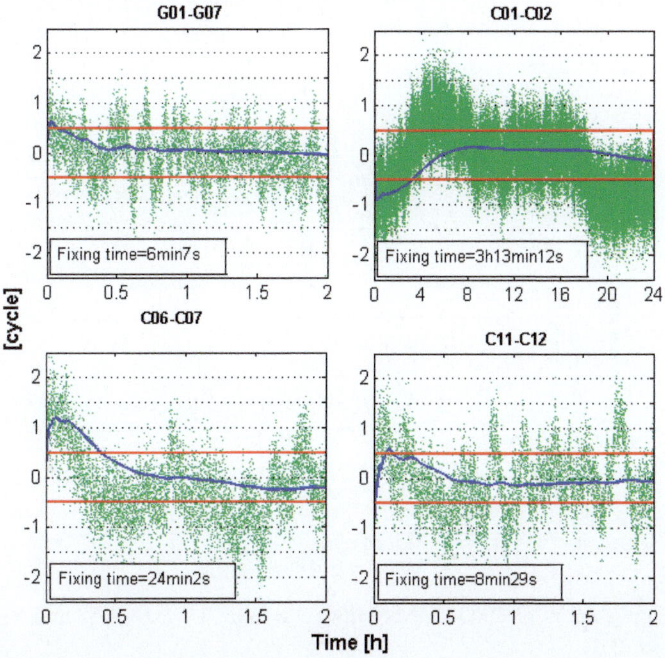

Figure 5. Averaged values of the HMW combination for baseline WHCD-WHHP. The red lines indicate the threshold of rounding. Green dots indicate the single-epoch values of the HMW combination. Blue lines indicate the averaged value of the HMW combination.

Table 9. Minimum time needed for reaching 100% fixing rate. The results of the WL combination are computed based on the HMW combination. The results of the EWL-I and EWL-II combinations are computed based on Equations (3) and (5), respectively.

Combinations	BDS GEO	BDS IGSO	BDS MEO
WL	24 h	1 h	45 min
EWL-I	1 s	3 s	2 s
EWL-II	2 s	25 s	25 s

6.2. Empirical Accuracy

Observational data of seven baselines listed in Table 6 are used for this test. The rounding threshold of the DFWL-DGNSS method is set as 0.2 cycles to guarantee reliable results. The ILS algorithm is applied for AR for TFWL-DGNSS method. Ratio test [44] is used to validate the ambiguity results. The threshold is set as 3.0 for the ambiguity validation of both EWL-I and EWL-II. Figure 6 shows the positioning errors of ORC-DGNSS, CSC-DGNSS, DFWL-DGNSS and TFWL-DGNSS obtained from BDS observations for baseline WHCD-WHHP as an example. As can be noticed, the positioning accuracy of ORC-DGNSS is much lower than those of the other three methods. The positioning accuracy of DFWL-DGNSS and TFWL-DGNSS are almost comparable to that of CSC-DGNSS. Theoretically, the accuracy of DFWL-DGNSS and TFWL-DGNSS will be equivalent each other once ambiguities are fixed. However, the errors of some epochs for DFWL-DGNSS are larger than that of TFWL-DGNSS, especially during the first several hours. This can be attributed to two reasons. Firstly, in some epochs, many WL observations are excluded because the fractional parts of their float ambiguities exceed the rounding threshold, thus the observations which can be used for positioning is few. The second reason is that there are some wrongly fixed wide-lane ambiguities in some epochs. We also can see that the positioning errors of CSC-DGNSS in the first several tens of epochs are relatively large. This is because the noises of the code measurements have not yet been smoothed to a rather low level.

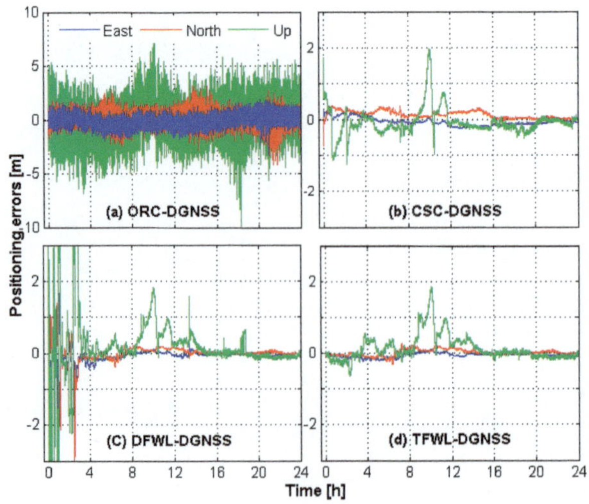

Figure 6. Positioning errors of ORC-DGNSS, CSC-DGNSS, DFWL-DGNSS and TFWL-DGNSS for baseline WHCD-WHHP.

Figure 7 shows the RMS of the positioning errors for the seven baselines. In order to figure out whether the positioning accuracy is related to the baseline length, the results are ordered by the baseline distance. For comparison, the CSC-DGNSS results obtained from GPS observations are also presented. It is shown that the positioning accuracy is somehow dependent on the baseline distance. For short baselines, the positioning accuracy is relatively high, especially for TFWL-DGNSS results. For medium baselines, the accuracy of most results is lower than that of the short baseline case. As the baseline length increases, the positioning accuracy shows a decreasing trend. Some abnormal cases are seen for BDS results, which may be caused by the systematic multipath errors of BDS code measurements [43]. Fortunately, the positioning accuracy of TFWL-DGNSS for all the six medium baselines is at the decimeter level.

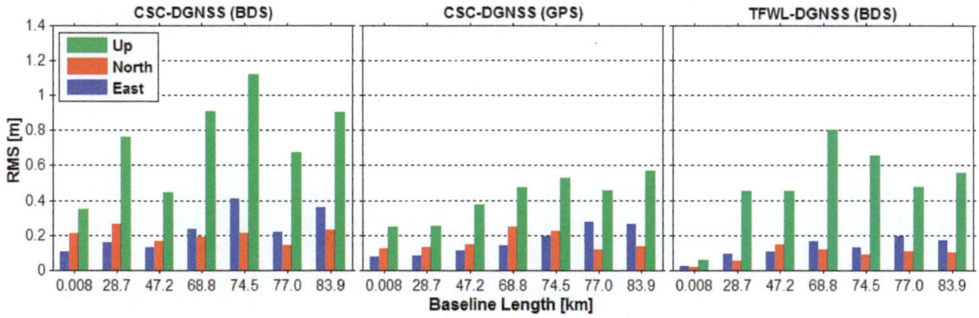

Figure 7. RMS of the positioning errors *versus* the baseline length. The left panel is for CSC-DGNSS results derived from BDS observations, the middle panel is for CSC-DGNSS results derived from GPS observations whereas the right panel is for TFWL-DGNSS results derived from BDS observations.

Table 10 presents the averaged RMS of the positioning results of ORC-DGNSS, CSC-DGNSS and TFWL-DGNSS for the six medium baselines. As can be seen, the accuracy of ORC-DGNSS is 5–7 dm in horizontal and lower than 18 dm in the up direction, much lower than those of the CSC-DGNSS and TFWL-DGNSS results. The accuracy of TFWL-DGNSS is 1.0–1.5 dm in horizontal and about 6 dm in the up direction. The accuracy of CSC-DGNSS for BDS observations and GPS observations is also different. The accuracy of CSC-DGNSS with BDS observations is 1.5–2.5 dm in horizontal and about 8 dm in the up direction, which is lower than that of TFWL-DGNSS in both the horizontal and up directions. The accuracy of CSC-DGNSS with GPS observations is about 1.7 dm in horizontal and about 4.5 dm in the up direction, which is a little lower than that of TFWL-DGNSS in the horizontal, but a little higher than that of TFWL-DGNSS in the up direction.

Table 10. Averaged RMS of CSC-DGNSS and TFWL-DGNSS. Units are in meters.

Method	East	North	Up	3D
ORC-DGNSS (BDS)	0.528	0.642	1.713	1.904
CSC-DGNSS (BDS)	0.248	0.187	0.790	0.852
CSC-DGNSS (GPS)	0.176	0.164	0.441	0.507
TFWL-DGNSS (BDS)	0.140	0.099	0.565	0.593

In order to evaluate the impact of the signal interruption to these methods, we processed 4 h of WHCD-WHHP data with simulated signal interruptions. The simulation is implemented by reinitializing all the estimating parameters for every 300 s during the processing. The results with no signal interruptions are also obtained for comparison. Figure 8 shows the positioning errors of CSC-DGNSS and TFWL-DGNSS in both the signal interruption case and no signal interruption case for baseline WHCD-WHHP. As can be noticed, the CSC-DGNSS results of both BDS and GPS are affected by signal interruptions. Relatively high accuracy can only be reliably achieved when certain time smoothing is implemented after signal interruption for CSC-DGNSS. TFWL-DGNSS, however, is almost not affected by signal interruption. Since single-epoch AR can be implemented with very high success rate, high accuracy positioning results can be obtained immediately after signal interruption.

By splitting the observational data into different time spans, ranging from several seconds to several minutes, we also computed the accuracy *versus* the time of CSC-DGNSS and TFWL-DGNSS. The results are presented in Figure 9. As can be seen, for CSC-DGNSS, the shorter the time span, the lower the accuracy is. Relatively high accuracy only can be achieved after 60–200 s of data are used. In contrary, the accuracy of TFWL-DGNSS remains stable regardless of the time span. These results display the advantage of TFWL-DGNSS for obtaining high accuracy positioning results rapidly.

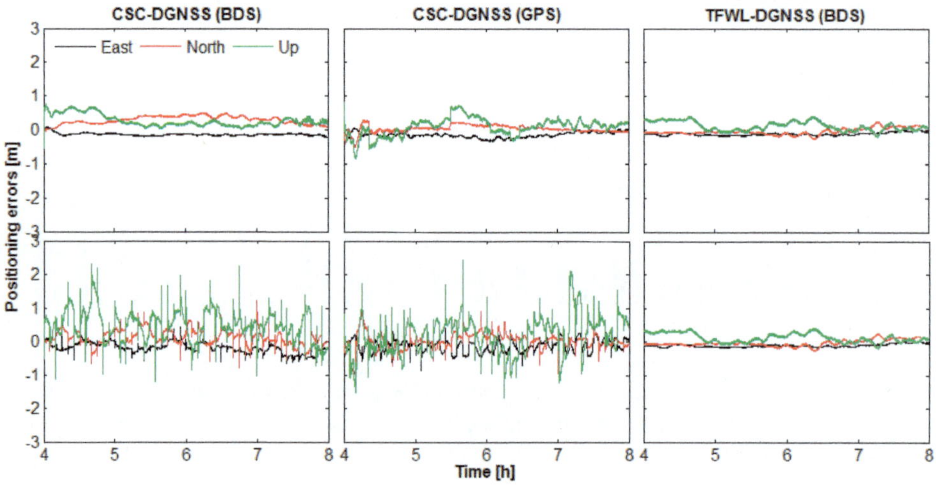

Figure 8. Positioning errors of the CSC-DGNSS and TFWL-DGNSS in both the signal interruption case and no signal interruption case for baseline WHCD-WHHP. The top panels are for the results without simulated signal interruption, whereas the bottom panels are for the results with simulated signal interruption.

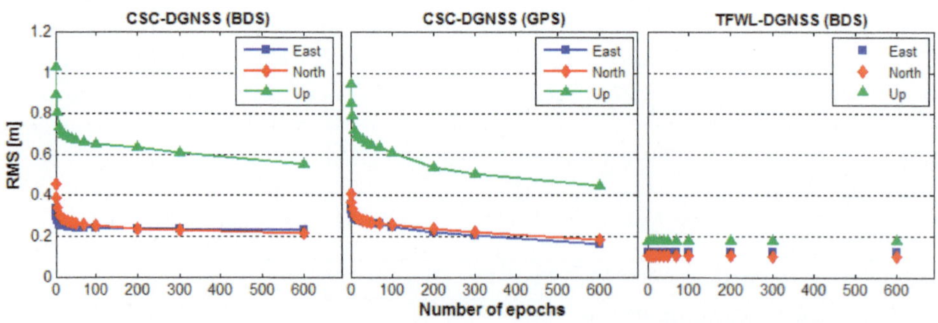

Figure 9. Positioning accuracy *versus* time for CSC-DGNSS and TFWL-DGNSS. The left panel is for CSC-DGNSS results derived from BDS observations, the middle panel is for CSC-DGNSS results derived from GPS observations whereas the right panel is for TFWL-DGNSS results derived from BDS observations.

7. Conclusions

In this paper we present a wide-lane positioning algorithm based on BDS triple-frequency signals, which can implement instantaneous real-time kinematic positioning with decimeter level positioning accuracy over medium distances. The fixing rate and positioning accuracy are carefully analyzed theoretically first, and then verified empirically by processing the data sets of seven baselines. The positioning performance is compared with that of carrier-smoothed code positioning.

The results indicates that the AR fixing rate of triple-frequency WL method with BDS observations can reach over 98%, which is much higher than that of DFWL-DGNSS. Decimeter level accuracy can be achieved by TFWL-DGNSS method, which is comparable to that of CSC-DGNSS method when using continuous observations. Signal interruption simulation experiments indicate that relatively high accuracy positioning results cannot be immediately obtained with the CSC-DGNSS method when signal interruption happens. However, they can be achieved using the TFWL-DGNSS method. The results imply that a high AR fixing rate and relatively high positioning accuracy can be achieved for the TFWL-DGNSS method with single-epoch or very short time observations, representing a significant advantage compared to the CSC-DGNSS method. The presented method can be an alternative method for the users requiring decimeter level rapid positioning results over medium distances. In future studies, the atmospheric errors should be considered to further improve the positioning accuracy. In addition, triple-frequency observations of the other GNSS systems can be added to improve the performance.

Acknowledgments: The authors gratefully acknowledge IGS and GNSS Research Centre of Curtin University of Technology (CUT) for providing BDS data and products. The BeiDou data sets collected from Wuhan CORS network (China) were used in this study, which is acknowledged. This study was supported by National Natural Science Foundation of China (Grant Nos. 41474025, 41204030 and 41304005), Natural Science Foundation for Distinguished Young Scholar of Hubei Province (No. 2015CFA039).

Author Contributions: Xiyang He, Xiaohong Zhang and Long Tang provided the initial idea and conception; He implemented the software and carried out the experimental validation; Xiyang He and Xiaohong Zhang wrote the main manuscript; Long Tang and Wanke Liu helped with the writing. All authors reviewed the manuscript.

Conflicts of Interest: The authors declare no conflict of interest.

References

1. Larson, K.; Bodin, P.; Gomberg, J. Using 1-Hz GPS data to measure deformations caused by the Denali fault earthquake. *Science* **2003**, *300*, 1421–1424. [CrossRef] [PubMed]
2. Bock, Y.; Prawirodirdjo, L.; Melbourne, T. Detection of arbitrarily large dynamic ground motions with a dense high-rate GPS network. *Geophys. Res. Lett.* **2004**, *31*, 177–182. [CrossRef]
3. Thuilot, B.; Cariou, C.; Martinet, P.; Berducat, M. Automatic guidance of a farm tractor relying on a single CP-DGPS. *Auton. Robot.* **2002**, *13*, 53–71. [CrossRef]
4. Logan, S.A.; Leahy, F.J.; Kealy, A. Integration of GPS carrier phase and other measurements for kinematic mapping. *J. Geod.* **2003**, *76*, 543–556. [CrossRef]
5. Satirapod, C.; Rizos, C.; Wang, J. GPS single point positioning with SA off: How accurate can we get? *Surv. Rev.* **2001**, *36*, 255–262. [CrossRef]
6. Zumberge, J.F.; Heflin, M.B.; Jefferson, D.C.; Watkins, M.M.; Webb, F.H. Precise point positioning for the efficient and robust analysis of GPS data from large networks. *J. Geophys. Res.* **1997**, *102*, 5005–5017. [CrossRef]
7. Ge, M.; Gendt, G.; Rothacher, M.; Shi, C.; Liu, J. Resolution of GPS carrier-phase ambiguities in precise point positioning (PPP) with daily observations. *J. Geod.* **2008**, *82*, 389–399. [CrossRef]
8. Bisnath, S.; Gao, Y. Current State of Precise Point Positioning and Future Prospects and Limitations. *Int. Assoc. Geod. Symp.* **2008**, *133*, 615–623.
9. Kleusberg, A. Kinematic relative positioning using GPS code and carrier beat phase observations. *Mar. Geod.* **1986**, *10*, 257–274. [CrossRef]
10. Cannon, M.E. *Kinematic Positioning Using GPS Pseudo-Range and Carrier Phase Observations*; (UCSE 20019); The University of Calgary: Calgary, AB, Canada, 1987.
11. Schwarz, K.P.; Cannon, M.E.; Wong, R.V.C. A comparison of GPS kinematic models for the determination of positioning and velocity along a trajectory. *Manuscr. Geod.* **1989**, *14*, 345–353.
12. Teunissen, P.J.G.; Odolinski, R.; Odijk, D. Instantaneous BeiDou + GPS RTK positioning with high cut-off elevation angles. *J. Geod.* **2014**, *88*, 335–350. [CrossRef]
13. Odolinski, R.; Teunissen, P.J.G.; Odijk, D. Combined GPS + BDS + Galileo + QZSS for Long Baseline RTK Positioning. In Proceedings of the 27th International Technical Meeting of the Satellite Division of the Institute of Navigation (ION GNSS + 2014), Tampa, FL, USA, 8–12 September 2014; pp. 2326–2340.

14. Han, S.; Rizos, C. GPS Network Design and Error Mitigation for Real-Time Continuous Array Monitoring Systems. In Proceedings of the ION GPS-1996, 9th International Technical Meeting of The Satellite Division of the Institute of Navigation, Kansas City, MO, USA, 17–20 September 1996; pp. 1827–1836.
15. Li, X.; Zhang, X.; Ge, M. Regional reference network augmented precise point positioning for instantaneous ambiguity resolution. *J. Geod.* **2011**, *85*, 151–158. [CrossRef]
16. Zhang, B.; Teunissen, P.J.G.; Odijk, D. A Novel Un-differenced PPP-RTK Concept. *J. Navig.* **2011**, *64*, S180–S191. [CrossRef]
17. Li, X.; Ge, M.; Zhang, H.; Wickert, J. A method for improving uncalibrated phase delay estimation and ambiguity-fixing in real-time precise point positioning. *J. Geod.* **2013**, *87*, 405–416. [CrossRef]
18. Blackwell, E.G. Overview of differential GPS methods. *Navigation* **1985**, *32*, 114–125. [CrossRef]
19. Hatch, R. The synergism of GPS code and carrier measurement. In Proceedings of the Third International Geodetic Symposium Satellite Doppler Positioning, La Cruces, NM, USA, 8–12 February 1982; pp. 1213–1232.
20. Goad, C. Optimal filtering of pseudo-ranges and phases from single-frequency GPS receivers. *Navigation* **1990**, *37*, 249–262. [CrossRef]
21. Rizos, C. Network RTK research and implementation—A geodetic perspective. *J. Glob. Position. Syst.* **2003**, *1*, 144–150. [CrossRef]
22. Li, X.; Zhang, X.; Ren, X.; Fritsche, M.; Wickert, J.; Schuh, H. Precise positioning with current multi-constellation Global Navigation Satellite Systems: GPS, GLONASS, Galileo and BeiDou. *Sci Rep.* **2015**, *5*. [CrossRef] [PubMed]
23. Harris, R.A. *Direct Resolution of Carrier Phase Ambiguity by "Bridging the Wavelength Gap"*; ESA Publisher: Paris, France, 1997.
24. Forssell, B.; Martin-Neira, M.; Harris, R.A. Carrier Phase Ambiguity Resolution in GNSS-2. In Proceedings of the ION GPS-1997, Institute of Navigation, Kansas City, MO, USA, 16–19 September 1997; pp. 1727–1736.
25. Jung, J. High integrity carrier phase navigation for future LAAS using multiple civilian GPS signals. In Proceedings of the ION GPS-1999, Institute of Navigation, Nashville, TN, Alexandria, USA, 14–17 September 1999; pp. 14–17.
26. Feng, Y. GNSS three carrier ambiguity resolution using ionosphere-reduced virtual signals. *J. Geod.* **2008**, *82*, 847–862. [CrossRef]
27. Feng, Y.; Li, B. Wide area real time kinematic decimetre positioning with multiple carrier GNSS signals. *Sci. China Earth Sci.* **2010**, *53*, 731–740. [CrossRef]
28. Tang, W.; Deng, C.; Shi, C.; Liu, J. Triple-frequency carrier ambiguity resolution for Beidou navigation satellite system. *GPS solut.* **2014**, *18*, 335–344. [CrossRef]
29. Zhang, X.; He, X. Performance analysis of triple-frequency ambiguity resolution with BeiDou observations. *GPS Solut.* **2015**. [CrossRef]
30. Melbourne, W. The case for ranging in GPS-based geodetic systems. In Proceedings of the First International Symposium on Precise Positioning with the Global Positioning System, Rockville, MD, USA, 15–19 April 1985; pp. 373–386.
31. Wübbena, G. Software developments for geodetic positioning with GPS using TI-4100 code and carrier measurements. In Proceedings of the First International Symposium on Precise Positioning with the Global Positioning System, Rockville, MD, USA, 15–19 April 1985; pp. 403–412.
32. Cocard, M.; Bourgon, S.; Kamali, O.; Collins, P. A systematic investigation of optimal carrier-phase combinations for modernized triple-Frequency GPS. *J. Geod.* **2008**, *82*, 555–564. [CrossRef]
33. Zhang, X.; He, X. BDS triple-frequency carrier-phase linear combination models and their characteristics. *Sci. China: Earth Sci.* **2015**, *58*, 896–905. [CrossRef]
34. Teunissen, P.J.G. The least squares ambiguity decorrelation adjustment: A method for fast GPS integer estimation. *J. Geod.* **1995**, *70*, 65–82. [CrossRef]
35. Chang, X.; Yang, X.; Zhou, T. MLAMBDA: A modified LAMBDA method for integer least-squares estimation. *J. Geod.* **2005**, *79*, 552–565. [CrossRef]
36. Xu, P. Mixed integer geodetic observation models and integer programming with applications to GPS ambiguity resolution. *J. Geod. Soc. Jpn.* **1998**, *44*, 169–187.
37. Blewitt, G. Carrier phase ambiguity resolution for the global positioning system applied to geodetic baselines up to 2000 km. *J. Geophys. Res.* **1989**, *94*, 135–151. [CrossRef]

38. McGraw, G. How can dual frequency code and carrier measurements be optimally combined to enhance position solution accuracy. *Inside GNSS* **2006**, *1*, 17–19.
39. Teunissen, P.J.G. Integer estimation in the presence of biases. *J. Geod.* **2001**, *75*, 399–407. [CrossRef]
40. Teunissen, P.J.G. An optimality property of the integer least-squares estimator. *J. Geod.* **1999**, *73*, 587–593. [CrossRef]
41. Zhang, B.; Teunissen, P.J.G. Characterization of multi-GNSS between-receiver differential code biases using zero and short baselines. *Sci. Bull.* **2015**, *60*, 1840–1849. [CrossRef]
42. Wang, G.; Jong, K.; Zhao, Q.; Hu, Z.; Guo, J. Multipath analysis of code measurements for BeiDou geostationary satellites. *GPS Solut.* **2014**, *19*, 129–139. [CrossRef]
43. He, X.; Zhang, X. Characteristics Analysis of BeiDou Melbourne-Wübbena Combination. In Proceedings of the China Satellite Navigation Conference (CSNC) 2015, Xi'an, China, 12–14 May 2015; pp. 587–593.
44. Euler, H.; Schaffrin, B. On a measure for the discernibility between different ambiguity solutions in the static-kinematic GPS-mode. In *Kinematic Systems in Geodesy, and Surveying, and Remote Sensing*; Springer: New York, NY, USA, 1991; pp. 285–295.

2

Parallax-Robust Surveillance Video Stitching

Botao He [1,*] and Shaohua Yu [2]

Academic Editor: Gonzalo Pajares Martinsanz

1 School of Optical and Electronic Information, Huazhong University of Science and Technology, Wuhan 430074, China
2 Wuhan Research Institute of Posts and Telecommunications, Wuhan 430074, China; shyu@fhzz.com.cn
* Correspondence: hebotao@hust.edu.cn

Abstract: This paper presents a parallax-robust video stitching technique for timely synchronized surveillance video. An efficient two-stage video stitching procedure is proposed in this paper to build wide Field-of-View (FOV) videos for surveillance applications. In the stitching model calculation stage, we develop a layered warping algorithm to align the background scenes, which is location-dependent and turned out to be more robust to parallax than the traditional global projective warping methods. On the selective seam updating stage, we propose a change-detection based optimal seam selection approach to avert ghosting and artifacts caused by moving foregrounds. Experimental results demonstrate that our procedure can efficiently stitch multi-view videos into a wide FOV video output without ghosting and noticeable seams.

Keywords: video stitching; video surveillance; layered warping; parallax

1. Introduction

Image stitching, also called image mosaicing or panorama stitching, has received a great deal of attention in computer vision [1–8]. After decades of development, the fundamentals of image stitching are well studied and relatively mature now. There are many research works on image stitching [1–8], and it is typically solved by estimating a global 2D projective warp to align the input images. A 2D projective warp uses a homography parameterized by 3×3 matrices [1–3,9], which can preserve global image structures, but cannot handle parallax. It is correct only if the scene is planar or if the views differ purely by rotation. However, in practice, such conditions are usually hard to satisfy, thus ghosting and seams yield (see Figure 1). If there is parallax in input images, no global homograhpy exists that can be used to align these images. When a global warp is used to stitch these images, ghosting like Figure 1a would appear. Some advanced image composition techniques such as seam cutting [10–12] can be used to relieve these artifacts. However, if there are moving objects across the seams, another kind of ghosting like Figure 1b would yield.

Previous research indicates that one of the most challenging problems to create seamless and drift-free panoramas is performing a correct image alignment rather than using a simple global projective model and then fix the alignment error [6,8,9]. Thus, some recent image stitching methods focus on using spatially-varying warping algorithms to align the images [6–8], these methods can handle parallax and allow for local deviation to some extent but require more computation.

With wide applications in robotics, industrial inspection, surveillance and navigation, video stitching faces all the problems as image stitching does and can be more challenging due to moving objects in videos. Some researchers tried to build panoramic images by aligning video sequences [13,14], which is panoramic image generation rather than expansion of the FOV of dynamic videos. Other works focus on freely moving devices [15–18], especially mobile devices, which include techniques such as efficient computation of temporal varying homography [15], optimal seam selection for blending [16], and so on. However, due to complex computation and low resolution, they may not be suitable for surveillance application.

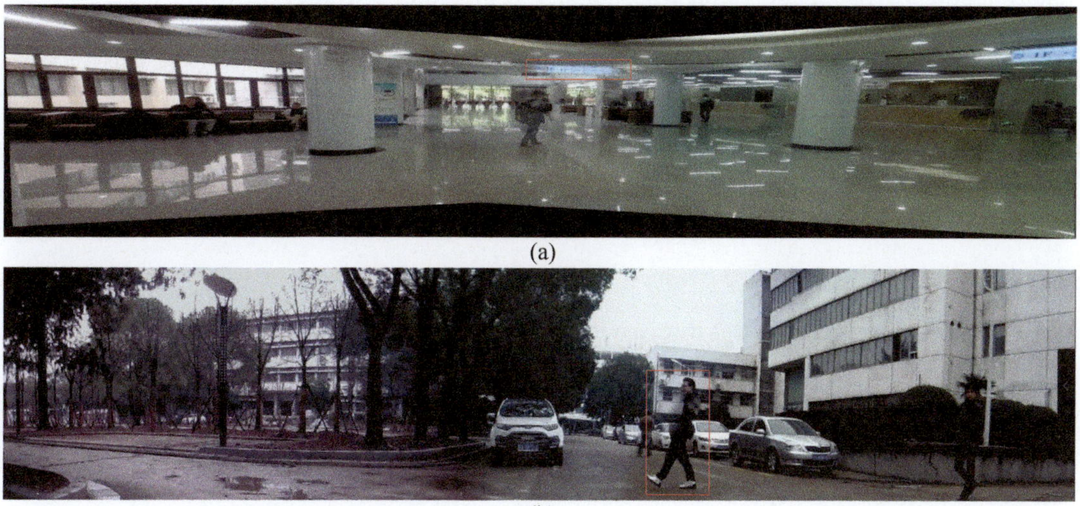

(a)

(b)

Figure 1. The main reason for ghosting comes from misalignment error. Here we show two types of ghosting in red boxes: (**a**) ghosting caused by using a global projective warping; (**b**) ghosting caused by persons moving across seams.

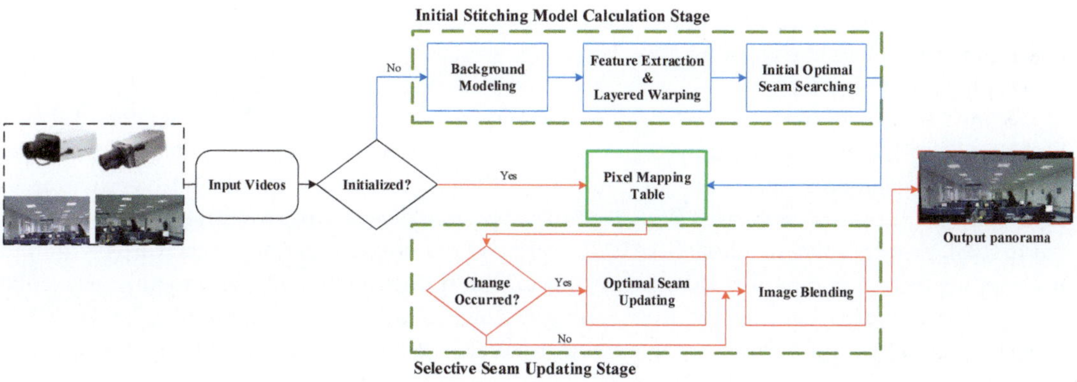

Figure 2. Outline of the proposed video stitching procedure. Our method consists of two stages: initial stitching model calculation stage and selective seam updating stage. In the initial stitching model calculation stage, we first use background modeling algorithm to generate still background of each input source video, then utilize layered warping algorithm to align background images, finally, we perform optimal seam searching and image blending to generate panorama backgrounds. The resulting stitching model is a mapping table in which each entry indicates the correspondence between the pixel index of source images and that of panorama image. To relieve the ghosting effect caused by moving objects, at selective seam updating stage, we perform seam updating according to whether there are objects moving across previous seams or not. Our method reaches a balance between suppressing ghosting artifacts and real-time requirement.

In this paper, we present an efficient parallax-robust surveillance video stitching procedure that combines layered warping and the change-detection based seam updating approach. As the alignment validity of video stitching is still crucial as it is in seamless image stitching, we use a layered warping method for video registration instead of a simple global projective warp.

Xiao *et al.* [19] provides a similar layer-based video registration algorithm, but it aims at aligning a single mission video sequence to the reference image via layer mosaics and region expansion, rather than building a dynamic panoramic video for motion monitoring. Through dividing matched feature pairs into multiple layers (or planes) and local alignment based on these layers, the layered warping method seems to be more robust to parallax. Moreover, the warping data for fixed surveillance videos are stored in an index table for "recycling-use" in subsequent frames to avoid repeated registration and interpolations. This index table is referred to as the initial stitching model in this paper. Aside from layered warping, a local change-detection based seam updating method for overlapping regions is performed to disambiguate the ghosting caused by moving foregrounds. Figure 2 shows the video stitching process presented in this paper.

2. Related Works

Recently, video stitching has drawn a lot of attentions [11,20,21] due to its wide usage in public security. Generally speaking, surveillance video stitching can be regarded as image stitching for every individual frame since the camera positions are always fixed. Different from image stitching technologies, video stitching requires more strict real-time processing ability, and large parallax and dynamic foregrounds must be carefully considered to obtain consistent wide field of view videos. Image stitching is relatively a well studied problem in computer vision [1,9,22]. Several freewares and commercial softwares are also available for performing image stitching, like AutoStitch [23], Microsoft's Image Compositing Editor [24], and Adobe's Photoshop CS5 [25] mosaicing feature. However, these approaches all work under the assumption that the input images contains little or no parallax, which implies that the scene is either sufficiently far away from the camera to be considered planar, or that the images have been taken from a camera carefully rotated about its center of projection. This assumption is too strict to be satisfied in real surveillance scenarios. Thus, misalignment artifacts like ghosting or broken image structures will make the final panorama visually unacceptable (see Figure 1).

Some advanced image composition techniques, such as seam cutting [10–12] and blending [26,27], have been employed to reduce the artifacts. However, these methods alone still cannot handle significant parallax. In this paper, we also use seam cutting and blending as the final steps to suppress artifacts. Recent studies on spatially-varying warping are another way out [6,7,28]. As-projective-as-possible (APAP) warps [7] employed local projective warps within the overlapping regions and performed moving direct linear transformation [29] to smoothly extrapolate local projective warps into the non-overlapping regions. Shape-Preserving Half-Projective (SPHP) warp [28] spatially combines a projective transformation and a similarity transformation and has the strengths of both. However, instead of improving alignment accuracy, its main concern is to decrease distortion of non-overlapping area caused by the projective transformation. So even if it introduces APAP [7] into their warp, they cannot solve the problem of structure distortion in the overlapping area in APAP [7]. Gao *et al.* [6] proposed to uses a dual homography warp (DHW) algorithm for scenes containing two dominant planes (ground plane and distant plane). While it performs well if the required setting is true, it may fail when there are more than two planes in the source images. Inspired by DHW [6], we propose to use a layered warping algorithm to align the background scenes, which is location-dependent and turned out to be more robust to parallax than the traditional global projective warping methods and more flexible than DHW [6] which can only process images with two planes.

Apart from parallax, moving foregrounds are another reason for ghosting in video stitching. Although we propose to use layered warping to align images as accurate as possible and to utilize seam cutting to composite source images, some artifacts may still exist when objects move across the seam. Liu *et al.* [20] only used the stitching model calculated with first few frames to stitch following frames and didn't consider the moving foregrounds. In contrast, Tennoe *et al.* [30] and Hu *et al.* [11] update the seam in every frame, which is very time-consuming. To balance between suppressing the artifacts and the real-time requirement, we propose to first detect changes around the previous seams, and only perform seam update when there are moving objects across seams. Since the price of change detection around seams is much lower than that of updating seams, artifacts caused by moving foregrounds can be suppressed with acceptable time consumption by our method.

3. Initial Stitching Model Calculation

Since our focus is on improving image alignment accuracy and reducing artifacts caused by moving objects, we do not change the conventional pipeline [9] of image stitching with different number of input sources. For ease of illustration, in the following text, we only describe the layered warping algorithm and selective seam updating algorithm with two input videos. In the experiment section, we provide stitching results with both two and more than two input videos.

3.1. Background Image Generation and Feature Extraction

For fixed surveillance cameras, we only perform alignment at the stitching model calculation stage because the computation of temporal varying homography may result in palpable jitter of background scenes in the panoramic video. Background modeling is essential to avoid volatile foreground [31,32]. We take the first N_{gmm} frames of input videos to establish the background frame utilizing the Gaussian Mixture Model (GMM) [33]. SIFT [3,34] features of the background frame is extracted and matched into pairs through Best-Bin-First (BBF) algorithm [35].

3.2. Layered Warping

Inspired by DHW [6], we assume that different objects in a scene usually lie in different depth layers, the objects in the same layer (plane) shall be consistent with each other in spatial transformation. Compared to warping using a global homography or warping using dual homography, layered warping may be more adequate and robust for abundant scenes.

Layer registration. We denote the input images as I_1 and I_2 respectively, and the matching feature pairs as $\mathcal{F} = \{(\mathbf{p}_i^1, \mathbf{p}_i^2)\}_{i=1}^N$, where $\mathbf{p}_i^k$ is the coordinate of the i-th matching point from I_k $(k = 1, 2)$. Given the matching feature pairs, we first utilize Random Sample Consensus (RANSAC) algorithm [36] to robustly group the feature pairs into different layers, then estimate the homography for each layer. Denote the consistent matching feature pairs of layer k as $\mathcal{L}_k$, the number of matches in $\mathcal{L}_k$ as $|\mathcal{L}_k|$ and its corresponding homography as $\mathbf{H}_k$. To guarantee the grouped layer to be representative, we introduce a threshold N_{min} which denotes the minimum number of matching pairs in $|\mathcal{L}_k|$. Layers whose number of matching pairs is smaller than N_{min} are simply dropped. The detailed layer registration process is presented in Algorithm 1.

Algorithm 1 Layer registration utilizing multiple-layer RANSAC

Input: Initial pair set $\mathcal{F}_1 = \mathcal{F}$, threshold N_{min} and iteration index $k = 0$;
Output: Each layer's matching pair set $\mathcal{L}_k$ and its corresponding homography $\mathbf{H}_k$;
repeat
  $k \leftarrow k + 1$
  RANSAC in pair set $\mathcal{F}_k$ for model $\mathbf{p}_{k_n}^1 \times \mathbf{H}\mathbf{p}_{k_n}^2 = 0$, where $(\mathbf{p}_{k_n}^1, \mathbf{p}_{k_n}^2) \in \mathcal{F}_k$;
  Divide outliers $\mathbf{V}_{out}$ and inliers $\mathbf{V}_{in}$ according to $\mathbf{H}$;
  if $|\mathbf{V}_{in}| \geq N_{min}$ **then**
    Set matching pair set of the k-th layer as $\mathcal{L}_k = \mathbf{V}_{in}$;
    Set homography of the k-th layer as $\mathbf{H}_k = \mathbf{H}$;
  end if
  Set the pair set of next iteration as $\mathcal{F}_{k+1} = \mathbf{V}_{out}$;
until $|\mathcal{F}_{k+1}| < N_{\text{min}}$

Through the layer registration process, we divide the matched feature pairs to multiple layers, each of which contains a set of feature pairs that are consistent with a common homography. Figure 3 shows two examples of layer registration, in which feature points are illustrated as circles and points with the same color are from the same layer. From Figure 3 we can see that the grouped matching pairs of the same layer are almost from the same plane or with the same depth, which is in accordance with our expectation.

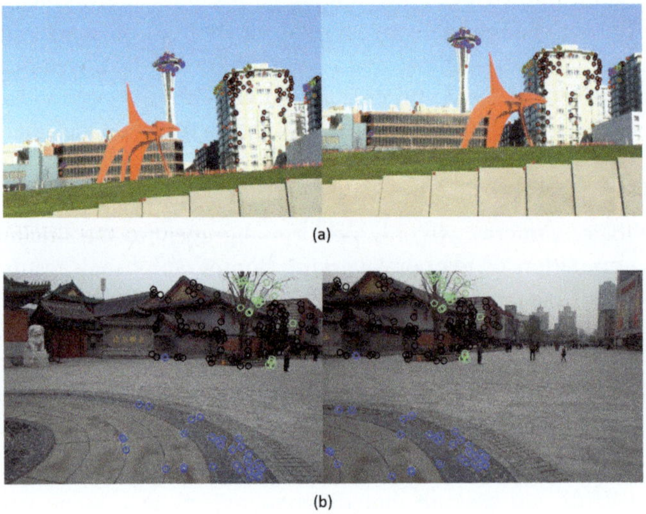

Figure 3. Examples of matched feature points in multiple layers, different color indicate different layers: (**a**) multiple-layer matched feature pairs of images with large parallax; (**b**) multiple-layer matched feature pairs of images with a distant plane and ground plane

Local alignment. To simplify calculations of local alignment, the source image is divided into $M \times N$ grids. Since a feature point is not usually coincident with any grid vertex, we use the distances between the grid center and its nearest neighbors in different layers to vote on the warp of the due grid. A grid g_j is represented by its center point $\mathbf{c}_j$. The homography of the grid g_j, denoted by $\mathbf{H}_j^*$, is computed by fusing $\mathbf{H}_k$ of multiple layers using a weight w_k^j by Equation (1)

$$\mathbf{H}_j^* = \sum_k w_k^j \mathbf{H}_k \tag{1}$$

where $w_k^j = a_k^j / \sum_i a_i^j$ and a_k^j is a position dependent Gaussian weight:

$$a_k^j = \exp(-\frac{||\mathbf{c}_j - \mathbf{p}_k^*||^2}{\sigma^2}) \tag{2}$$

Here $\mathbf{p}_k^*$ denotes the nearest neighboring feature point of $\mathbf{c}_j$ in layer k and σ is a scale constant.

After deriving the local homography for each grid, the target pixel position $\mathbf{p}'$ in the reference image for the source pixel at position $\mathbf{p}$ in grid g_j can be easily obtained by Equation (3):

$$\mathbf{p}' = \mathbf{H}_j^* \mathbf{p} \tag{3}$$

This process is referred to as forward mapping [9]. To accelerate the computation, we only perform forward mapping once, and store the correspondence between source pixel positions and target pixel positions in the pixel mapping table. This pixel mapping table is exactly the stitching model. The index tables are stored so that the warped image can be obtained by looking up each corresponding pixel in the source image instead of repeated transformation when new frames arrive. Figure 4a,b shows the panorama images with global projective warp and layered warp respectively. It is clear that the building in Figure 4b is better aligned than that in Figure 4a.

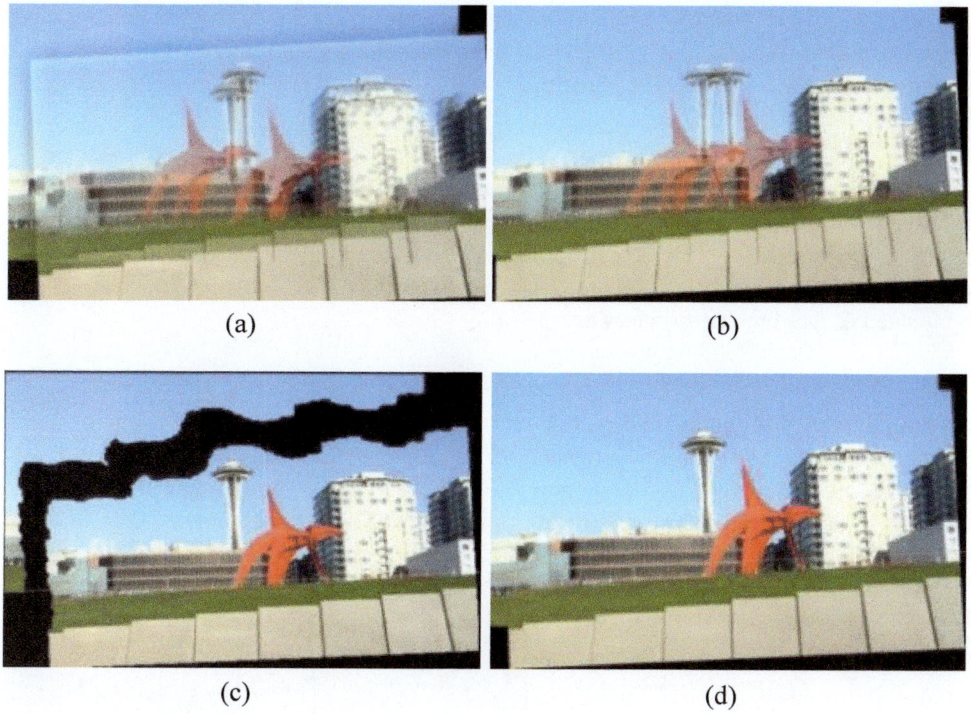

Figure 4. Examples of panoramas with (**a**) global projective warp and (**b**) layered warp, and seam selection results in (**c**). (**d**) is the final fused image according to selected seam in (**c**). The seam selection result (**c**) is based on stitched image (**b**) with layered warp.

3.3. *Optimal Seam Cutting*

Though layered warping is parallax-robust, it is applied only in the stitching model calculation stage on the extracted background scene. We perform the seam selection method to disambiguate the ghosting caused by the moving foreground.

Optimal seam searching method, also called optimal seam selection [5] or seam cutting, is to search for an optimal seam path which is a pixel-formed continuous curve in the overlapping area to connect pairwise warped images.

The seam should neither introduce inconsistent scene elements nor intensity differences. Therefore, two criteria are applied in this paper to form the difference map of overlaps: the intensity energy E_{ij}^{C} and gradient energy E_{ij}^{g} which are defined as:

$$\begin{cases} E_{ij}^{C} & = \frac{||I_A(i,j)-I_B(i,j)||}{\max(I_A(i,j),I_B(i,j))} \\ E_{ij}^{g} & = ||\nabla I_A(i,j) - \nabla I_B(i,j)||^2 \end{cases} \quad (4)$$

Here $I_A(i,j)$, $I_B(i,j)$, $\nabla I_A(i,j)$ and $\nabla I_B(i,j)$ are the intensity and gradient of pixel (i,j) in image A and B respectively. Finding the optimal stitching seam is an energy minimization problem (see Equation (5)) and can be converted to a binary Markov Random Field (MRF) labeling problem [37]:

$$\arg\min_{ij} \sum_{ij} (E_{ij}^{C} + \lambda E_{ij}^{g}) \tag{5}$$

To accelerate the computing process, the warped images and background images are down sampled before seam selection and restored to the original size after seam selection in our procedure. Figure 4c and d show an example of seam selection. From Figure 4c we can see, the selected seam mainly crosses flat areas with little gradients or intensity differences, thus the resulted panorama is visually consistent with no noticeable ghosting.

4. Selective Seam Updating

After initial stitching model calculation, the videos are overall pre-aligned. However, for moving foregrounds, the previous seam may lose its optimality or even miss information. Since the seam cutting algorithm requires complex computation even on the down-sampled frames, it is difficult to be used to update video frames in real-time. So we perform a change detection based seam updating method instead of real-time seam selection.

4.1. Change Detection around Previous Seams

First of all, we resize the new warped frames to a smaller scale as we did in the optimal seam selection section. Even if the images have been scaled down, direct calculation of the gradient of the two images and evaluation of the change may be expensive. However, we observed that compared with changes in non-overlapping area, those in the overlapping area are more likely to violate the optimal seam. Furthermore, only changes across the optimal seam may result in the failure of it (see Figure 5), which cam be measured by gradient difference.

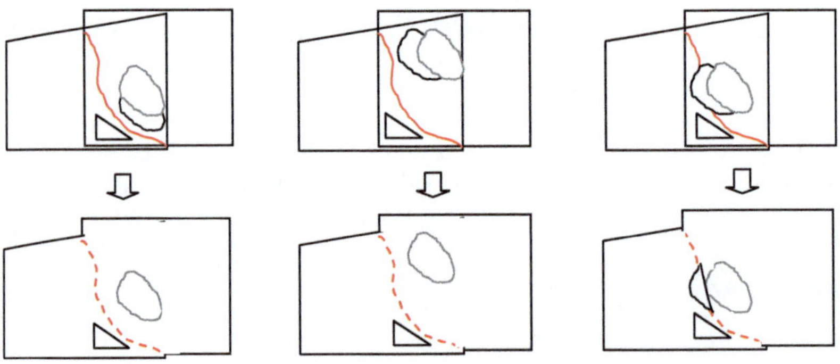

(a) (b) (c)

Figure 5. (**a**) The original aligned frame with a stitching line; (**b**) newly aligned frame with changes not cross the line; (**c**) frame with changes cross the line.

As an optimal seam is searched, the original gradient value set $\tilde{G}_0$ of the seam is stored. We set $\tilde{g}_{i0}$ as the original gradient of pixel p_i in the present seam, and $\tilde{g}_{it}$ as the gradient in time t. To calculate the number of pixels that have large gradient variations, we use the following rule to judge if changes occured at pixel p_i or not:

$$C_t = \{p_i | \frac{\tilde{g}_{it} - \tilde{g}_{i0}}{\tilde{g}_{i0}} > \delta\} \tag{6}$$

If the total pixel number in C_t is bigger than N_{cd}, we consider that there are new moving objects in the overlapping area and the optimal seam shall be updated. In Equation (6), δ is a constant chosen empirically, whereas N_{cd} depends on $\tilde{N}$, which is the total pixel number of the seam. We set it as $0.3\tilde{N}$ in our experiments.

4.2. Seam Updating

For each new warped frame, change detection, as described in the previous section, is performed to see if an alteration of the seam is needed, if so, we select a new seam by the seam selection algorithm presented in Section 3.3, otherwise, we continue to use the previous seam and move on to the image blending process.

4.3. Blending

Seam cutting can eliminate ghosting, but it provides images without overlaps which may result in noticeable seams. We expand the seamed image with a spherical dilating kernel, and use a simple weighted linear blending method [9] to blend the images.

5. Experiments

To evaluate the performance of our video stitching procedure, we conduct some experiments on both still images with parallax and actual surveillance videos.

5.1. Experimental Settings

There are several empirical parameters which should be manually tuned for different cases. At the background modeling stage, we use the initial 20 frames to construct a GMM model with 5 components for each pixel. The frame number may be set larger if the scenes are more cluttered. In our experiment, N_{min} is set as 12, which is the minimum number of matching pairs for each layer, and the source 720P image is divided into 80×45 grids for layered warping. We provide stitching results with both two, three and four channels as input.

5.2. Stitching Still Images

To evaluate the proposed layered warp algorithm, we first conduct experiments on still images with parallax and compare our results with other image stitching methods in Figures 6 and 7.

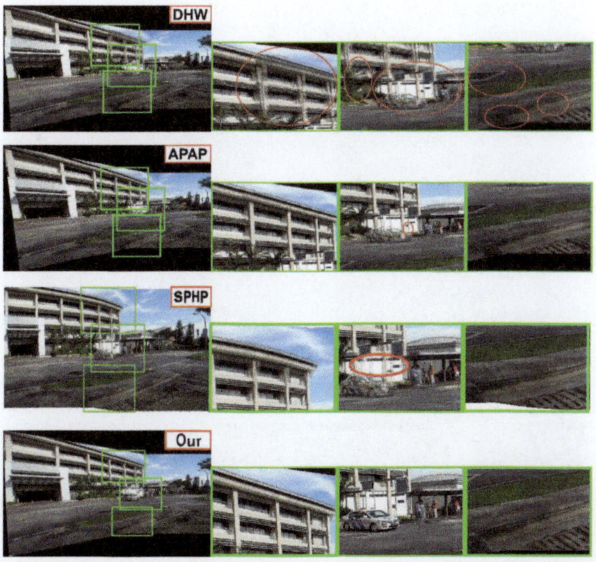

Figure 6. Comparisons among dual homography warp (DHW) [6], as-projective-as-possible (APAP) [18], Shape-Preserving Half-Projective (SPHP) [28] and Our algorithm, Red circles highlight errors.

DHW [6] only considers two layers (a distant plane and a ground plane), so it fails when the scene has multiple depth layers or contains complex objects. Our algorithm is more adequate and robust than DHW for abundant scenes (see Figure 6). Figure 7 shows the comparison among our algorithm and some state-of-art algorithms on a tough scene from [8]. As APAP tries to align two images over the whole overlapping region, it distorts the salient image structure, such as the pillar indicated by the red rectangle. SPHP also fails when the overlapping area covers most of the image and there is large parallax in it (see Figure 7c). From Figures 6 and 7 we can see, the stitching results of our algorithm contain no noticeable distortions and are more visually acceptable than other methods.

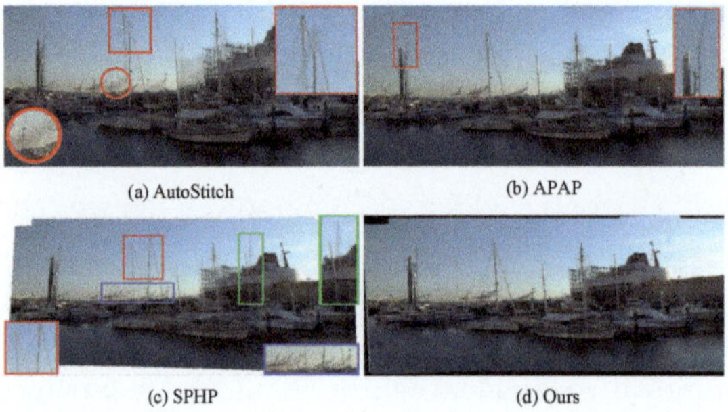

Figure 7. Comparisons among various stitching methods. colored rectangles and their zoom-in rectangles illustrate the false misalignment.

5.3. *Stitching Fixed Surveillance Videos*

To evaluate the effectiveness of selective seam updating strategy, we perform extensive experiments on fixed surveillance videos, and compare our results with that of without seam updating. The results are shown in Figure 8, from which we can see, when we do not perform seam updating in the video stitching stage, some obvious artifacts caused by moving objects would make the stitched video visually unacceptable. The comparison results indicate that the change detection based seam updating approach is rather helpful to avert ghosting and mis-alignments caused by moving targets, especially when there are moving objects across previous seams.

Figure 8. Frames taken from the output Field-of-View (FOV) videos: (**a1**–**d1**) are taken from the output videos using initialized seam without updating. (**a2**–**d2**) are taken from the output video using change-detection based seam updating approach. (**a1**), (**a2**), (**b1**) and (**b2**) are the stitching results of two channels of 720P videos. (**c1**) and (**c2**) are the stitching results of three channels of 720P videos. While for (**d1**) and (**d2**), they are the stitching results of four 720P videos.

5.4. Time Analysis

Apart from surpressing ghosting caused by parallax and moving targets, the real-time requirement of video stitching task should also be seriously considered when developing algorithms. To evaluate the speed of the proposed method, we take two videos as input and test the time consumption for different video resolutions. To make a fair comparison, the speed of stitching with temporal varying homography, layered warping without selective seam updating and layered warping with selective seam updating are all listed in Table 1. These experiments are all conducted on a PC with Intel Core i3 2.33 GHz CPU with two cores and 2GB RAM. All tested algorithms are implemented with C++.

From Table 1 we can see, as the index table has already been established according to the registration result in the stitching model calculation stage, frames can be projected to the panoramic frame by directly indexing instead of alignment and mapping for every new frame, thus the stitching efficiency of the proposed method is greatly improved. Table 1 also shows that the selective seam updating process slows down the frame stitching process to some extent.

Table 1. Comparisons of the speed among different video stitching methods on input-videos with different resolutions.

Resolution	Stitching with Temporal Varying Homography	Proposed Algorithm without Seam Updating	Proposed Algorithm with Seam Updating
720p: 1280×720	4.764 s	0.051 s	0.083 s
480P: 720×480	2.326 s	0.035 s	0.045 s
CIF: 352×288	1.025 s	0.021 s	0.032 s

6. Conclusions

This paper presents an efficient parallax-robust stitching algorithm for fixed surveillance videos. The proposed method consists of two parts: alignment work done at the stitching model calculation stage and change detection based frame updating at the video stitching stage. The algorithm uses layered warping method to pre-align the background scene which is robust to scene parallax. As for each new frame, an efficient change detection based seam updating method is adopted to avert ghosting and artifacts caused by moving foregrounds. Thus, the algorithm can provide good stitching performance with no ghosting and artifacts for dynamic scenes efficiently.

Author Contributions: The work was realized with the collaboration of all of the authors. Botao He and Shaohua Yu contributed to the algorithm design and code implementation. Botao He contributed to the data collection and baseline algorithm implementation and drafted the early version of the manuscript. Shaohua Yu organized the work and provided the funding, supervised the research and critically reviewed the draft of the paper. All authors discussed the results and implications, commented on the manuscript at all stages and approved the final version.

Conflicts of Interest: The authors declare no conflict of interest.

References

1. Brown, M.; Lowe, D. Recognising Panoramas. In Proceedings of the Ninth IEEE International Conference on Computer Vision, Nice, France, 13–16 October 2003; pp. 1218–1225.
2. Szeliski, R. Video mosaics for virtual environments. *IEEE Comput. Graph. Appl.* **1996**, *16*, 22–30.
3. Eden, A.; Uyttendaele, M.; Szeliski, R. Seamless image stitching of scenes with large motions and exposure differences. *Comput. Vis. Pattern Recognit.* **2006**, doi:10.1109/CVPR.2006.268.
4. Levin, A.; Zomet, A.; Peleg, S.; Weiss, Y. *Seamless Image Stitching in the Gradient Domain;* Springer Berlin Heidelberg: Berlin, Germany, 2004; pp. 377–389.
5. Mills, A.; Dudek, G. Image stitching with dynamic elements. *Image Vis. Comput.* **2009**, *27*, 1593–1602.

6. Gao, J.; Kim, S.J.; Brown, M.S. Constructing Image Panoramas Using Dual-Homography Warping. In Proceedings of the EEE Conference on Computer Vision and Pattern Recognition (CVPR), Providence, RI, USA, 20–25 June 2011; pp. 49–56.
7. Zaragoza, J.; Chin, T.J.; Tran, Q.H.; Brown, M.S.; Suter, D. As-projective-as-possible image stitching with moving DLT. *IEEE Trans. Pattern Anal. Mach. Intell.* **2014**, *36*, 1285–1298.
8. Zhang, F.; Liu, F. Parallax-Tolerant Image Stitching. In Proceedings of the IEEE Conference on Computer Vision and Pattern Recognition (CVPR), Columbus, OH, USA, 23–28 June 2014; pp. 3262–3269.
9. Szeliski, R. Image alignment and stitching: A tutorial. *Found. Trends Comput. Graph. Vis.* **2006**, *2*, 1–104.
10. Agarwala, A.; Dontcheva, M.; Agrawala, M.; Drucker, S.; Colburn, A.; Curless, B.; Salesin, D.; Cohen, M. Interactive digital photomontage. *ACM Trans. Graph.* **2004**, *23*, 294–302.
11. Hu, J.; Zhang, D.Q.; Yu, H.; Chen, C.W. Discontinuous Seam Cutting for Enhanced Video Stitching. In Proceedings of the IEEE International Conference on Multimedia and Expo (ICME), 29 June–3 July 2015; pp. 1–6.
12. Kwatra, V.; Schödl, A.; Essa, I.; Turk, G.; Bobick, A. Graphcut textures: Image and video synthesis using graph cuts. *ACM Trans. Graph.* **2003**, *22*, 277–286.
13. Steedly, D.; Pal, C.; Szeliski, R. Efficiently Registering Video into Panoramic Mosaics. In Proceedings of the Tenth IEEE International Conference on Computer Vision, Beijing, China, 17–21 October 2005; pp. 1300–1307.
14. Hsu, C.T.; Tsan, Y.C. Mosaics of video sequences with moving objects. *Signal Process. Image Comm.* **2004**, *19*, 81–98.
15. El-Saban, M.; Izz, M.; Kaheel, A. Fast Stitching of Videos Captured from Freely Moving Devices by Exploiting Temporal Redundancy. In Proceedings of the 17th IEEE International Conference on Image Processing (ICIP), Hong Kong, China, 26–29 September 2010; pp. 1193–1196.
16. El-Saban, M.; Izz, M.; Kaheel, A.; Refaat, M. Improved Optimal Seam Selection Blending for Fast Video Stitching of Videos Captured from Freely Moving Devices. In Proceedings of 18th IEEE International Conference on the Image Processing (ICIP), Brussels, Belgium, 11–14 September 2011; pp. 1481–1484.
17. Okumura, K.I.; Raut, S.; Gu, Q.; Aoyama, T.; Takaki, T.; Ishii, I. Real-Time Feature-Based Video Mosaicing at 500 fps. In Proceedings of the IEEE/RSJ International Conference on Intelligent Robots and Systems (IROS), Tokyo, Japan, 3–7 November 2013; pp. 2665–2670.
18. Au, A.; Liang, J. Ztitch: A Mobile Phone Application for Immersive Panorama Creation, Navigation, and Social Sharing. In Proceedings of the IEEE 14th International Workshop on Multimedia Signal (MMSP), Banff, AB, USA, 17–19 September 2012; pp. 13–18.
19. Xiao, J.; Shah, M. Layer-based video registration. *Mach. Vis. Appl.* **2005**, *16*, 75–84.
20. Liu, H.; Tang, C.; Wu, S.; Wang, H. Real-Time Video Surveillance for Large Scenes. In Proceedings of the International Conference on Wireless Communications and Signal Processing (WCSP), Nanjing, China, 9–11 November 2011; pp. 1–4.
21. Zeng, W.; Zhang, H. Depth Adaptive Video Stitching. In Proceedings of the Eighth IEEE/ACIS International Conference on Computer and Information Science, Shanghai, China, 1–3 June 2009; pp. 1100–1105.
22. Brown, M.; Lowe, D.G. Automatic panoramic image stitching using invariant features. *Int. J. Comput. Vis.* **2007**, *74*, 59–73.
23. AutoStitch. Available online: http://cvlab.epfl.ch/ brown/autostitch/autostitch.html (accessed on 18 December 2015).
24. Microsoft Image Composite Editor. Available online: http://research.microsoft.com/en-us/um/redmond/groups/ivm/ICE/ (accessed on 18 December 2015).
25. Adobe Photoshop CS5. Available online: http://www.adobe.com/products/photoshop (accessed on 18 December 2015).
26. Burt, P.J.; Adelson, E.H. A multiresolution spline with application to image mosaics. *ACM Trans. Graph.* **1983**, *2*, 217–236.
27. Pérez, P.; Gangnet, M.; Blake, A. Poisson image editing. *ACM Trans. Graph.* **2003**, *22*, 313–318.
28. Chang, C.H.; Sato, Y.; Chuang, Y.Y. Shape-Preserving Half-Projective Warps for Image Stitching. In Proceedings of the IEEE Conference on Computer Vision and Pattern Recognition (CVPR), Columbus, OH, USA, 23–28 June 2014; pp. 3254–3261.

29. Alexa, M.; Behr, J.; Cohen-Or, D.; Fleishman, S.; Levin, D.; Silva, C.T. Computing and rendering point set surfaces. *IEEE Trans. Vis. Comput. Graph.* **2003**, *9*, 3–15.
30. Tennoe, M.; Helgedagsrud, E.; Næss, M.; Alstad, H.K.; Stensland, H.K.; Gaddam, V.R.; Johansen, D.; Griwodz, C.; Halvorsen, P. Efficient Implementation and Processing of a Real-Time Panorama Video Pipeline. In Proceedings of the IEEE International Symposium on Multimedia (ISM), Anaheim, CA, USA, 9–11 December 2013; pp. 76–83.
31. Kumar, P.; Dick, A.; Brooks, M.J. Integrated Bayesian Multi-Cue Tracker for Objects Observed from Moving Cameras. In Proceedings of the 23rd International Conference on Image and Vision Computing New Zealand, Christchurch, New Zealand, 26–28 November 2008; pp. 1–6.
32. Kumar, P.; Dick, A.; Brooks, M.J. Multiple Target Tracking with an Efficient Compact Colour Correlogram. In Proceedings of the 10th International Conference on Control, Automation, Robotics and Vision, Hanoi, Vietnam, 17–20 December 2008; pp. 699–704.
33. Zivkovic, Z. Improved adaptive Gaussian mixture model for background subtraction. *Proc. Int. Conf. ICPR Pattern Recognit.* **2004**, *2*, 28–31.
34. Kumar, P.; Ranganath, S.; Huang, W. Queue Based Fast Background Modelling and Fast Hysteresis Thresholding for Better Foreground Segmentation. In Proceedings of the Fourth Pacific Rim Conference on Multimedia Information, Communications and Signal, Singapore, 15–18 December 2003; pp. 743–747.
35. Beis, J.S.; Lowe, D.G. Shape Indexing Using Approximate Nearest-Neighbour Search in High-Dimensional Spaces. In Proceedings of the IEEE Computer Society Conference on Computer Vision and Pattern Recognition, San Juan, Territory, 17–19 June 1997; pp. 1000–1006.
36. Fischler, M.A.; Bolles, R.C. Random sample consensus: A paradigm for model fitting with applications to image analysis and automated cartography. *Comm. ACM* **1981**, *24*, 381–395.
37. Szeliski, R.; Zabih, R.; Scharstein, D.; Veksler, O.; Kolmogorov, V.; Agarwala, A.; Tappen, M.; Rother, C. A comparative study of energy minimization methods for markov random fields with smoothness-based priors. *IEEE Trans. Pattern Anal. Mach. Intell.* **2008**, *30*, 1068–1080.

3

Robust Foot Clearance Estimation Based on the Integration of Foot-Mounted IMU Acceleration Data

Mourad Benoussaad [1,*,†], Benoît Sijobert [1], Katja Mombaur [2] and Christine Azevedo Coste [1]

Academic Editor: Gert F. Trommer

1 INRIA, LIRMM, Rue Saint Priest, Montpellier 34000, France; benoit.sijobert@inria.fr (B.S.); christine.azevedo@inria.fr (C.A.C.)
2 ORB, IWR, University of Heidelberg, Heidelberg 69117, Germany; katja.mombaur@iwr.uni-heidelberg.de
* Correspondence: mourad.benoussaad@enit.fr
† Current Address: LGP Lab, Tarbes 65000, France

Abstract: This paper introduces a method for the robust estimation of foot clearance during walking, using a single inertial measurement unit (IMU) placed on the subject's foot. The proposed solution is based on double integration and drift cancellation of foot acceleration signals. The method is insensitive to misalignment of IMU axes with respect to foot axes. Details are provided regarding calibration and signal processing procedures. Experimental validation was performed on 10 healthy subjects under three walking conditions: normal, fast and with obstacles. Foot clearance estimation results were compared to measurements from an optical motion capture system. The mean error between them is significantly less than 15% under the various walking conditions.

Keywords: foot clearance estimation; inertial measurement unit; 3D transformation; driftless integration

1. Introduction

Foot clearance, defined as the foot elevation during the stride phase, is an important indicator of gait quality and safety in subjects with disabilities [1]. The minimum foot clearance is one of common gait deviations in people with stroke, which increase the risk of unsuccessful foot clearance (trips) and then of falls [2,3]. In a previous study, we showed that this parameter could be used to adapt the applied functional electrical stimulation in the context of drop foot correction in subjects with post-stroke hemiplegia [4]. In the present study, foot clearance (FC) is defined as the continuous heel height relative to the ground, whereas in other previous studies, maximum or minimum FC values during the stride phase were used instead [5].

Optical motion capture systems (OMCS) are the most commonly-used devices for measuring foot elevation during walking, providing high accuracy and direct measurement of the position of reflective markers located on body landmarks. However, these devices present multiple drawbacks, such as limited operating space and high cost, making their clinical use technically and financially difficult to implement.

The use of wearable devices, such as inertial measurement units (IMU) presents a promising alternative for gait analysis in outdoor and clinical contexts [6,7]. Such devices typically include three-axis accelerometers, three-axis gyroscopes and three-axis magnetometer sensors.

Raw data provided by IMU sensors requires complex post-processing, creating a source of errors in the estimation of IMU displacement and orientation. A major IMU raw data processing issue is the integration of acceleration signals and angular velocity signals (provided by gyroscopes). It is well known that the numerical integration of such data suffers from a high drift error rate, introducing significant errors in the estimation of IMU position and orientation. Magnetometers, on the other hand, provide an absolute orientation and do not require numerical integration. However, such systems can be disturbed by ferromagnetic materials, thus making this technique not reliable everywhere.

In the past few years, several teams have focused on the use and processing of IMU data for human motion analysis. Some authors have estimated IMU orientation based only on gyroscope data [8,9], or in combination with acceleration data [10], or acceleration and magnetometer data [10,11], using data fusion algorithms to overcome integration drift problems. To achieve driftless integration using gyroscope data alone, a weighted Fourier linear combiner (WFLC) algorithm [8] and an empirical mode decomposition (EMD) method [9] have been proposed for estimating the 3D orientation of the lower trunk during walking. Other authors have developed human biomechanical models to estimate joint angles [12] or walking parameters, such as stride length [13], based on IMU orientation estimates.

Only a few authors have focused on the estimation of IMU position or displacement based on acceleration data, mainly because such data require not single, but double integration, which makes the integration drift problem even more significant. In an effort to overcome this issue, an extended Kalman filter has been applied to fuse inertial (accelerometer and gyroscope) sensor data and magnetic sensor data for body position and orientation tracking [14].

Acceleration-based estimation of foot displacement during walking appears to be very complex due to the high accelerations involved. Therefore, only a few authors have investigated this approach. A simple biomechanical foot model (heel-toe distance) has been used to estimate maximum heel clearance and minimum toe clearance values with a mean error of approximately 4.1 cm and 1.3 cm, respectively [15]. These results were obtained with 12 healthy participants.

Only a few techniques not involving the use of a biomechanical foot model have been explored to avoid increased drift errors during walking [5,16–20]. Such a method has been applied to estimate maximum foot clearance values from double integration of acceleration data [16]. These authors used an optimally-filtered direct and reverse integration (OFDRI) technique based on frequency domain analysis of signals to be integrated [21]. Another published method is based on time domain analysis and the application of the zero velocity update (ZVU) principle. This method relies on the detection of foot-flat periods (*i.e.*, foot in contact with the ground) to set the velocity to a known constant value [17] or zero [18]. Such events can be detected using a foot-switch sensor [22] or based on the analysis of accelerometer and gyroscope data [23,24].

Based on the ZVU principle, integrated acceleration can be corrected using so-called "de-drifted" integration techniques [5,17]. Linear de-drifted integration has been applied in the case of five participants walking on a treadmill [17], whereas de-drifted integration based on a p-chip interpolation function has been used to estimate maximum foot clearance values with a mean error of approximately 1.9 cm [5]. ZVU-based integration correction based on extended Kalman filtering has also been proposed [17,18].

According to another proposed ZVU-based correction method [19,20], the sensor error was modeled as a constant acceleration bias in the sensor coordinate system and corrected backwards based on the estimated error observed at each detected event. In these two studies, ZVU-based correction was applied to estimate driftless velocities and thereby calculate the walking speed over the stride. However, displacements were not estimated.

All of these cited works do not go further to get the foot clearance, directly integrated the velocity without any correction or estimated only the maximal foot clearance instead of a curve of foot clearance, which is required in our work. In the present study, we focused on the double integration of foot acceleration data from a foot-mounted IMU. We therefore used the ZVU principle to limit the increase in integration drift between two successive strides and then applied the above-mentioned acceleration correction method [19,20] to each stride. For displacement estimation purposes (particularly foot clearance estimation), we adapted the same principle to vertical velocity integration and correction, by generalizing to a zero displacement update (ZDU) principle, where walking is assumed as occurring on flat and horizontal ground. In accordance with our previous work and with clinical context requirements [4], our goal was to obtain a foot clearance estimation error of less than 2 cm (minimum foot clearance [3]) and to make our method robust to misalignment of IMU axes and foot axes. The proposed IMU-based foot clearance estimation algorithm is detailed in Section 2 below. Experimental setup and validation methods are described in Section 3. Estimation results and the performance evaluation are presented and discussed in Section 4. Conclusions and future prospects are given in Section 5.

2. IMU-Based Foot Clearance Estimation

Figure 1 shows sagittal and frontal views of IMU placement for foot clearance estimation (FCE). Possible misalignment of the IMU with respect to the vertical axis is also illustrated (Figure 1b). As mentioned earlier, we defined foot clearance as the continuous heel clearance (Figure 1a). Estimation of ankle joint clearance (where the IMU is placed) is adequate, since there is only one bias to be added, corresponding to the distance between IMU-based estimated clearance and heel clearance. Moreover, this ankle joint clearance is most appropriate in the context of drop foot correction [4].

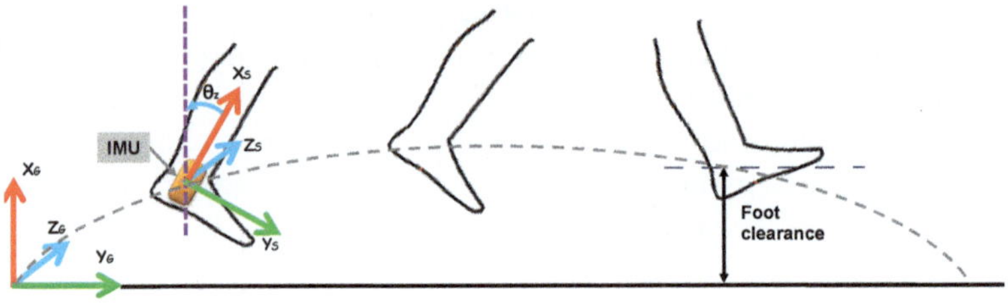

(**a**)

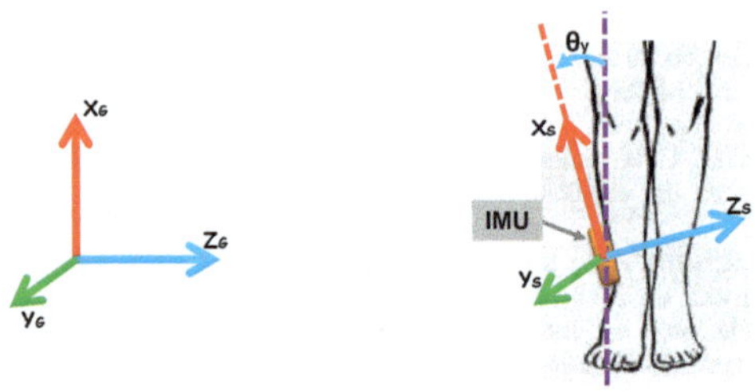

(**b**)

Figure 1. IMU placement and frame transformation in (**a**) the sagittal view and (**b**) the frontal view.

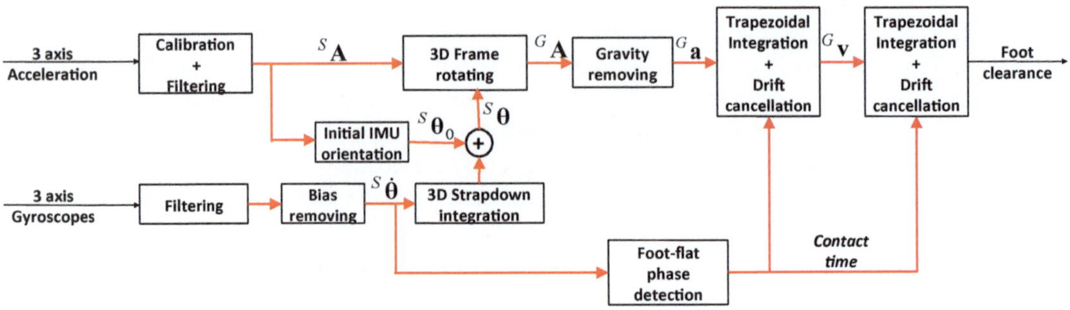

Figure 2. IMU-based foot clearance estimation algorithm.

The proposed FCE algorithm (illustrated in Figure 2) uses only acceleration in the global frame. However, since acceleration is measured in the sensor frame, gyroscope-based angular rate data are used to estimate foot acceleration in the global frame. Each step of the method is detailed below.

2.1. Accelerometer Calibration

An efficient calibration procedure is essential for ensuring accurate accelerometer-based acceleration measurement. The proposed calibration procedure is based on a previous study [25] and consists of optimizing the three-axis accelerometer output, such that, under stationary conditions, the norm of the acceleration vector is equal to the norm of the gravity vector (*i.e.*, 9.81 m/s^2). For this purpose, the IMU sensor was placed in N (>20) different random orientations, and each corresponding acceleration vector was measured under static conditions. The mathematical model of calibrated accelerometer output is described as follows [25]:

$$\mathbf{A} = \mathbf{S}(\mathbf{A}_{\mathbf{m}} - \mathbf{B}) \tag{1}$$

where $\mathbf{A}$ is the calibrated acceleration, $\mathbf{A}_{\mathbf{m}}$ is the measured acceleration prior to calibration and $\mathbf{S}$ and $\mathbf{B}$ are, respectively, the scale matrix and bias vector identified during the calibration procedure. $\mathbf{S}$ and $\mathbf{B}$ were identified using a Newton iterative optimization algorithm that minimizes quadratic errors between the norm of the calibrated acceleration vector and the norm of the gravity vector for each static orientation. This calibration was performed once, and the resulting calibration matrices ($\mathbf{S}$, $\mathbf{B}$) were used to correct measured acceleration data (Figure 2) using Equation (1).

2.2. Data Filtering and Bias Removing

Since IMU data were post-processed offline, there were no real-time constraints. We therefore chose a data acquisition frequency of 200 Hz. However, frequency sampling should be optimized for future real-time applications, so as to ensure sufficient measurement quality and compliance with real-time constraints.

Three-axis gyroscope data were band-pass filtered between 0.001 Hz and 5 Hz using a first-order Butterworth filter to remove constant and high frequency components. The angular rate in the stationary state (subjects were asked to remain static at the start of each trial) was considered as a bias and, therefore, automatically detected, quantified and removed from the entire gyroscope signal. For three-axis acceleration data, a first-order low-pass Butterworth filter with a 10-Hz cutoff frequency was applied to remove high-frequency noise. However, the bias was not subtracted from the acceleration signal at this stage, since it corresponds to the gravity acceleration, which changes according to IMU orientation.

2.3. Gyroscopic Data Integration

As illustrated in Figure 2, processing of gyroscopic data is crucial for processing of gravity-free acceleration (due to the movement only), since the measured acceleration also includes the gravity effect. Therefore, proper processing and integration of gyroscopic data was required to expect proper acceleration-based displacement estimation (Section 2.4).

The filtered angular rate ${}^{S}\dot{\theta}$ (see Figure 2) was then integrated using a 3D strapdown integration method [11,26]. This integration method has the advantage of considering rotations around all three axes, since IMU rotation exclusively in the sagittal plane cannot be ensured during walking. This is based on a recursive equation (Equation (2)) that updates the 3D orientation matrix of the IMU:

$${}^{G}\mathbf{R}_i = {}^{G}\mathbf{R}_{i-1} + {}^{G}\mathbf{R}_{i-1}[\Delta t \cdot {}^{S}\dot{\boldsymbol{\theta}}_i \times] \tag{2}$$

where ${}^{G}\mathbf{R}_i$ is the 3D orientation matrix in the global frame at sample i, the initial transformation matrix ${}^{G}\mathbf{R}_0$ is the identity matrix (I), Δt is the time interval between two successive samples ($i-1, i$) and ${}^{S}\dot{\boldsymbol{\theta}}_i$ is the angular rate vector at sample i. $[\mathbf{a}\times]$ is the skew-symmetric matrix representing the cross-product operator, such that:

$$[\mathbf{a}\times] = \begin{bmatrix} 0 & -a_z & a_y \\ a_z & 0 & -a_x \\ -a_y & a_x & 0 \end{bmatrix} \quad (3)$$

where a_x, a_y, a_z are the elements of $\mathbf{a}$.

At each time instant, the transformation matrix ${}^{G}\mathbf{R}_i$ represents the IMU frame transformation with respect to the initial IMU configuration. In practice, IMU orientation is never absolutely vertical in the initial position, and as a result, the initial frame is not the global frame. Therefore, in order to define the transformation matrix with respect to a global frame independently of initial placement, we made the initial matrix ${}^{G}\mathbf{R}_0$ equal to a transformation matrix between the initial IMU position and the global frame (vertical orientation).

This initial transformation matrix consists of a rotation about the Z-axis (pitch) with an angle of θ_{z0}, followed by another rotation about the Y-axis (roll) with an angle of θ_{y0}, where the sign of rotation is as shown in Figure 1. This initial transformation matrix is therefore defined as follows:

$${}^{G}\mathbf{R}_0 = \begin{bmatrix} cos(\theta_{z0}) & sin(\theta_{z0}) & 0 \\ -sin(\theta_{z0}) & cos(\theta_{z0}) & 0 \\ 0 & 0 & 1 \end{bmatrix} \cdot \begin{bmatrix} cos(\theta_{y0}) & 0 & sin(\theta_{y0}) \\ 0 & 1 & 0 \\ -sin(\theta_{y0}) & 0 & cos(\theta_{y0}) \end{bmatrix} \quad (4)$$

where θ_{z0} and θ_{y0} were estimated based on the projection of the gravity vector on the three axes [11], measured by accelerometers during the stationary period, *i.e.*, prior to the start of movement. Considering the order of rotation defined above (Equation (4)), these initial angles (tilt angles) were calculated using the following equations:

$$\theta_{z0} = \arctan\left(\frac{{}^{S}G_{Y}}{{}^{S}G_{X}}\right) \qquad \theta_{y0} = \arctan\left(\frac{{}^{S}G_{Z}}{\sqrt{{}^{S}G_{X}^{2} +{}^{S} G_{Y}^{2}}}\right) \quad (5)$$

where ${}^{S}G = \left({}^{S}G_X \quad {}^{S}G_Y \quad {}^{S}G_Z\right)$ is the gravity vector measured on the three IMU axes during the stationary period. Based on our measurements, the drift in gyroscopic data integration was neglected, and as a result, no drift cancellation was required at this level.

2.4. Displacement Estimation

The proposed displacement estimation method is based on double integration of foot acceleration data on the global frame axes. To achieve this objective and thereby overcome the integration drift problem, the following steps are applied.

2.4.1. 3D Frame Transformation and Gravity Removing

It is well known that IMU-based acceleration measurements correspond to the acceleration in the sensor frame, whereas the useful acceleration should be the acceleration with respect the global frame. Moreover, the measured acceleration includes gravitational acceleration, which should be removed to obtain gravity-free acceleration due to movement only. Therefore, the acceleration in the sensor frame ${}^{S}\mathbf{A}$ was first transformed to the global frame (${}^{G}\mathbf{A}$ in Figure 2) using the strapdown transformation matrix (Equation (2)):

$${}^{G}\mathbf{A} ={}^{G} \mathbf{R}_i *{}^{S} \mathbf{A} \quad (6)$$

Gravitational acceleration was then removed to obtain the gravity-free acceleration in the global frame ${}^{G}\mathbf{a}$:

$$ {}^{G}\mathbf{a} = {}^{G}\mathbf{A} - \begin{pmatrix} 9.81 \\ 0 \\ 0 \end{pmatrix} \tag{7} $$

To obtain the vertical foot displacement (foot clearance), the vertical component of the gravity-free acceleration ${}^{G}\mathbf{a}$ needs to be integrated twice (Equation (7)).

2.4.2. Foot-Flat Phase Detection

One of the main problems with the integration of acceleration data is the significant integration drift, which is further increased by double integration. The proposed solution to overcome this problem is an adaptation of the ZVU principle [27]. The proposed ZDU principle consists of detecting when the foot is flat on the ground and setting the vertical velocity and displacement to zero. These particular events are detected by the foot-flat phase detection algorithm (illustrated in Figure 2), which uses the angular rate about the Z-axis (${}^{S}\dot{\theta}_z$) to automatically detect when the foot is flat on the ground [20,28]. Such events are detected based on the detection of the minimum angular velocity at each stride cycle. This foot-flat phase corresponds to the period of time between heel-strike and heel-off.

2.4.3. Drift Cancellation

The foot-flat phase detection algorithm segments walking into several strides. This segmentation avoids the accumulation of drift error between different strides. However, at the end of each stride, a local drift (occurring only during a current stride) appears as an error between integrated and theoretical data, corresponding to the foot-flat phase. In the proposed method, we applied a drift cancellation on the vertical foot velocity for each stride, based on the error accumulated at the end of each stride and the zero vertical velocity assumption. This correction used the following model [19]:

$$ V_{x\_corrected}(t) = V_x(t) - \frac{V_x(T)}{T}t \tag{8} $$

where $V_x(t)$ is the vertical foot velocity obtained using trapezoidal integration of gravity-free acceleration on the X-axis (${}^{G}a_x$), $V_x(T)$ is the calculated velocity at the end of the stride phase, t is the time instant of the current stride, T is the entire duration of the current stride and $V_{x\_corrected}(t)$ is the vertical foot velocity after correction for the entire stride phase.

To estimate the vertical foot displacement (foot clearance), we first integrated the driftless corrected vertical velocity ($V_{x\_corrected}$) and then applied the same drift cancellation on this vertical foot displacement, assuming zero foot clearance at the end of the stride (foot-flat phase). This second correction used the following model:

$$ D_{x\_corrected}(t) = D_x(t) - \frac{D_x(T)}{T}t \tag{9} $$

where $D_x(t)$ is the vertical foot displacement obtained by trapezoidal integration of the corrected velocity $V_{x\_corrected}(t)$, $D_x(T)$ is the calculated displacement at the end of the stride phase and $D_{x\_corrected}(t)$ is the vertical foot displacement after correction during the stride phase, corresponding to the estimated foot clearance.

Since the drift cancellation was applied during the swing phase, the foot clearance was set to zero during the foot-flat phase in agreement with our assumptions.

3. Experimental Validation

For validation of our foot clearance estimation method, an experimental procedure was applied on both feet of ten healthy participants whose characteristics are summarized in Table 1 below.

Table 1. Characteristics of study participants.

Subject Number	Age (Years)	Height (m)	Sex (M/F)
1	35	1.85	M
2	24	1.89	M
3	40	1.60	F
4	23	1.55	F
5	23	1.71	F
6	29	1.93	M
7	26	1.72	M
8	26	1.62	F
9	24	1.76	M
10	23	1.58	F

Two IMU sensors (HikoB Fox©Villeurbanne, France) were strapped to both feet of each subject (at the ankle joint), as shown in Figure 3.

Each subject stood upright, remained still for a few seconds and then walked a distance of approximately 10 m on flat horizontal ground. Each subject covered this distance four times walking normally, four times walking fast and two times crossing obstacles. The fast walk was to test a higher level of acceleration and then to increase the drift level. The purpose of walking with obstacles was to test the estimation of a higher level of clearance and nonsymmetrical and nonperiodic walking. The obstacle heights range from 20 cm to 50 cm and were placed to influence normal walking, increase the foot clearance and test the robustness of our estimation method. As illustrated in Figure 3, foot clearance estimation was validated with a BONITA VICON motion capture system, using reflective markers located on the IMUs.

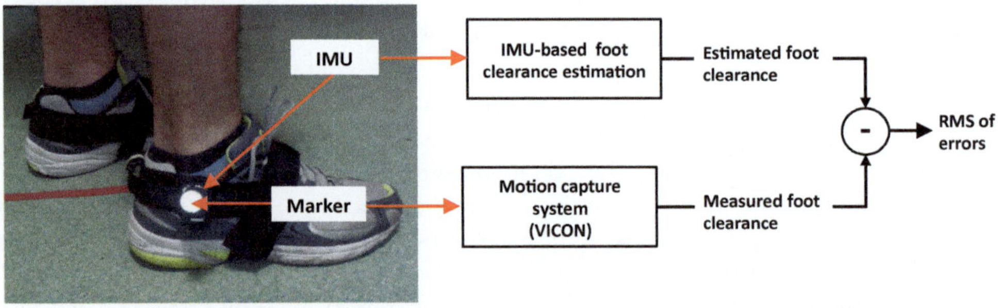

Figure 3. Experimental setup and validation of IMU-based foot clearance estimation.

Measured and estimated peak foot clearance values were automatically detected and used to synchronize the two datasets. Foot clearance data for the entire movement, measured with the motion capture system and estimated using the proposed method, was then compared (Figure 3). Subsequently, the RMS value of foot clearance errors ($RMSE$) was quantified for the evaluation of foot clearance estimation performance, using the following criterion:

$$RMSE = \sqrt{\frac{1}{N}\sum_{i=1}^{N}(X_m - X_e)^2} \tag{10}$$

where X_m and X_e are respectively the measured and estimated foot clearance, discretized to Nsamples throughout the walking cycle duration. This number of samples (N) was associated with the walking cycle duration and discretization (sampling) frequency, *i.e.*, approximately 200 Hz for both VICON and IMU signals.

To compare the foot clearance estimation performance under different walking conditions, it was necessary to calculate another error criterion, since foot clearance values obtained during walking with obstacles were three-times higher than during normal walking. A normalized RMSE percentage value was therefore calculated for each subject and walking type by normalizing the RMSE with respect to the maximum foot clearance during the current walking test as follows:

$$NRMSE = \frac{RMSE}{max(X_m)} \times 100 \tag{11}$$

4. Presentation and Discussion of Results

4.1. Results

Figure 4 shows a typical set of foot clearance measurements obtained with a VICON motion capture system and compared to estimated values generated by the proposed algorithm (Section 2). Normal and fast walking data (Figure 4a,b) was obtained on the left and right foot, respectively, of Subject 1 (35 years old, 1.85 m), and walking-with-obstacles data (Figure 4c) were obtained on the right foot of Subject 10 (23 years old, 1.58 m).

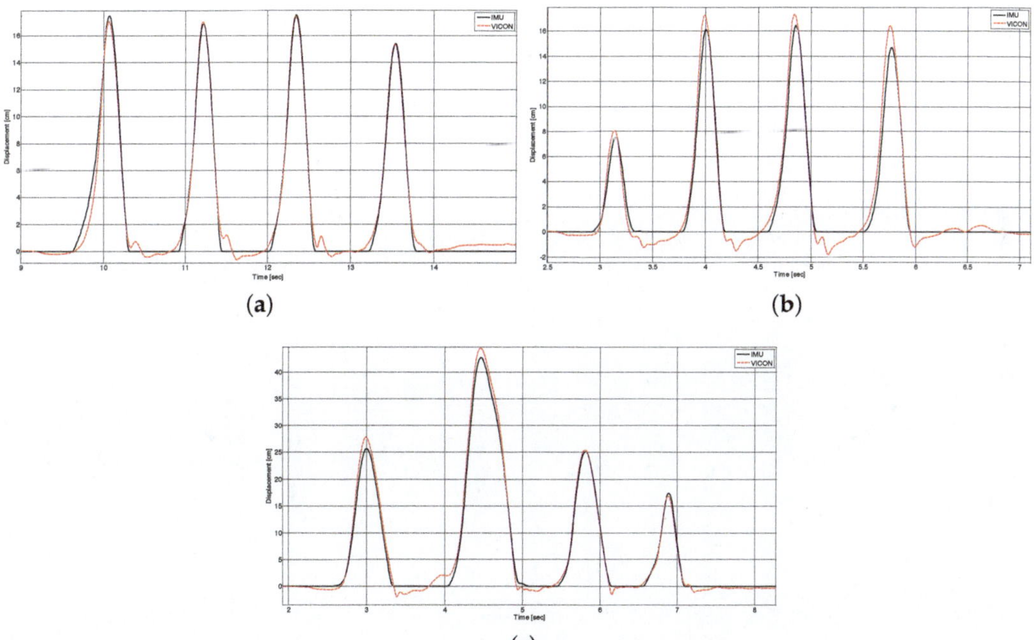

(**a**) (**b**) (**c**)

Figure 4. Foot clearance values measured by the VICON system and estimated by the IMU during: (**a**) Normal walking (Subject 1, left foot); (**b**) Fast walking (Subject 1, right foot); and (**c**) Walking with obstacles (Subject 10, right foot).

Foot clearance measurements obtained with the VICON system and estimated using the IMU-based algorithm for both feet of all 10 participating subjects are shown as histograms in Figure 5.

These three histograms correspond to results obtained during normal walking (Figure 5a), fast walking (Figure 5b) and walking with obstacles (Figure 5c). In each histogram, the height of each bar corresponds to the maximum foot clearance measured with the VICON system on each foot of each subject. The bracket centered at the top of each bar represents the RMS value of foot clearance errors (Equation (10)) between VICON measurements and the proposed algorithm.

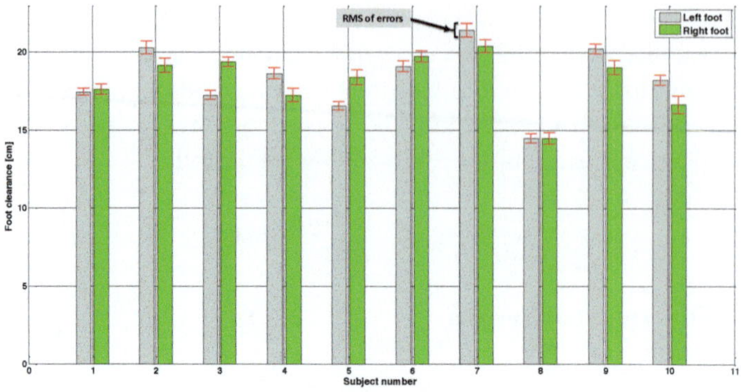

(a)

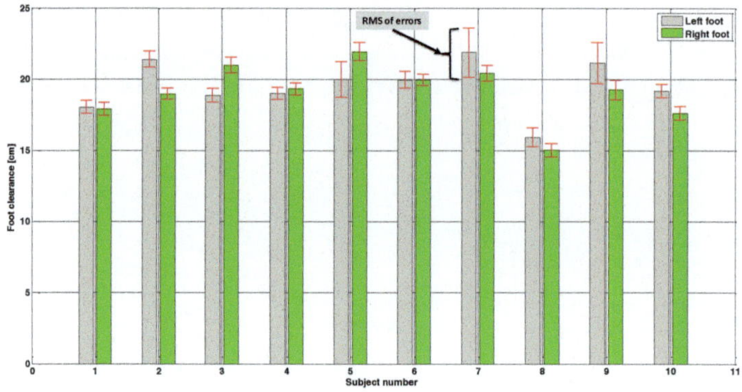

(b)

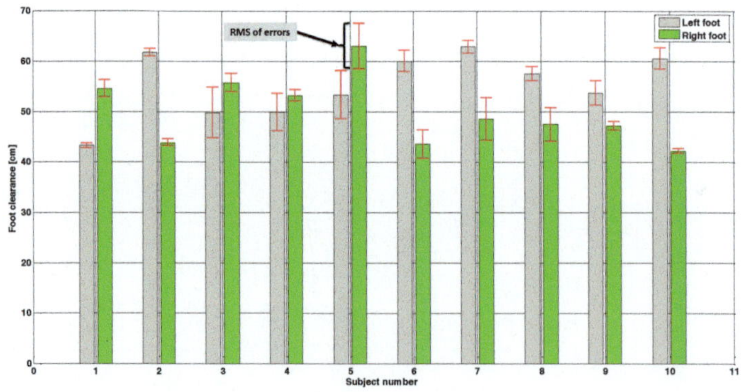

(c)

Figure 5. Maximum foot clearance and RMSE between VICON measurements and IMU-based estimates for both feet of each subject during: (**a**) Normal walking; (**b**) Fast walking; and (**c**) Walking with obstacles.

The individual step error rate between VICON measurements and IMU-based foot height estimates was analyzed using a Student *t*-test. Benjamini–Hochberg correction was applied to limit the false positive rate. The mean error is significantly (p-value < 0.05) lower than 15% under the various walking conditions. A Spearman test showed that walking speed affects the error-rate.

The results shown in Figure 4 exhibit very good correspondence between foot clearance values estimated with the proposed method and those measured with the VICON motion capture system. In these particular examples, the good results obtained are illustrated by a low RMS of foot clearance estimation errors, *i.e.*, approximately 0.35 cm during normal walking (Figure 4a), approximately 0.62 cm during fast walking (Figure 4b) and approximately 0.89 cm during walking with obstacles (Figure 4c).

Results obtained in all subjects and summarized in Figure 5 show on the whole a good estimation of foot clearance compared to a measured one with VICON motion capture, which is considered as a reference due to its high accuracy. Indeed, the average of all normalized errors (NRMSE) is approximately 6.5%, and in 96% of all experiment cases, these normalized errors are less than 15%. In addition, in 83% of all experiment cases, these normalized errors are less than 10%, which correspond to 2 cm if the maximum foot clearance is approximately 20 cm.

The results obtained during the normal walking experiments, summarized by the histogram in Figure 5a, show that all RMS values of foot clearance estimation errors are less than 1.13 cm, whereas the average value of all of these errors is approximately 0.74 cm. This shows very good foot clearance estimation using the proposed method during normal walking for various subjects with different heights.

The estimation results declined during the fast walking experiments (Figure 5b), since the average of RMS errors increase to 1.3 cm, whereas there maximum value is approximately 3.5 cm (left foot, Subject 7). However, even in these experiment sets, the results are good, since in 85% of the experiments, the RMS errors remained less than 1.5 cm with an average of RMS errors around 1 cm.

Walking with obstacles is experimented on here to test the limits and the strength of our foot clearance estimation method. The results of these experiments for all subjects, summarized in Figure 5c, show larger RMS values of foot clearance estimation, which reached 10 cm (right foot, Subject 3). However, these results in cm cannot be compared to the previous results, since the maximum foot clearances are not comparable. Therefore, normalized RMS errors (Equation (11)) are used hereinafter for the purposes of comparison and discussion.

Table 2 summarizes the subject average speed during each walking type using VICON data measurements. This brings numerical information about speed and, thereby, acceleration levels involved during each walking type.

Table 2. Subject average speed during each walking type.

Subject Number	Walking Speed (m/s)		
	Normal	Fast	With Obstacles
1	1.05	1.71	1.07
2	1.21	1.83	1.09
3	0.95	1.66	0.70
4	1.21	1.61	1.06
5	0.97	1.84	1.21
6	0.94	1.33	1.02
7	1.05	1.50	1.04
8	0.93	1.71	0.85
9	1.00	1.81	1.18
10	1.01	1.42	0.98

4.2. Discussion

From the results, in walking with obstacles, 85% of the normalized RMS errors are lower than 15%. The average value of NRMSE in these 85% is approximately 6.9%, which correspond to 1.4 cm if the maximum foot clearance is approximately 20 cm. This result is therefore comparable to previous results and shows better foot clearance estimation performance if we compare it to the average RMS errors obtained in the fast walking case (1.3 cm). The estimation performance decrease in this case cannot be related to the speed difference with normal walking, since their speeds are close (see Table 2). This performance degradation is mainly due to the foot-flat phase detection algorithm (Figure 2), which presents some limits when the walking gait is not very natural, as in the case of walking with obstacles. Indeed, the accuracy of walking segmentation has an important impact on the foot clearance estimation, which decreased the estimation performances in the case of walking with obstacles.

In the foot-flat phase detection algorithm, the detection of the first events can be a source of errors in the case of the foot, which stands at the beginning, when the other foot begins the swing phase. Indeed, the motion start of this foot, detected from gyroscope data, is not the real start, since the ankle rotates around the heel before heel-off. To overcome this problem, we have combined the segmentation method with the linear acceleration start to make the detection of this first event more robust. Assuming that each subject starts walking with the same foot for all experiments, it seems interesting to compare for each subject the difference between the RMS errors obtained on the right and left foot. The average of all subjects of absolute variation between right and left foot estimation error is approximately 0.19 cm during normal walking and approximately 0.63 cm during fast walking. This small variation shows the robustness of the proposed segmentation to the choice of starting foot during walking. This comparison cannot be made in the case of walking with obstacles, since obstacles make the gait nonsymmetrical, and thus, the error produced in the right foot and left foot cannot be compared.

In addition to its good accuracy, the originality of the proposed foot clearance estimation method lies in its robustness to errors in IMU sensor placement on the foot around the Y-axis and Z-axis (Figure 1). This was facilitated by initial tilt angles (Equation (5)) measured under initial stationary conditions and included in the gravity-free acceleration calculation.

As mentioned in the experimental validation section (Section 3), data obtained from the IMU sensors cannot be used online. This IMU sensors technology was however enough to validate our method and quantify its performance by comparing results to measurements with the VICON motion capture system, which is known for its high accuracy.

As in the current work, we assumed walking occurs on a flat and horizontal floor, some modifications should be made in the future in order to consider other types of walking, like climbing stairs. In this case, the height of each stair should be known and included in the correction model (Equation (9)) Another future prospect consists of the online estimation of the foot clearance. This requires using another IMU technology.

The difference in accuracy estimation between normal and fast walking can be due to the limit of the acquisition frequency in the presence of a higher acceleration, related to a higher walking speed (see Table 2). A study comparing the acquisition frequency and its effect on different walking speeds can be interesting to explore and is planned in future works.

5. Conclusions

In the present work, a single mounted inertial measurement unit (IMU) was used to estimate the foot clearance, from the double integration and de-drift correction of acceleration. An original algorithm of foot clearance estimation is proposed, which is robust to misalignment of IMU axes with respect to the foot axes. Our foot clearance estimation method was validated on both feet of 10 healthy subjects, with different sizes and gaits (normal walking, fast walking and walking with obstacles). For this experimental validation, an optical measurement of IMU position through a motion capture system was used as the reference. Then, the deviation between the IMU-based foot clearance estimation and the one measured with the motion capture system was quantified and discussed.

Experimental validation results obtained for all subjects showed good foot clearance estimation performance, where the average of all normalized RMSE values, including all walking types, is approximately 6.5%, which corresponds to 1.3 cm, if the maximum foot clearance is approximately 20 cm.

Detailed results showed the best estimation performance during normal walking, with an average of RMS error around 0.74 cm, where this average of RMS errors is around 1.3 cm in the case of the fast walking. This performance degradation can be due to the limitation of acquisition frequency in the presence of a higher acceleration. The results obtained during walking with obstacles show a decrease in estimation performance, even when normalized with a higher foot clearance than for other walking types. However, even in this case, the average of normalized RMS errors in 85% of experiments is approximately 6.9%, which presents a good estimation performance. This performance loss can be due to a decrease of segmentation walking accuracy, since this walking type is not as usual and periodic as normal and fast walking types.

Our foot clearance estimation method was performed offline, since IMU data were acquired locally on the SD card of the sensor. Regarding potential prospects for IMU wireless data transmission, an adaptation of this method will be required for an online version of the proposed foot clearance estimation algorithm.

Some of the proposed algorithms are included in a Python toolkit devoted to developments regarding acquisition, analysis and visualization of IMU-based motion capture. Open access to this toolkit is available via the link [29].

Acknowledgments: The authors wish to thank all volunteers who participated to the experiments. The study was supported by INRIA SENSBIO ADT. We also warmly thank Christelle Reynes from "Laboratoire de Physique Industrielle et Traitement de l'Information", Montpellier France for her help and expertise in statistics.

Author Contributions: Mourad Benoussaad and B.S. developed and tested the algorithms; Mourad Benoussaad, Benoît Sijobert and Christine Azevedo Coste designed and performed the experiments; Mourad Benoussaad, Benoît Sijobert, Christine Azevedo Coste and Katja Mombaur analyzed the data and results; Mourad Benoussaad wrote the paper; Benoît Sijobert, Christine Azevedo Coste and Katja Mombaur contributed in the paper correction and improvements.

Conflicts of Interest: The authors declare no conflict of interest.

References

1. Barrett, R.; Mills, P.; Begg, R. A systematic review of the effect of ageing and falls history on minimum foot clearance characteristics during level walking. *Gait Posture* **2010**, *32*, 429–435.
2. Burpee, J.L.; Lewek, M.D. Biomechanical gait characteristics of naturally occurring unsuccessful foot clearance during swing in individuals with chronic stroke. *Clin. Biomech.* **2015**, *30*, 1102–1107.
3. Begg, R.; Best, R.; Dell'Oro, L.; Taylor, S. Minimum foot clearance during walking: Strategies for the minimisation of trip-related falls. *Gait Posture* **2007**, *25*, 191–198.
4. Benoussaad, M.; Mombaur, K.; Azevedo-Coste, C. Nonlinear model predictive control of joint ankle by electrical stimulation for drop foot correction. In Proceeding of the 2013 IEEE/RSJ International Conference on Intelligent Robots and Systems, IEEE, Tokyo, Japan, 3–7 November 2013; pp. 983–989.
5. Mariani, B.; Hoskovec, C.; Rochat, S.; Büla, C.; Penders, J.; Aminian, K. 3D gait assessment in young and elderly subjects using foot-worn inertial sensors. *J. Biomech.* **2010**, *43*, 2999–3006.
6. Chelius, G.; Braillon, C.; Pasquier, M.; Horvais, N.; Pissard-Gibollet, R.; Espiau, B.; Azevedo Coste, C. A Wearable Sensor Network for Gait Analysis: A 6-Day Experiment of Running Through the Desert. *IEEE/ASME Trans. Mechatron.* **2011**, *16*, 878–883.
7. Azevedo Coste, C.; Sijobert, B.; Pissard-Gibollet, R.; Pasquier, M.; Espiau, B.; Geny, C. Detection of freezing of gait in Parkinson disease: Preliminary results. *Sensors* **2014**, *14*, 6819–6827.
8. Bonnet, V.; McCamley, J.; Mazza, C.; Cappozzo, A. Trunk orientation estimate during walking using gyroscope sensors. In Proceeding of the 2012 4th IEEE RAS EMBS International Conference on Biomedical Robotics and Biomechatronics (BioRob), Rome, Italy, 24–27 June 2012; pp. 367–372.
9. Bonnet, V.; Ramdani, S.; Azevedo Coste, C.; Fraisse, P.; Mazzà, C.; Cappozzo, A. Integration of human walking gyroscopic data using empirical mode decomposition. *Sensors* **2013**, *14*, 370–381.
10. Madgwick, S.; Harrison, A.; Vaidyanathan, R. Estimation of IMU and MARG orientation using a gradient descent algorithm. In Proceeding of the 2011 IEEE International Conference on Rehabilitation Robotics (ICORR), Zurich, Switzerland, 29 June–1 July 2011; pp. 1–7.
11. Luinge, H.J.; Veltink, P.H. Measuring orientation of human body segments using miniature gyroscopes and accelerometers. *Med. Biol. Eng. Comput.* **2005**, *43*, 273–282.
12. Seel, T.; Raisch, J.; Schauer, T. IMU-Based Joint Angle Measurement for Gait Analysis. *Sensors* **2014**, *14*, 6891–6909.

13. Aminian, K.; Najafi, B.; Büla, C.; Leyvraz, P.F.; Robert, P. Spatio-temporal parameters of gait measured by an ambulatory system using miniature gyroscopes. *J. Biomech.* **2002**, *35*, 689–699.
14. Schepers, H.M.; Roetenberg, D.; Veltink, P.H. Ambulatory human motion tracking by fusion of inertial and magnetic sensing with adaptive actuation. *Med. Biol. Eng. Comput.* **2010**, *48*, 27–37.
15. Mariani, B.; Rochat, S.; Büla, C.J.; Aminian, K. Heel and toe clearance estimation for gait analysis using wireless inertial sensors. *IEEE Trans. Bio-Med. Eng.* **2012**, *59*, 3162–3168.
16. Trojaniello, D.; Cereatti, A.; Valeri, N.; Ravaschio, A.; Croce, U.D. Foot clearance estimation during overground walking and obstacle passing using shank-worn MIMU in healthy elderly and Parkinson's disease subjects. *Gait Posture* **2015**, *42*, doi:10.1016/j.gaitpost.2015.07.052.
17. Bailey, G.; Harle, R. Assessment of Foot Kinematics During Steady State Running Using a Foot-Mounted IMU. *Proced. Eng.* **2014**, *72*, 32–37.
18. Foxlin, E. Pedestrian Tracking with Shoe-Mounted Inertial Sensors. *IEEE Comput. Graph. Appl.* **2005**, *25*, 38–46.
19. Yang, S.; Mohr, C.; Li, Q. Ambulatory running speed estimation using an inertial sensor. *Gait Posture* **2011**, *34*, 462–466.
20. Laudanski, A.; Yang, S.; Li, Q. A concurrent comparison of inertia sensor-based walking speed estimation methods. *Conf. IEEE Eng. Med. Biol. Soc.* **2011**, *2011*, 3484–3487.
21. Zok, M.; Mazzà, C.; Della Croce, U. Total body centre of mass displacement estimated using ground reactions during transitory motor tasks: application to step ascent. *Med. Eng. Phys.* **2004**, *26*, 791–798.
22. Veltink, P.H.; Slycke, P.; Hemssems, J.; Buschman, R.; Bultstra, G.; Hermens, H. Three dimensional inertial sensing of foot movements for automatic tuning of a two-channel implantable drop-foot stimulator. *Med. Eng. Phys.* **2003**, *25*, 21–28.
23. Bird, J.; Arden, D. Indoor navigation with foot-mounted strapdown inertial navigation and magnetic sensors [Emerging Opportunities for Localization and Tracking]. *IEEE Wirel. Commun.* **2011**, *18*, 28–35.
24. Kotiadis, D.; Hermens, H.J.; Veltink, P.H. Inertial Gait Phase Detection for control of a drop foot stimulator: Inertial sensing for gait phase detection. *Med. Eng. Amp Phys.* **2010**, *32*, 287–297.
25. Frosio, I.; Pedersini, F.; Borghese, N.A.; Model, A.S. Autocalibration of MEMS Accelerometers. *IEEE Trans. Instrum. Meas.* **2009**, *58*, 2034–2041.
26. Ignagni, M.B. Optimal Strapdown Attitude Integration Algorithms. *J. Guid. Control Dyn.* **1990**, *13*, 363–369.
27. Schepers, H.M.; Koopman, H.; Veltink, P.H. Ambulatory Assessment of Ankle and Foot Dynamics. *IEEE Trans. Biomed. Eng.* **2007**, *54*, 895–902.
28. Sijobert, B.; Benoussaad, M.; Denys, J.; Pissard-gibollet, R. Implementation and Validation of a Stride Length Estimation Algorithm, Using a Single Basic Inertial Sensor on Healthy Subjects and Patients Suffering from Parkinson's Disease. *Health* **2015**, *7*, 704–714.
29. GitHub. Avaliable online: https://github.com/sensbio/sensbiot (accessed on 2 December 2015).

4

Application of Carbon-Microsphere-Modified Electrodes for Electrochemistry of Hemoglobin and Electrocatalytic Sensing of Trichloroacetic Acid

Wen-Cheng Wang, Li-Jun Yan, Fan Shi, Xue-Liang Niu, Guo-Lei Huang, Cai-Juan Zheng * and Wei Sun

Academic Editors: Yu Lei, Ashutosh Tiwari and Hongyun Liu

Key Laboratory of Tropical Medicinal Plant Chemistry of Ministry of Education,
College of Chemistry and Chemical Engineering, Hainan Normal University, Haikou 571158, China;
Sdwangwencheng@126.com (W.-C.W.); 18289751071@163.com (L.-J.Y.); shifan0802@126.com (F.S.);
xueliangniu@163.com (X.-L.N.); huangguolei1982@163.com (G.-L.H.); sunwei@qust.edu.cn (W.S.)
* Correspondence: caijuan2002@163.com

Abstract: By using the hydrothermal method, carbon microspheres (CMS) were fabricated and used for electrode modification. The characteristics of CMS were investigated using various techniques. The biocompatible sensing platform was built by immobilizing hemoglobin (Hb) on the micrometer-sized CMS-modified electrode with a layer of chitosan membrane. On the cyclic voltammogram, a couple of quasi-reversible cathodic and anodic peaks appeared, showing that direct electrochemistry of Hb with the working electrode was achieved. The catalytic reduction peak currents of the bioelectrode to trichloroacetic acid was established in the linear range of 2.0~70.0 mmol$\cdot$ L^{-1} accompanied by a detection limit of 0.30 mmol$\cdot$ L^{-1} (3σ). The modified electrode displayed favorable sensitivity, good reproducibility and stability, which suggests that CMS is promising for fabricating third-generation bioelectrochemical sensors.

Keywords: hemoglobin; carbon microsphere; direct electrochemistry; trichloroacetic acid

1. Introduction

Recently, the electrochemical behavior of proteins has roused great interest, and the results can be applied to the study of electron transfer mechanisms in biosystems and the construction of third-generation electrochemical biosensors or biofuel cells [1,2]. However, the electroactive centers are often buried in the polypeptide chains of redox proteins, which make electron transfer of proteins in a conventional biosensor more difficult [3]. Therefore, various protein-based biosensors have been fabricated for the realization of electrochemical behavior with the usage of multifarious modifiers such as polymers, surfactants and nanosized materials [4,5]. The existence of modifiers can preserve the original structure of redox proteins and their enzymatic activity, which offer a suitable microenvironment for electron transfer between electrode and proteins.

Nanomaterials with multifarious morphologies and unique properties such as excellent biocompatibility and large surface area have been applied to the electrochemistry of protein [6]. Among them, carbon nanomaterials are commonly used due to their excellent electrical conductivity. Various carbon materials have been applied to the investigation of electrochemical behavior of proteins, such as mesoporous carbon [7], carbon nanotubes (CNT) [8], and graphene (GR) [9,10]. Carbon microspheres (CMS) are a kind of carbon material that has been widely investigated. Shin *et al.* proposed a hydrothermal technique for the fabrication of colloidal spheres from aqueous cyclodextrin solution [11]. Sun *et al.* prepared core-shell structures of colloidal carbon spheres with the loading of different nobel-metal nanoparticles [12]. Chen *et al.* synthesized monodispersed carbon spheres from glucose for the supercapacitor [13]. Jin *et al.* proposed a direct hydrocarbon pyrolysis technique for the large-scale synthesis of carbon spheres [14]. The synthesized carbon spheres can be used for the loading of other nanomaterials and applied in different fields such as magnetism, catalysis and biosensor [15–17]. However, few reports about the application of CMS in the field of protein electrochemistry have been found.

In the present study, CMS was synthesized from glucose using a hydrothermal method and was further applied to the protein electrochemistry. CMS-hemoglobin (Hb) composite was prepared and cast on the ionic liquid modified carbon paste electrode (CILE) surface. Then, CTS were cast for the immobilization of the composite on the electrode surface. The fabrication procedure of this Hb modified electrode is shown in Scheme 1. CILE has exhibited many advantages, such as wide electrochemical windows, high conductivity, good antifouling capability and inherent electrocatalytic ability, which is reported in electroanalysis and electrochemical sensors [18,19]. CTS is an abundant natural cationic biopolymer that is composed of structural repeating units of N-acetyl-glucosamine and glucosamine, which offers a biocompatible microenvironment for the immobilized redox proteins. Therefore, CTS-modified electrodes have been commonly used in electrochemical biosensors [20]. The synthesized CMS were checked using different techniques and exhibited large surface area with porous structure. On CTS/CMS-Hb/CILE the direct electrochemistry of Hb was carried out and the electrocatalysis of trichloroacetic acid (TCA) was achieved, demonstrating the potential applications of this electrochemical sensor.

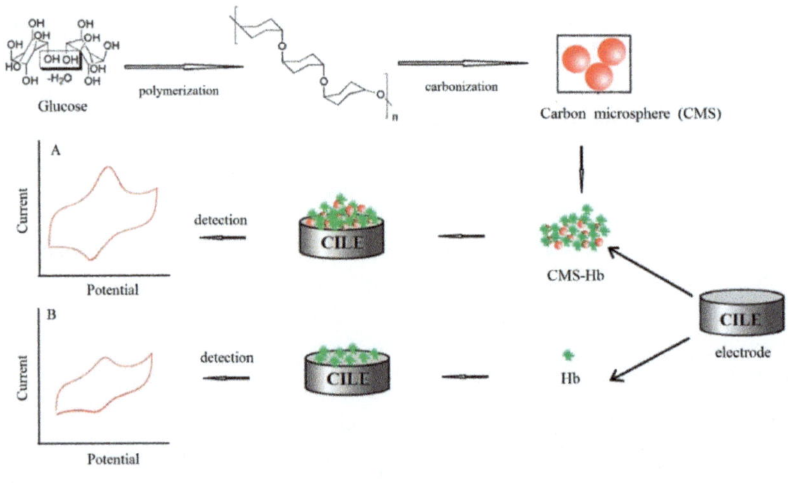

Scheme 1. The fabrication process of this Hb modified electrode.

2. Experimental

2.1. Reagents

Glucose (Beijing Chem. Reagent Factory, Beijing, China), Hb (Sinopharm. Chem. Reagent, Shanghai, China), 1-butylpyridinium hexafluorophosphate ($BPPF_6$ > 99%, Lanzhou Greenchem. CAS., Lanzhou, China), chitosan (CTS, Dalian Xindie Ltd., Dalian, China) graphite powder (30 μm, Shanghai Colloid Chem., Shanghai, China) and TCA (Tianjin Kemiou Chem. Ltd., Tianjin, China).The experiments were conducted in 0.1 $mol \cdot L^{-1}$ phosphate buffer saline (PBS) at room temperature (25 ± 1 °C), which was bubbled with pure N_2 for half hour before experiments to deoxygenate and keep under N_2 atmosphere during the electrochemical measurements. Ultrapure water and other chemicals (analytical reagent grade) were used in the experiments.

2.2. Apparatus

CHI 660D workstation (Shanghai CH Instrument, Shanghai, China); Nicolet 6700 FT-IR spectrophotometer (Thermo Fisher Scientific Inc., Waltham, Massachusetts, USA); TU-1901 double beam UV-Visible spectrophotometer (Beijing General Instrument Ltd. Co., Beijing, China); Renishaw InVia Raman microspectrometer using 532 nm lasers (Renishaw Plc., London, UK); JSM-7100F scanning electron microscope (Japan Electron Co., Tokyo, Japan); FEI Tecnai G2 F20 microscope (FEI, Hillsborough, Oregon, USA) with a field-emission gun operating at 200 kV; D8 advance X-ray diffractometer (Germany Bruker Co., Karlsruhe, Germany). A three-electrode system was used with a modified electrode (the working electrode), platinum wire (the counter electrode) and saturated calomel electrode (SCE, the reference electrode).

2.3. Synthesis of CMS

CMS was fabricated on the basis of the previous report [12]. In general, 8.0 g of glucose was added to 40 mL water with ultrasonic agitation for 2 min to get a colorless solution, which was positioned in teflon-sealed autoclave and kept for 4 h in 180 °C. The brown or black product was segregated by centrifugal separation, cleaned by 3 cycles of centrifugal separation/washing/re-dispersion in water and alcohol, and dehydrated at 80 °C for over 4 h to get black powder. A 1.0 $mg\cdot mL^{-1}$ CMS liquor was produced by redispersing CMS into the water with ultrasonic agitation for 3 h to get a homogeneous suspension solution.

2.4. Electrode Fabrication

CILE was manufactured as described previously [21]. In brief, graphite powder were mixed thoroughly with ionic liquid $BPPF_6$ at 3/1 (w/w) in a mortar, then the paste was inserted into a glass tube ($\phi = 4.0 \times 10^{-3}$ m). Prior to use, CILE was smoothed to get a mirror-like surface.

The step of fabricating the modified electrode was as follows. A 6.0 μL of 0.5 $mg\cdot mL^{-1}$ CMS and 15.0 $mg\cdot mL^{-1}$ Hb mixture solution was directly cast on the CILE surface with a 10.0 μL microsyringe. Then, the working electrode (CTS/CMS-Hb/CILE) was fabricated by spreading 5.0 μL of 1.0 $mg\cdot mL^{-1}$ chitosan (in 1.0% HOAC) solution evenly onto the CMS-Hb/CILE surface. A uniform film on the modified electrode was formed by covering a beaker to alleviate the evaporated solution. The preparation processes of CTS/Hb/CILE, CTS/CMS/CILE, CTS/CILE *etc.*, were parallel to that of CTS/CMS-Hb/CILE.

3. Results and Discussion

3.1. Morphological and Structural Characterization

SEM images of CMS with different magnification are shown in Figure 1A,B. It can be seen that the synthesized CMS had an average diameter of 600 nm, which was in good agreement with the reference [12]. TEM images of CMS at different magnitude were shown as Figure 1C,D, which exhibited that the solid structure of the surface had many nanosized pores. These pores on the surface of CMS could be attributed to the release of unreacted organic compounds such as oligosaccharides that been washed by water and alcohol. The existence of pores could offer large surface area and more reactive species for aiding in penetration and adsorption. Based on the reference [12], this synthesis process of CMS is a completely environmental-friendly procedure without the employment of any poisonous reagents, surfactants or organic reagent. Therefore the as-prepared CMS are nontoxic, with potential applications in biosensors or bioelectrochemistry.

The graphitization degrees of CMS were further checked by XRD and Raman spectroscopy. As shown in Figure 1E, two peaks at 22.5° and 42.8° appeared on XRD, which could be specified as the typical graphitic (002) and (100) planes. The broadening of these two peaks suggests that the degree of graphitization was low and the amorphous carbon was possibly subsistent [14]. Figure 1F shows the Raman spectrum of CMS, which had one significant peak at 1574.0 cm^{-1}. This peak corresponds to the G-band that is attributed to the ordered graphite structure [22].

The functional groups on CMS were further characterized by an FT-IR spectrum with the result shown in Figure 1G. The O–H bond stretching vibration led to the strong characteristic peak at 3434.7 cm^{-1}. The C–H bond stretching vibration led to the weak peaks at 3172.4 cm^{-1} and 2927.5 cm^{-1}. The absorption bands at 1398.2 cm^{-1} corresponded to C-C stretching vibration and that of 1625.7 cm^{-1} was the carboxyl groups (C=O) stretching vibration [13]. Therefore, on the surface of CMS, many functional groups were present, which resulted in the hydrophilicity and stability of CMS in aqueous solution. As shown in Figure 1H, the CMS solution could remain stable for 2 days without congregation, indicating that CMS was stable in water and could be used for bioapplications.

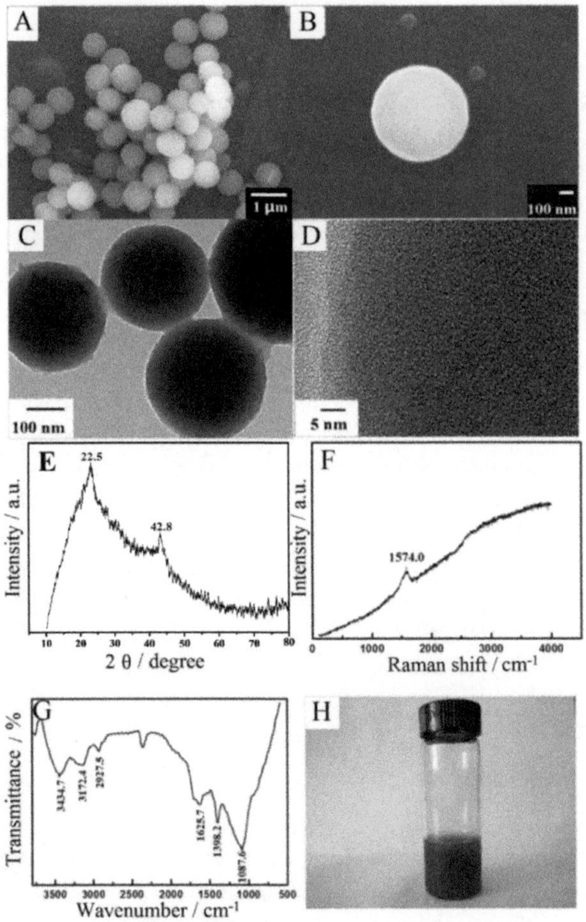

Figure 1. (**A,B**) SEM; (**C,D**) HRTEM; (**E**) XRD; (**F**) Raman spectrum; (**G**) FT-IR spectrum of CMS; (**H**) Photo of 1.0 mg· mL^{-1} CMS solution that kept for 2 days.

3.2. Spectroscopic Results

FT-IR spectroscopy was employed to examine the integrity of the structure and structural changes of the proteins [23]. Detailed data of the polypeptide chain could be determined from the shapes of amide I and II [24]. The amide I (1650.8 cm^{-1}) and II band (1542.8 cm^{-1}) of Hb after being mixed with CMS is shown in Figure 2Ab, which had less difference compared with amide I (1647.0 cm^{-1}) and II (1533.2 cm^{-1}) bands of the native Hb (Figure 2Aa). The results indicated that the original structure of Hb after being mixed with CMS was unchanged. UV-Vis adsorption spectroscopy is another useful tool to monitor the conformation change of heme proteins [25]. The Hb molecules had a characteristic band (406.0 nm) in pH 3.0 PBS (Figure 2Ba), which was the same as that of CMS-Hb (Figure 2Bb), meaning Hb molecules retained their original conformation in the CMS-Hb solution. All the spectroscopic research indicated that the biocompatibility of CMS and Hb kept the fundamental active conformation of the original structure after being mixed with CMS.

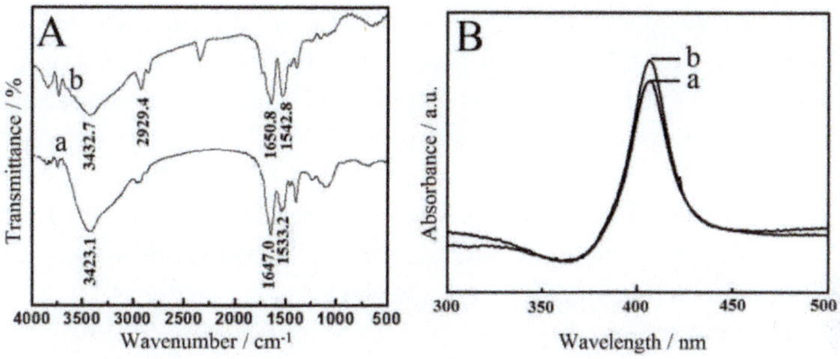

Figure 2. (**A**) FT-IR spectra of Hb (**a**) and CMS-Hb (**b**); (**B**) UV-Vis spectra of Hb (**a**) and CMS-Hb (**b**).

3.3. Electrochemical Characterization

Electrochemical behaviors were verified by cyclic voltammetry in the 0.1 mol$\cdot$L^{-1} KCl and 10.0 mmol$\cdot$L^{-1} $[Fe(CN)_6]^{3-/4-}$ mixture solution and the results are shown in Figure 3A,B. On CILE (Figure 3Aa) a couple of symmetric redox peaks was shown and it was the representative response of CILE. The electrochemical response of CTS/CILE (Figure 3Ab) was weaker than that of CILE, which was attributed to the existence of unconductive CTS on the surface of electrode impeding the electron transfer. The smallest redox peak currents appeared on CTS/Hb/CILE (Figure 3Bc), proving that the existence of Hb molecules on the electrode surface further impeded the electron transfer. However, on CTS/CMS-Hb/CILE (Figure 3Bd) the redox peak currents increased with the highest value, which was ascribed to the existence of CMS that promoted the electron transfer rate of $[Fe(CN)_6]^{3-/4-}$ with increased responses.

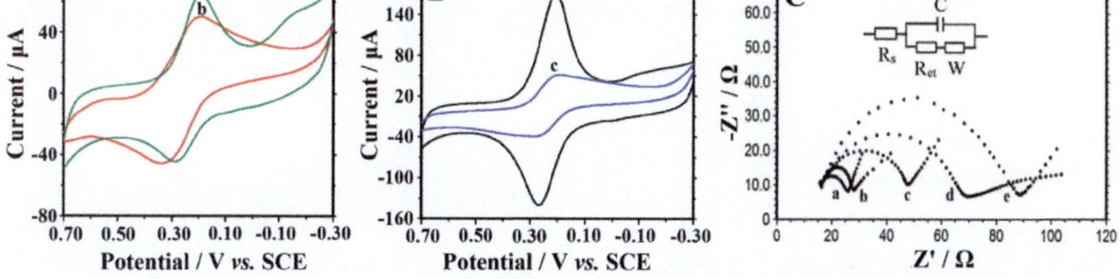

Figure 3. CV of (**A**) CILE (a), CTS/CILE (b); (**B**) CTS/Hb/CILE (c), CTS/CMS-Hb/CILE (d) in a 10.0 mmol$\cdot$L^{-1} $[Fe(CN)_6]^{3-/4-}$ and 0.1 mol$\cdot$L^{-1} KCl solution, scan rate:0.1 V$\cdot$s^{-1}; (**C**) EIS of (**a**) CTS/CMS/CILE; (**b**) CTS/CMS-Hb/CILE; (**c**) CILE; (**d**) CTS/CILE and (**e**) CTS/Hb/CILE with the frequencies ranging from 10^5 to 10^{-1} Hz. (Inset is the Randles circuit model in the cell).

Figure 3C shows the electrochemical impedance spectroscopy (EIS) of different electrodes, which was employed to investigate the interfacial information [26]. The electron transfer resistance (R_{et}) value of CILE was found to be 37.97 Ω (curve c) and that of CTS/CILE increased to 54.75 Ω (curve d), proving that the interfacial resistance was increased with the existence of unconductive CTS film. On CTS/Hb/CILE (curve e) the R_{et} was 82.83 Ω, which could be ascribed to the existence of Hb further increasing the interfacial resistance. While on CTS/CMS/CILE (curve a) and CTS/CMS-Hb/CILE (curve b) the R_{et} decreased to 10.76 Ω and 15.25 Ω, respectively, showing that the existence of CMS decreased the resistance. CMS is a carbon material with good conductivity that can quicken the electron transfer.

3.4. Direct Electrochemistry of Hb

Figure 4 showed the electrochemical behaviors of different modified electrodes in PBS (pH 3.0). No redox peaks were found on CTS/CILE (curve a) and CTS/CMS/CILE (curve b). With a layer of IL present on the surface, CILE exhibits good conductivity and a biocompatible surface [9], and a couple of asymmetric redox peaks were found on CTS/Hb/CILE (curve c), showing that direct electron transfer between Hb and CILE was realized. On CTS/CMS-Hb/CILE the redox peaks increased greatly (curve d), which remained nearly unchanged at multi-scan cyclic voltammogram. The mixture of CMS with Hb on the electrode can form a biocomposite with good stability, which is suitable for accelerating electron transfer from the electroactive center of Hb to the electrode. The cathodic (Epc) and the anodic (Epa) peak potential were found to be −0.140 V and −0.227 V with the ΔE_p as 87 mV. The formal peak potential ($E^{0'}$) was calculated as −0.184 V (*vs.* SCE). The ratio of the cathodic (Ipc) and the anodic (Ipa) peak current was found to be 1.06. To explore the best CV responses of Hb on CTS/CMS-Hb/CILE, the amount of CMS cast on the electrode was optimized in the control experiments with a concentration range of 0.05 to 5.0 $mg \cdot mL^{-1}$. The highest redox currents appeared at 0.5 $mg \cdot mL^{-1}$ CMS, which was used for electrode modification.

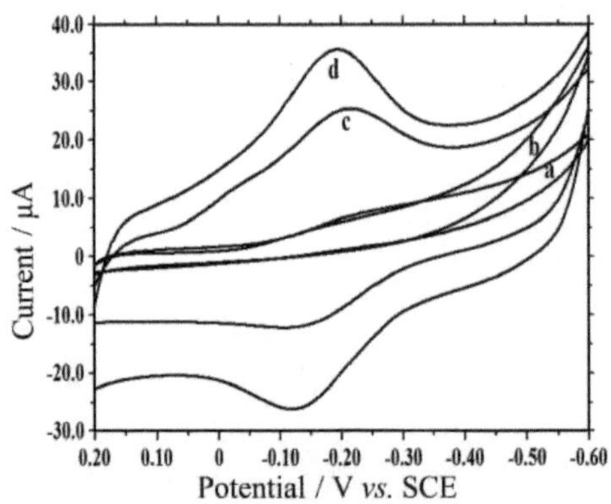

Figure 4. CV of (**a**) CILE; (**b**) CTS/CMS/CILE; (**c**) CTS/Hb/CILE and (**d**) CTS/CMS-Hb/CILE in pH 3.0 PBS, scan rate: 0.1 $V \cdot s^{-1}$.

3.5. Electrochemical Investigation

As shown in Figure 5A, the influence of scan rate (υ) on voltammetric responses of CTS/CMS-Hb/CILE was checked. A couple of redox peak appeared at different scan rates with the regression equations of peak currents and υ as Ipc(μA) = 92.5·υ ($V \cdot s^{-1}$) + 6.9 (n = 10, γ = 0.999) and Ipa(μA) = −78.0 υ· ($V \cdot s^{-1}$) − 10.0 (n = 10, γ = 0.999) (Figure 5B), indicating a surface-controlled thin-layer electrode behavior. The integration of the redox peak in cyclic voltammograms can count the surface coverage (Γ*) of electroactive Hb by the formula (Γ* = Q/nAF). By subtracting the background current, the peak currents of Hb were integrated with the Γ* value found to be 9.06×10^{-10} $mol \cdot cm^{-2}$, which was larger than the theoretical monolayer coverage (1.89×10^{-11} $mol \cdot cm^{-2}$) [27]. The fraction of electroactive Hb among the total Hb (1.20×10^{-8} $mol \cdot cm^{-2}$) was calculated as 7.6%. As shown in Figure 5C, the increase in υ resulted in the change of the redox peak potentials and the relationships of Ep with lnυ at high scan rate range were built. Two regression equations were got as Epc(V) = −0.03lnυ − 0.23 (n = 7, γ = 0.997) and

Epa(V) = 0.02lnυ − 0.10 (n = 7, γ = 0.993). Based on the Laviron's method [28], the kinetic of the redox reaction could be calculated with the electron transfer coefficient (α) and electron transfer rate constant (k_s) as 0.444 and 0.946 s^{-1}. The k_s value was larger than those of Nafion/Hb-graphene oxide-IL/CILE (0.92 s^{-1}) [29], GR/Fe_3O_4/Hb/GCE (0.30 s^{-1}) [30], Nafion/GR-TiO_2-Hb/CILE (0.65 s^{-1}) [31] and Hb-IL-MWCNT-CPE (0.84 s^{-1}) [32], indicating a relatively high rate of electron transfer. Also the *ks* value of CTS/Hb/CILE was calculated as 0.389 s^{-1} with the same method, which was less than that of CTS/CMS-Hb/CILE (0.946 s^{-1}). Therefore the presence of CMS with high conductivity provides an enhanced electron transfer reaction for Hb.

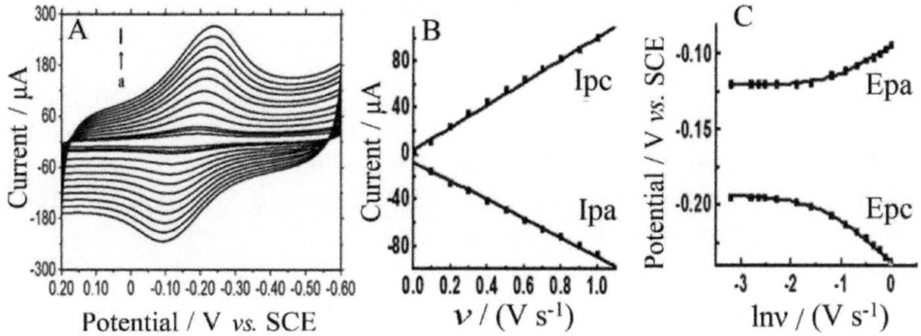

Figure 5. (**A**) Effect of scan rates (υ) (from a→l as 50, 80, 100, 200, 300, 400, 500, 600, 700, 800, 900, 1000 mV·s^{-1}) on electrochemical responses of CTS/CMS-Hb/CILE in pH 3.0 PBS; (**B**) Plot of the redox peak currents against υ; (**C**) plot of the redox peak potentials against lnυ.

The influence of buffer pH on electrochemical behaviors of CTS/CMS-Hb/CILE was studied in different PBS and the results are displayed in Figure 6A. The linear equation between $E^{0'}$ and pH was found to be $E^{0'}$(mV) = −44.25·pH − 54.73 (γ = 0.998). The slope value (−44.25 mV·pH^{-1}) was smaller than the theoretical value (−59.0 mV·pH^{-1}) for a one-proton and one-electron transfer process [33,34]. At pH 3.0 buffer solution, the largest redox peak currents appeared, and it was selected for the electrochemical experiments.

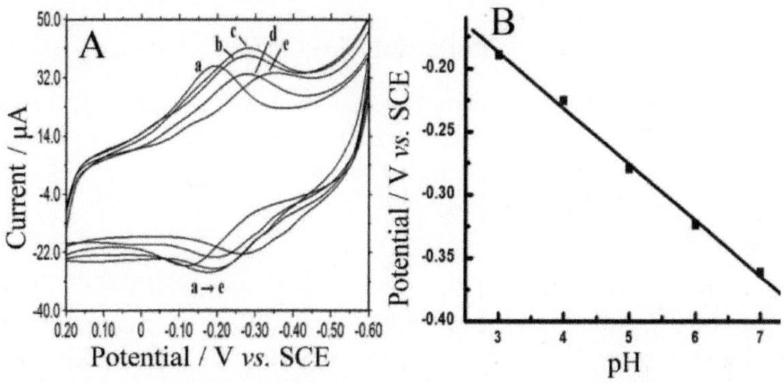

Figure 6. (**A**) CV of CTS/CMS-Hb/CILE in 0.1 mol·L^{-1} different pH PBS (from a→e: 3.0, 4.0, 5.0, 6.0, 7.0), scan rate: 0.1 V·s^{-1}; (**B**) the relationship of $E^{0'}$ with pH.

3.6. Electrocatalysis

The redox-protein-based biosensor exhibits excellent electrocatalytic ability for TCA, which is an important analytical target in biochemistry and environmental chemistry. As shown in Figure 7, when various concentrations of TCA were analyzed by CTS/CMS-Hb/CILE, the reduction peak current increased at −0.243 V without oxidation peak (curves a-h). With the further increase of the TCA concentration another reduction peak appeared at −0.515 V, indicated that di- and mono-chloroacetic acid might be dechlorinated by the formation of Hb [Hb Fe(I)] after TCA dechlorination with Hb Fe(II) [35]. On CTS/CMS/CILE direct electroreduction of TCA was studied with the potential negatively than −0.8 V (curves i, j). Therefore the reduction potential of TCA was reduced due to the existence of Hb. The reaction of electrocatalysis could be inferred from the following equations [35]:

$$\begin{aligned}&\mathrm{HbFe(III)} + e \rightarrow \mathrm{HbFe(II)}\\&\mathrm{2HbFe(II)} + H^+ + \mathrm{Cl_3CCOOH} \rightarrow \mathrm{2HbFe(III)} + \mathrm{Cl^-} + \mathrm{Cl_2CHCOOH}\\&\mathrm{HbFe(II)} + e \rightarrow \mathrm{HbFe(I)}\\&\mathrm{2HbFe(I)} + H^+ + \mathrm{Cl_2CHCOOH} \rightarrow \mathrm{2HbFe(II)} + \mathrm{Cl^-} + \mathrm{ClCH_2COOH}\\&\mathrm{2HbFe(I)} + H^+ + \mathrm{ClCH_2COOH} \rightarrow \mathrm{2Hb\ Fe(II)} + \mathrm{Cl^-} + \mathrm{CH_3COOH}\end{aligned}$$

The catalytic cathodic peak current at −0.243 V depended linearly on the TCA concentration with the equation of Iss(μA) = 3.04 C (mmol· L^{-1}) − 4.89 (n = 13, $\gamma = 0.999$). The linear range and the detection limit were obtained as 2.0 ~ 70.0 mmol· L^{-1} and 0.30 mmol· L^{-1} (3σ), respectively. The cathodic peak current reached a stable value when the TCA concentration exceeded 70.0 mmol· L^{-1}, indicating a typical Michaelis-Menten kinetic process. The Lineweaver-Burk equation of $1/I_{ss} = (1/I_{max})\ (1 + K_M{}^{app}/C)$ [36] was used to calculate the apparent Michaelis-Menten constant ($K_M{}^{app}$) at a value of 1.60 mmol· L^{-1}, which was smaller than those of published values [27,37–39]. The low KMapp value indicated that Hb entrapped on the electrode retained its bioactivity and had a much higher biological affinity to TCA.

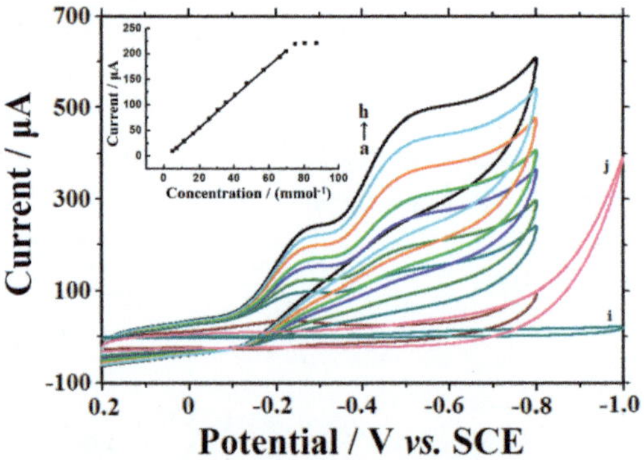

Figure 7. CV of CTS/CMS-Hb/CILE with 0, 15.0, 25.0, 35.0, 45.0, 55.0, 65.0, 70.0 mmol· L^{-1} TCA (curves a→h) and of CTS/CMS/CILE with of 0, 80.0 mmol· L^{-1} TCA (curves i and j), scan rate: 0.1 V· s^{-1} (Inset was the linearity of catalytic reduction currents and TCA concentration).

3.7. *Analytical Application*

To verify its application, CTS/CMS-Hb/CILE was used for determining of TCA in lab water with the standard addition method. As shown in Table 1, no TCA were found in the water specimens and the recovery was in the range of 97.00%–104.00%. Therefore this Hb-modified electrode could be used for water specimen detection.

Table 1. Detection results of TCA in the lab water specimen (n = 3).

Specimen	Found (mmol· L^{-1})	Added (mmol· L^{-1})	Total (mmol· L^{-1})	Recovery (%)	RSD (%)
		2.00	2.08	104.00	3.06
water	0	4.00	3.88	97.00	3.10
		6.00	6.12	102.00	2.98

3.8. *Stability and Reproducibility*

The reproducibility was studied by applying six Hb-modified electrodes to the determination of 10.0 mmol· L^{-1} TCA independently. An acceptable relative standard deviation (RSD) of 2.15% was determined. CTS/CMS-Hb/CILE was put into a 4 °C refrigerator for a certain period to check the storing stability. Every 5 days, the peak response of CTS/CMS-Hb/CILE to 10.0 mmol· L^{-1} TCA was tested, which decreased by 3.1% after 10 days and 7.5% for 25-days storage, proving the relative good stability of CTS/CMS-Hb/CILE.

4. Conclusions

CMS was prepared using a hydrothermal method and direct electrochemistry of Hb was carried out on CMS-modified CILE. A couple of well-defined redox peaks could be seen on CTS/CMS-Hb/CILE, indicating the acceleration of direct electron transfer of Hb due to high conductivity, large surface area and good biocompatibility of CMS. The immobilized Hb molecules kept their original structure and exerted good electrocatalytic ability for the reduction of TCA. As compared with other types of redox-protein-modified electrodes for TCA detection [29,31,37–39], this Hb-based electrode exhibited advantages such as favorable sensitivity, linear range, detection limit, stability and reproducibility. Hence, CMS has potential for the construction of third-generation redox-protein-based bioelectrochemical sensors.

Acknowledgments: We acknowledge the financial support of the National Natural Science Foundation of China (Nos. 21365010 and 51363008), the Nature Science Foundation of Hainan Province (20152016), the International Science and Technology Cooperation Project of Hainan Province (KJHZ2015-13), the Marine Science and Technology Program of Hainan Province (2015XH06) and Graduate Student Innovation Research Projects of Hainan Normal University (Hsyx2015-47).

Author Contributions: All authors contributed extensively to the work in this paper. Wen-Cheng Wang synthesized the sensing material, Li-Jun Yan preformed the electrocatalysis, Fan Shi prepared the modified electrode, Xue-liang Niu conducted the sensor analysis, Guo-Lei Huang characterized the carbon microsphere, Cai-juan Zheng conducted the characterization of protein, and Wei Sun investigated the electrochemical performances.

Conflicts of Interest: The authors declare no conflict of interest.

Abbreviations

The abbreviations involved in the paper.

Sigillum	Full Caption	Sigillum	Full Caption
$BPPF_6$	1-butylpyridinium hexafluorophosphate	HOAC	acetic acid
CILE	carbon ionic liquid electrode	IL	ionic liquid
CPE	carbon paste electrode	Ipa	anodic peak current
CMS	carbon microsphere	Ipc	cathodic peak current
CTS	chitosan	$K_M{}^{app}$	apparent michaelis-menten constant
CV	cyclic voltammograms	MWCNT	multi-walled carbon nanotube
EIS	electrochemical impedance spectroscopy	PBS	phosphate buffer solutions
$E^{0'}$	formal peak potential	Ret	the electron transfer resistance
Epa	anodic peak potential	RSD	the relative standard deviation
Epc	cathodic peak potential	SEM	scanning electron microscopy
GOD	glucose oxidase	TCA	trichloroacetic acid
GR	graphene	TEM	transmission electron microscopy
GCE	glass carbon electrode	XRD	X-ray diffraction
Hb	hemoglobin	Γ^*	surface coverage

References

1. Armstrong, F.A.; Hill, H.A.O.; Walton, N.J. Direct electrochemistry of redox proteins. *Acc. Chem. Res.* **1988**, *21*, 407–413. [CrossRef]
2. Ramanavicius, A.; Ramanaviciene, A. Hemoproteins in design of biofuel cells. *Fuel Cells* **2009**, *1*, 25–36. [CrossRef]
3. Armstrong, F.A.; Wilson, G.S. Recent developments in faradaic bioelectrochemistry. *Electrochim. Acta* **2000**, *45*, 2623–2645. [CrossRef]
4. Prakash, P.A.; Yogeswaran, U.; Chen, S.M. A review on direct electrochemistry of catalase for electrochemical sensors. *Sensors* **2009**, *9*, 1821–1844. [CrossRef] [PubMed]
5. Blanford, C.F. The birth of protein electrochemistry. *Chem. Commun.* **2013**, *49*, 11130–11132. [CrossRef] [PubMed]
6. Mani, V.; Devadas, B.; Chen, S.M.; Li, Y. Immobilization of enzymes and redox proteins and their electrochemical biosensor applications. *ECS Trans.* **2013**, *50*, 35–41. [CrossRef]
7. Sun, W.; Guo, C.X.; Zhu, Z.H.; Li, C.M. Ionic liquid/mesoporous carbon/protein composite microelectrode and its biosensing application. *Electrochem. Commun.* **2009**, *11*, 2105–2108. [CrossRef]
8. Wang, J.; Li, M.; Shi, Z.; Li, N.; Gu, Z. Direct electrochemistry of cytochrome c at a glassy carbon electrode modified with single-wall carbon nanotubes. *Anal. Chem.* **2002**, *74*, 1993–1997. [CrossRef] [PubMed]
9. Li, M.G.; Xu, S.D.; Tang, M.; Liu, L.; Gao, F.; Wang, Y.L. Direct electrochemistry of horseradish peroxidase on graphene-modified electrode for electrocatalytic reduction towards H_2O_2. *Electrochim. Acta* **2011**, *56*, 1144–1149. [CrossRef]
10. Kang, X.H.; Wang, J.; Wu, H.; Aksay, I.A.; Liu, J.; Lin, Y.H. Glucose oxidase–graphene–chitosan modified electrode for direct electrochemistry and glucose sensing. *Biosens. Bioelectron.* **2009**, *25*, 901–905. [CrossRef] [PubMed]
11. Shin, Y.S.; Wang, L.Q.; Bae, I.T.; Arey, B.W.; Exarhos, G.J. Hydrothermal syntheses of colloidal carbon spheres from cyclodextrins. *J. Phys. Chem. C* **2008**, *112*, 14236–14240. [CrossRef]
12. Sun, X.M.; Li, Y.D. Colloidal carbon spheres and their core/shell structures with noble-metal nanoparticles. *Angew. Chem. Int. Ed.* **2004**, *43*, 597–601. [CrossRef] [PubMed]
13. Chen, J.X.; Xia, N.N.; Zhou, T.X.; Tan, S.X.; Jiang, F.P.; Yuan, D.S. Mesoporous carbon spheres: Synthesis, characterization and supercapacitance. *Int. J. Electrochem. Sci.* **2009**, *4*, 1063–1073.
14. Jin, Y.Z.; Gao, C.; Hsu, W.K.; Zhu, Y.Q.; Huczko, A.; Bystrzejewski, M.; Roe, M.; Lee, C.Y.; Acquah, S.; Kroto, H.; *et al.* Large-scale synthesis and characterization of carbon spheres prepared by direct pyrolysis of hydrocarbons. *Carbon* **2005**, *43*, 1944–1953. [CrossRef]
15. Auer, E.; Freund, A.; Pietsch, J.; Tacke, T. Carbons as supports for industrial precious metal catalysts. *Appl. Catal. A* **1998**, *173*, 259–271. [CrossRef]
16. Wang, Y.; Su, F.; Wood, C.D.; Lee, J.Y.; Zhao, X.S. Preparation and characterization of carbon nanospheres as anode materials in lithium-ion secondary batteries. *Ind. Eng. Chem. Res.* **2008**, *47*, 2294–2300. [CrossRef]
17. Yan, X.B.; Xu, T.; Xu, S.; Chen, G.; Liu, H.W.; Yang, S.R. Fabrication of carbon spheres on a C: H films by heat-treatment of a polymer precursor. *Carbon* **2004**, *42*, 2769–2771. [CrossRef]
18. Shiddiky, M.J.A.; Torriero, A.A.J. Application of ionic liquids in electrochemical sensing systems. *Biosens. Bioelectron.* **2011**, *26*, 1775–1787. [CrossRef] [PubMed]
19. Sun, W.; Gao, R.F.; Jiao, K. Research and application of ionic liquids in analytical chemistry. *Chin. J. Anal. Chem.* **2007**, *35*, 1813–1819.
20. Sheng, M.L.; Gao, Y.; Sun, J.Y.; Gao, F. Carbon nanodots–chitosan composite film: a platform for protein immobilization, direct electrochemistry and bioelectrocatalysis. *Biosens. Bioelectron.* **2014**, *58*, 351–358. [CrossRef] [PubMed]
21. Sun, W.; Yang, M.X.; Jiao, K. Electrocatalytic oxidation of dopamine at an ionic liquid modified carbon paste electrode and its analytical application. *Anal. Bioanal. Chem.* **2007**, *389*, 1283–1291. [CrossRef] [PubMed]
22. Xu, X.; Shi, B.X.; Zheng, W.; Xia, W.; Guang, H. Photo-responsive behaviors and structural evolution of carbon-nanotube-supported energetic materials under a photoflash. *Mater. Lett.* **2012**, *88*, 27–29. [CrossRef]
23. Byler, D.M.; Susi, H. Examination of the secondary structure of proteins by deconvolved FTIR spectra. *Biopolymers* **1986**, *25*, 469–487. [CrossRef] [PubMed]

24. Kauppinen, J.K.; Moffat, D.J.; Mantsch, H.H.; Cameron, D.G. Fourier self-deconvolution: A method for resolving intrinsically overlapped ban. *Appl. Spectrosc.* **1981**, *35*, 271–276. [CrossRef]
25. Rusling, J.F.; Nassar, A.E.F. Enhanced electron transfer for myoglobin in surfactant films on electrodes. *J. Am. Chem. Soc.* **1993**, *115*, 11891–11897. [CrossRef]
26. Orazem, M.E.; Tribollet, B. An integrated approach to electrochemical impedance spectroscopy. *Electrochim. Acta* **2008**, *53*, 7360–7366. [CrossRef]
27. Wang, S.F.; Chen, T.; Zhang, Z.L.; Shen, X.C.; Lu, Z.X.; Pang, D.W.; Wong, K.Y. Direct electrochemistry and electrocatalysis of heme proteins entrapped in agarose hydrogel films in room-temperature ionic liquids. *Langmuir* **2005**, *21*, 9260–9266. [CrossRef] [PubMed]
28. Laviron, E. General expression of the linear potential sweep voltammogram in the case of diffusionless electrochemical systems. *J. Electroanal. Chem.* **1979**, *101*, 19–28. [CrossRef]
29. Sun, W.; Gong, S.X.; Shi, F.; Cao, L.L.; Ling, L.Y.; Zheng, W.Z.; Wang, W.C. Direct electrochemistry and electrocatalysis of hemoglobin in graphene oxide and ionic liquid composite film. *Mater. Sci. Eng. C* **2014**, *40*, 235–241. [CrossRef] [PubMed]
30. Wang, Y.Q.; Zhang, H.J.; Yao, D.; Pu, J.J.; Zhang, Y.; Gao, X.R.; Sun, Y.M. Direct electrochemistry of hemoglobin on graphene/Fe_3O_4 nanocomposite-modified glass carbon electrode and its sensitive detection for hydrogen peroxide. *J. Solid State Electrochem.* **2013**, *17*, 881–887. [CrossRef]
31. Sun, W.; Guo, Y.Q.; Ju, X.M.; Zhang, Y.Y.; Wang, X.Z.; Sun, Z.F. Direct electrochemistry of hemoglobin on graphene and titanium dioxide nanorods composite modified electrode and its electrocatalysis. *Biosens. Bioelectron.* **2013**, *42*, 207–213. [CrossRef] [PubMed]
32. Wei, W.; Jin, H.H.; Zhao, G.C. A reagentless nitrite biosensor based on direct electron transfer of hemoglobin on a room temperature ionic liquid/carbon nanotube-modified electrode. *Microchim. Acta* **2009**, *164*, 167–171. [CrossRef]
33. Yamazaki, I.; Araiso, T.; Hayashi, Y.; Yamada, H.; Makino, R. Analysis of acid-base properties of peroxidase and myoglobin. *Adv. Biophys.* **1978**, *11*, 249–281. [PubMed]
34. Liu, H.H.; Tian, Z.Q.; Lu, Z.X.; Zhang, Z.L.; Zhang, M.; Pang, D.W. Direct electrochemistry and electrocatalysis of heme-proteins entrapped in agarose hydrogel films. *Biosens. Bioelectron.* **2004**, *20*, 294–304. [CrossRef] [PubMed]
35. Fan, C.H.; Zhuang, Y.; Li, G.X.; Zhu, J.Q.; Zhu, D.X. Direct electrochemistry and enhanced catalytic activity for hemoglobin in a sodium montmorillonite film. *Electroanal* **2000**, *12*, 1156–1158. [CrossRef]
36. Kamin, R.A.; Wilson, G.S. Rotating ring-disk enzyme electrode for biocatalysis kinetic studies and characterization of the immobilized enzyme layer. *Anal. Chem.* **1980**, *52*, 1198–1205. [CrossRef]
37. Sun, W.; Li, X.Q.; Wang, Y.; Zhao, R.J.; Jiao, K. Electrochemistry and electrocatalysis of hemoglobin on multi-walled carbon nanotubes modified carbon ionic liquid electrode with hydrophilic EMIMBF4 as modifier. *Electrochim. Acta* **2009**, *54*, 4141–4148. [CrossRef]
38. Zhang, Z.Q.; Wu, J.; Sun, W.; Jiao, K. Direct electrochemistry of hemoglobin and its electrocatalysis based on a carbon nanotube paste electrode. *J. Chin. Chem. Soc.* **2009**, *56*, 561–567.
39. Sun, W.; Guo, Y.Q.; Lu, Y.P.; Hu, A.H.; Shi, F.; Li, T.T.; Sun, Z.F. Electrochemical biosensor based on graphene, Mg_2Al layered double hydroxide and hemoglobin composite. *Electrochim. Acta* **2013**, *91*, 130–136. [CrossRef]

5

The Performance Analysis of the Map-Aided Fuzzy Decision Tree Based on the Pedestrian Dead Reckoning Algorithm in an Indoor Environment

Kai-Wei Chiang †, Jhen-Kai Liao †, Guang-Je Tsai and Hsiu-Wen Chang *

Academic Editors: Kourosh Khoshelham and Sisi Zlatanova

Department of Geomatics, National Cheng-Kung University, 1 University Road, Tainan 701, Taiwan; kwchiang@mail.ncku.edu.tw (K.W.C.); cacalut1690@gmail.com (J.K.L.); tpp1114@gmail.com (G.J.T.)

* Correspondence: seawing1984@gmail.com

† These authors contributed equally to this work.

Abstract: Hardware sensors embedded in a smartphone allow the device to become an excellent mobile navigator. A smartphone is ideal for this task because its great international popularity has led to increased phone power and since most of the necessary infrastructure is already in place. However, using a smartphone for indoor pedestrian navigation can be problematic due to the low accuracy of sensors, imprecise predictability of pedestrian motion, and inaccessibility of the Global Navigation Satellite System (GNSS) in some indoor environments. Pedestrian Dead Reckoning (PDR) is one of the most common technologies used for pedestrian navigation, but in its present form, various errors tend to accumulate. This study introduces a fuzzy decision tree (FDT) aided by map information to improve the accuracy and stability of PDR with less dependency on infrastructure. First, the map is quickly surveyed by the Indoor Mobile Mapping System (IMMS). Next, Bluetooth beacons are implemented to enable the initializing of any position. Finally, map-aided FDT can estimate navigation solutions in real time. The experiments were conducted in different fields using a variety of smartphones and users in order to verify stability. The contrast PDR system demonstrates low stability for each case without pre-calibration and post-processing, but the proposed low-complexity FDT algorithm shows good stability and accuracy under the same conditions.

Keywords: pedestrian dead reckoning; map aided; fuzzy logic; smartphone; indoor navigation

---

1. Introduction

Over the past twenty years, several kinds of indoor navigation technologies have been developed based on various methods and theories [1]. This technological trend has changed and evolved quickly due to the growth of smart and wearable devices as well as advanced access points (AP). The ubiquitous smartphone has many kinds of internal micro-electro mechanical system (MEMS) sensors, such as Global Navigation Satellite System (GNSS chips), accelerometers, gyros, magnetometers, barometers and cameras. Those sensors are the basic components of some mainstream indoor navigation technologies such as pedestrian dead reckoning (PDR), image-aided PDR and radio frequency (RF)-aided PDR. Thus, the smartphone has the natural potential to become the ideal personal and mobile navigator. The number of people using their smartphones' navigational services have increased significantly, spurring the development of location-based service (LBS) and precise personal marketing. However, the problems associated with using smartphones for indoor pedestrian navigation, such as the low accuracy of MEMS sensors, non-repeatable and unstable MEMS error characteristics, GNSS-denied environments, and the imprecise predictability of human motion, have limited the effectiveness and subsequent popularity.

PDR is one of the most common methods for using inertial sensors to estimate two-dimensional positions for pedestrian navigation. Given an initial position, PDR uses a pedometer, step length model and heading sensor to obtain positions rather than constantly producing integral calculations like the inertial navigation system (INS). Because the model simplifies the 3D world into a 2D world, the impact of gravity projection on horizontal components in INS is eliminated in PDR. However, PDR errors accumulate with each step in a similar manner to the INS errors that accumulate over time. The estimated heading of PDR usually depends on the integration of the gyro angle rate or calculation from a magnetic field. Both of these methods, however, also have their inherent problems. It has long been understood that gyros' readings are strongly affected by human motion because of the vibration noise and dynamic misalignment between sensor body frame and pedestrian frame, causing error drifts quickly after integration. In regard to headings based on magnetic fields, magnetic materials and environmental obstacles degrade the magnetometer's reading, negatively influencing the accuracy of the calculated azimuth. In addition to the accuracy of the heading, most of the step length estimations derived from empirical formulas fail to accurately meet users' various individual characteristics. For example, step length can vary by as much as 40% with different pedestrians when they are walking at the same speed, and by up to 50% across the range of walking speeds of an individual [2]. Although tuning the parameters or calibration can enhance the accuracy in post-processing, it is inconvenient for real-time application. There is also the risk that the demonstration system performs better than the production system because of the uncertainty of the coefficients [3]. In conclusion, the PDR suffers error drift similar to that of the INS mechanization in terms of position or orientation because of the accumulated step-by-step impact of inaccurate step length and heading estimation. Therefore, in order for the PDR to maintain accuracy in individual cases, a PDR system requires pre-calibration or post-processing for tuning parameters similar to those in the Kalman Filter (KF) and step length model. However, this adjustment of parameters is time-consuming. Another method is using occasional position corrections from an external (absolute) positioning system to maintain stable performance [2]. Furthermore, PDR is a relative positioning system which needs an absolute coordinate for initialization.

Jeong-Min *et al.* propose an integrated system comprised of PDR and beacon-type AP to maintain stable performance [4]. The beacon-type APs' system generates a Wi-Fi signal within a few meters of each AP as footprints where these APs are installed at the crossroad. With footprints covering all directions of a corridor, proximity is taken into account to obtain better accuracy than fingerprinting; this is accomplished by setting the threshold value of the received signal for triggering location-based events. The beacon-type Wi-Fi AP can be replaced by a Bluetooth beacon which can also determine positions using the proximity method. The newest version is Bluetooth 4.0, which includes the technology of Bluetooth Low-Energy (BLE). The broadcast of BLE avoids the channels that Wi-Fi generally uses, thus reducing interference. As the name suggests, it has lower power consumption (usually battery powered), low cost, lower air data rate, small size and easy distribution in any indoor environment, unlike the larger WiFi AP which needs to be placed near a power source. Faragher and Harle analyzed the accuracy of WiFi and BLE fingerprinting in a large indoor space with a highly-accurate ultrasonic ground truth referencing system [5]. They also analyzed the errors caused by the measurement noise, distance, walls and human bodies and concluded that the measurement of distance using BLE Received Signal Strength Indicator (RSSI) is very accurate within one meter because the signal strength decreases as the inverse square of the distance to the source. Thus, the proximity method eliminates the influence of multipath disturbance, the requirements of database construction and has strong performance within one meter; the battery-powered beacon is easy to distribute and the Bluetooth signal mitigates the signal interference. For these reasons, we settled upon a method using proximity and bounding box (for some initial positions that are far away from the proximity-type beacon) to the Bluetooth beacon to initialize at any position and update measurements instead of using Wi-Fi positioning and fingerprinting.

At present, maps and geospatial data are often used as an aiding source for navigation because of their reliability. They generally consist of nodes, polylines, polygons, individual IDs and attribute data in the format of XML or computer-aided drawing (CAD) files. The map formats become more consistent because of the development of the map-matching algorithm and standards of the map database; however, the traditional generation of a numerical map is time consuming and the matching algorithm seems complex. In addition, for such a crowdsourcing map database like OpenStreetMaps (OSM), there may be inaccuracies and precision problems, since the editors are not necessarily professionals. Thus, instead of using the aforementioned maps, the proposed system uses the reliable map surveyed by the self-developed Indoor Mobile Mapping System (IMMS). The IMMS quickly obtains the indoor map in the WGS84 coordinate system with high accuracy, depending on the advanced hardware and algorithm. The map format references OSM with some self-defined attributes. The theoretical basis of the map-matching algorithm generally includes topological relation [6], probability map, *etc.* The matching algorithm first projects the navigation solution onto the passageway or corridor both to constrain the error drift and provide boundaries to avoid crossing a wall. They use the previous position, velocity and heading from the navigation solution in combination with geospatial data to perform the required mathematics, such as determining minimum distance and weighted algorithms. Once these processes have been completed, the map matching decides the most likely location of the user. However, those traditional algorithms and mathematics are more complex and time consuming when they achieve higher accuracy. Link *et al.* propose a novel footpath-matching algorithm that integrates their indoor maps with OSM which is thus similar to a rule-based system. To be specific, the algorithm implements the matching based on the footpath which is a route that consists of steps and corresponds with a step-by-step heading [7].

Therefore, in order to combine map information, Bluetooth beacon and PDR with lower complexity, some logic algorithms have been applied in the attempt to further improve navigation accuracy. In the real world, many decision-making tasks are too complex to be understood quantitatively, but humans succeed by flexibly using knowledge that is imprecise rather than precise [8]. Realizing this, Zadeh first proposed the concept of fuzzy logic in 1965, thus launching a system that often provides more appropriate and simplified expressions for interacting with applied systems than can be achieved by their more exact mathematical model counterparts. Fuzzy logic uses imprecise, abstract and subjective knowledge to generate decisions or to map general numerical ranges for specific systems; it behaves like human reasoning and cognition, generating useful and usable answers without the need for deep understanding, exact equations, and precise numerical values. In other words, it is suitable for systems with mathematical models that are ill defined and difficult to derive, and it allows decision making with estimated values using simple, nonlinear equations and rules with *a priori* qualitative, uncertain and expert knowledge. Specifically, fuzzy logic expresses the input crisp system in a fuzzy set with a membership function in the fuzzifier; then, using an inference engine, it obtains the output of a corresponding fuzzy set using defined, knowledge-based rules. Finally, the defuzzifier transforms the corresponding fuzzy set into crisp values. The basic concept of the fuzzy system is illustrated in Figure 1a.

The application of fuzzy logic has dramatically increased in the last two decades, being used extensively in control theory, artificial intelligence, finance and other decision-making problems. Trawinski applied fuzzy logic based on FURIA-based and J48G-based multi-classification systems (MCSs) to improve the accuracy of a topology-based Wi-Fi signal strength fingerprint approach, where FURIA represents Fuzzy Unordered Rule Induction Algorithm and J48G is a simple decision tree [9]. Lai *et al.* estimated walking frequency and strength per step based on fuzzy logic for step length estimation in their PDR system [10]. Shahram *et al.* further propose a complete fuzzy logic architecture for step length estimation with the membership functions such as step interval, human locomotion, user's height and trajectory curvature for the knowledge-based system (KBS) [11]. In looking at the literature, it becomes clear that fuzzy logic is used frequently for step length estimation and fingerprinting for pedestrian indoor navigation.

There are studies regarding fuzzy logic map matching for vehicle navigation. Haibin *et al.* propose an integrated map-matching algorithm based on fuzzy logic which combines a geometry map, probability map and the topology relation of the road network. Their fuzzy sets include three membership functions: projection distance, direction angle and comparability of trajectory [12]. Zhang and Gao have developed a fuzzy logic-based GPS system and map-matching integrated algorithm with fuzzy sets consisting of vehicle speed, heading error, perpendicular distance, gyro-rate and Horizontal Dilution of Precision (HDOP) for initial identification and tracking of the road link [13]. Working on similar problems, Wu has proposed a fuzzy sorting map-matching algorithm which generates road section candidates using a grid index, after which he uses a relativity function and fuzzy sorting method to determine the matching road section [14]. Ren and Karimi have even published a fuzzy logic map-matching algorithm for wheelchair navigation. They use fuzzy sets consisting of the perpendicular distance from a GPS point to a sidewalk and angular difference between the movement trajectories to build the necessary knowledge-based rules to produce the appropriate calculations [15]. Syed and Cannon have proposed a fuzzy logic map-matching algorithm with a high-sensitivity GPS receiver and a low-cost gyro for vehicle navigation in urban canyons. They designed two fuzzy inference systems which employ a variety of complete fuzzy sets to fix position and tracking [16]. Essentially, the strengths of fuzzy logic are its reduction of the depth and complexity of computations, improvement of accuracy and simplification of the positioning algorithm through the use of a relatively simple heuristic which determines a combination of the most appropriate rules, because an advanced indoor navigation system attempting to determine the necessary information through pure number crunching, with the required high computational complexity, would likely cause complete battery drain within a few hours. Keeping these problems in mind, this study focuses on developing a set of fuzzy rules based on the smartphone sensors and map knowledge for allowing individuals to successfully navigate an indoor environment in real time.

A decision tree is another low-complexity logic system for decision making; it consists of a hierarchical structure with a root, node and leaf, as shown in Figure 1b. Each internal node binary splits the dataset through the branches based upon the known conditions and features of the system; the classification will continue until a decision is produced at a leaf. The necessary conditions, or rules, can be obtained through a training process using sample datasets, and once generated, the new rules can be further used to make predictive estimations or system decisions. A decision tree based on Wi-Fi fingerprinting built during the off-line phase and used to determine a model for on-line application can be found in [17]. In addition, David *et al.* proposed methods based on the comparisons of RSSI and orientation information (magnetometer) to develop multiple weighted decision trees and data mining techniques for Wi-Fi fingerprinting [18]. There are also a variety of methods which use decision trees to aid in navigation, such as human behavior recognition [19] and Zero Velocity Update (ZUPT) detection. However, decision trees cannot solve problems of cognitive uncertainty or reduce classification ambiguity because of the need for exactly classified rules.

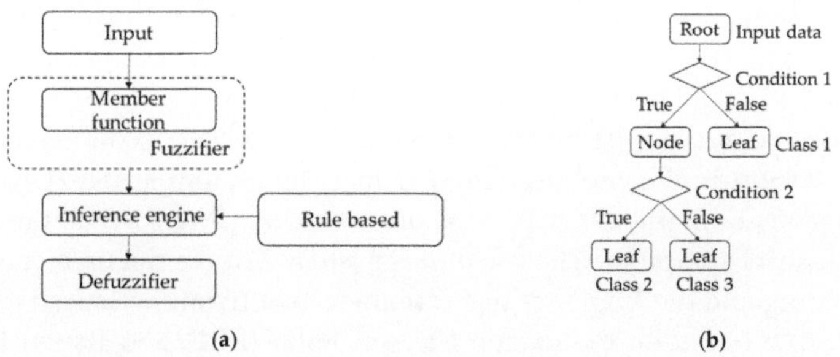

Figure 1. Two structures of decision making algorithms: **(a)** The fuzzy system includes fuzzifier, inference engine, expert rule and defuzzifer; **(b)** The binary decision tree includes the root, node and leaf.

Therefore, some studies combine fuzzy logic and decision tree to improve the performance of a rule-based system. Umano *et al.* have proposed an understandable fuzzy decision tree (FDT) based on the Iterative Dichotmiser 3 (ID3) algorithm and defined fuzzy sets for the analysis of gas in oil [20]. In trying to overcome the obstacles of decision tree caused by binary splits for classification, Yufei and Michael introduced an FDT which is based on the reduction of classification ambiguity through the use of fuzzy evidence for classified application; they use linguistic terms with soft boundaries to accommodate vagueness and ambiguity in human thinking and perception [21]. Instead of having one class (leaf) from binary logic, the major difference between FDT and traditional decision trees (binary logic) is that the results may be members of more than one class after nodes/branches are calculated with fuzzy logic. Some mathematically oriented fuzzy decision trees are introduced in [22,23]. Inspired by the discovery of these various applications in the literature, we have taken the step of constructing a map-aided FDT for PDR to estimate the location of a pedestrian through the use of hierarchical rules and probability-based fuzzy sets.

Although complicated algorithms, various integrations, mathematics and multi-sensor fusion do indeed significantly improve navigation accuracy, these high-complexity approaches are more time/power consuming and require higher hardware standards. In addition, external complementary signals suffer problems, such as environmental changes and maintenance costs. Other simpler algorithms usually need the adjustment of parameters based upon post/pre-processing to adapt to each user, device and environment which is inconvenient for real-time application. Therefore, developing a stable, accurate and low-complexity system is a much more attractive direction for smartphone and real-time indoor pedestrian applications. Based on this idea, a system using map-aided FDT is proposed. The system determines the user's location based on low-complexity rules, sensor data, fewer distributed Bluetooth beacons (depending on the floor plan's geometry). The FDT rules are trained once using experimental field data. Then these rules are fixed in the map-aided FDT system for all other new experimental field tests. In other words, after the map survey by IMMS and one experimental file is used as trained data, the proposed system has the generalization ability to work on various individuals and smartphones. The difference between the proposed system and the contrast PDR system is the additional logic rules related to the map information.

There are three main advantages to this approach: (1) based on a single time training, the proposed system is versatile regarding different users, smartphones and experimental fields. In other words, the proposed system has the ability to work for new devices and individuals; in a crowd or a new environment such as a plaza (where it is hard to apply general map matching); and in corridors and a magnetic-hostile environment; (2) the system can achieve better stability without time-consuming pre/post-processing; (3) compared to traditional PDR, a low-complexity method, the proposed method is also a relatively simple while maintaining a similar level of error acceptibility without the need for continuous tuning. It is worth mentioning that for traditional PDR to achieve asimilar performance, anew tuning process is required for each users, sensor and environment.

2. Experimental Section

2.1. *Indoor Map Production*

IMMS was used to survey the building to generate a precise geo-referenced map, which is an important component in the proposed algorithm. The system is INS/GNSS integration and consists of GNSS, navigation-grade Inertial Measurement Unit (IMU), a panoramic camera, a power supply and an industrial computer. The use of this system reduces the time-consuming procedure for mapping a building and the result has high accuracy. A software was used to precisely estimate the trajectory of IMMS based on the Extend Kalman Filter (EKF) and Rauch-Tung-Striebel (RTS) smoother. The Non-Holonomic Constraint (NHC) and Zero Velocity Update (ZUPT) are used in the EKF. Information regarding these can be found in our previous research [24]. Another software was used subsequently to digitize and geo-reference the floor plan and then produce the walkable area in a *.csv* format file with WGS84 coordinates with some self-defined attributes. In addition, It also measures the point of interest (POI) based on the theories of photogrammetry and direct geo-referencing (DG) for the position of the installed Bluetooth beacon and its subsequent application. The coordinate of POI can be further applied to some applications such as route planning and investigation of the user's hotspots. Those systems and software are part of an ongoing developed project at National Cheng Kung University (NCKU), as shown in Figure 2. Further details will not be discussed in this paper as a paper related to these systems will be published in the near future. Similar systems can be found in TIMMS [25].

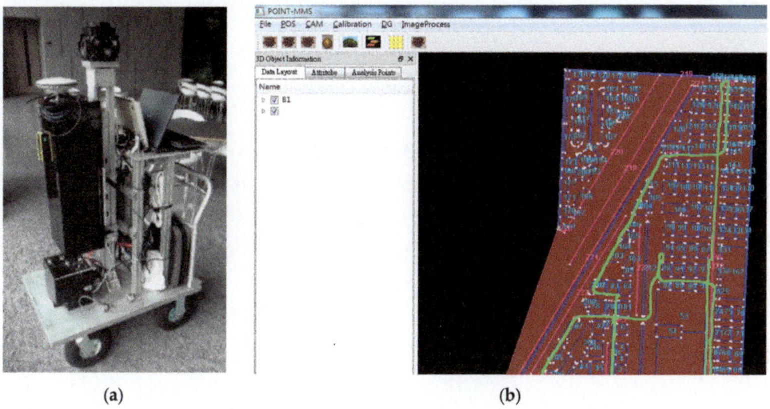

Figure 2. This figure shows the system and software for map production: (**a**) The hardware of IMMS; (**b**) The user interface of software for map production and measuring image.

2.2. Bluetooth Positioning

RF-based Bluetooth positioning is used for both initialization and update measurements during navigation. This experiment uses two methods, proximity and bounding box, implemented individually for each beacon with different setting IDs. Both of these methods are used to obtain the initial location, but only proximity is used for the update measurement. Some proximity-type beacons have heading constraint from map knowledge; the power level is one meter and is generally placed at the POI or corridor. When the system detects the signal from the beacons, the algorithm identifies their type based on their ID, and then decides on the next action. If the signal received is from the proximity-type beacon with an RSSI value larger than the threshold, then the algorithm obtains the coordinate and heading from the map database for initialization or updating. The signal strength distribution indoors is affected by the placement of the Bluetooth beacon and the environment [26]. To corroborate this idea, we have found that the proximity-type beacon may unevenly broadcast much longer compared to the setting power range (one meter in this study) because of the instability of RSSI, which may cause unexpected misdetection and a larger position error on the part of the proximity method, thereby rendering it unsuitable for placement in a plaza.

The bounding box method usually has a power level higher than ten meters, depending upon the field size, and beacons are placed in the corners of the plaza. Compared to the proximity method, initialization using the bounding box method is generally less accurate because of flaws in the RSSI-distance model, but it is more flexible in a large space. In addition, the heading used in the bounding box method relies on the magnetic heading because the initial location at the plaza has no corresponding heading in the map database. When the bounding box IDs are detected and the RSSI values exceed the defined threshold, the algorithm sorts their RSSI values and chooses the top three beacons, thus determining the coordinates according to their ID in map database for further position calculation. Each beacon produces a bounding box by the use of the beacon's location (x_i, y_i) as the square center with twice the distance (d) measured by RSSI-distance model as the edge length. In other words, the location of the vertex of the bottom left corner is ($x_i - d_i$,$y_i - d_i$), and the location of the upper right vertex is ($x_i + d_i$,$y_i + d_i$). The user's location, the intersection point of the squares, is defined by the following equations [27]:

$$x_{max} = \max(x_1 - d_1,\ x_2 - d_2,\ x_3 - d_3) \quad (1)$$

$$y_{max} = \max(y_1 - d_1,\ y_2 - d_2,\ y_3 - d_3) \quad (2)$$

$$x_{min} = \min(x_1 + d_1,\ x_2 + d_2,\ x_3 + d_3) \quad (3)$$

$$y_{min} = \min(y_1 + d_1,\ y_2 + d_2,\ y_3 + d_3) \quad (4)$$

$$x = x_{max} + (x_{min} - x_{max})/2 \quad (5)$$

$$y = y_{max} + (y_{min} - y_{max})/2 \quad (6)$$

where the equations are shown in a north–east coordinate system: (x_i, y_i) is the i-th beacon's coordinates as the center location of the square; d_i is the measured distance between the user and beacon determined by the RSSI value; (x,y) is the point of the intersection. The concepts of proximity and bounding box are shown in Figure 3. By using the proximity and bounding box methods, the proposed system can initialize the user's location almost anywhere, whether the user is in a large room or a narrow corridor; it also obtains some update constraints during navigation mode using a proximity-type beacon in conjunction with the map database.

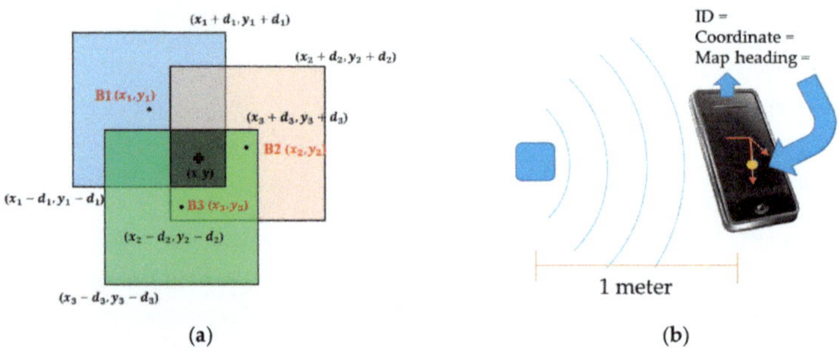

Figure 3. This figure shows the methods used in the proposed system for initialization: (**a**) The bounding box method; (**b**) The proximity method.

2.3. Pedestrian Dead Reckoning

The general PDR was developed to compare with the map-aided FDT algorithm. It has two Kalman Filters (KF) for heading and position estimation. The heading measurements are derived from gyro (A_g) and the magnetometer's readings (A_m). The characteristics of the gyro's heading are smooth and independent of the environment. Nevertheless, errors accumulate quickly over time. In addition, the magnetometer provides the initial heading for the gyro's heading as well as the heading derived from the magnetic field. The advantage of the magnetic heading is that it is independent of error drift. However, the magnetic heading is affected by the environment, which can cause unsmooth or even discontinuous measurements. The first KF estimates the heading as well as the gyro bias by fusing headings derived from the gyro and magnetometer. The fusion combines the advantages of both to obtain a smooth heading and reduce the error drift. The predictive stage of the first filter is shown in the following:

$$x_k = \Phi_{k-1}x_{k-1} = \begin{bmatrix} \delta\psi_k \\ \delta b_{\psi,\,k} \end{bmatrix} = \begin{bmatrix} 1 & \Delta t \\ 0 & 1 \end{bmatrix} \begin{bmatrix} \delta\psi_{k-1} \\ \delta b_{\psi,\,k-1} \end{bmatrix} + w_k,\ w_k \sim N(0,\ Q_k) \quad (7)$$

$$P_k = \Phi_{k-1}P_{k-1}\Phi_{k-1}^T + Q_k \quad (8)$$

where x is the state vector; $\delta\psi$ is the heading error; δb is the gyro bias; w_k is system noise assumed to be Gaussian distribution; Φ is the transition matrix which represents the relationship between the states at k and k−1 epochs; P is a covariance matrix of the state vector; Q is the covariance matrix of the system noise. The update stage is expressed as follows:

$$K_k = P_k H^T \left(\mathrm{H} P_k H^T + R_k\right)^{-1} \tag{9}$$

$$\hat{x}_k = x_k + K_k (z_k - H x_k) \tag{10}$$

$$\hat{P}_k = (\mathrm{I} - K_k \mathrm{H})\, P_k (I - K_k H)^T + K_k R_k K_k^T \tag{11}$$

where K is the Kalman gain; H is the design matrix for the measurement; R_k is the covariance matrix of the measurements at k epoch; $\hat{x}_k$ is the updated state vector at k epoch; z is a measurement of the difference between the two kinds of headings in this case. The second filter uses the coordinates of the update measurement broadcast by the Bluetooth beacon (proximity type) to update the PDR position which is calculated based on the fused heading from the first KF and step length. The state vector and transition matrix are shown in the following equations:

$$x_k = \begin{bmatrix} E_k & N_k & L_k & b_{E,\,k} & b_{N,k} & b_{L,k} \end{bmatrix} \tag{12}$$

$$\Phi_{k-1} = \begin{bmatrix} 1 & 0 & \vartheta & 1 & 0 & 0 \\ 0 & 1 & \eta & 0 & 1 & 0 \\ 0 & 0 & 1 & 0 & 0 & 1 \\ 0 & 0 & 0 & 1 & 0 & 0 \\ 0 & 0 & 0 & 0 & 1 & 0 \\ 0 & 0 & 0 & 0 & 0 & 1 \end{bmatrix} \tag{13}$$

where E is the local coordinate of the eastern direction; N is the local coordinate of north; L is the step length; ϑ and η are the second-order of Taylor series expansion for sine and cosine function; $b_{E,\,k}$ is the offset of positional error of east direction; $b_{N,\,k}$ is the offset of positional error in the northern direction; $b_{L,\,k}$ is the offset of step length. A detailed step length model is used from Ruizhi *et al.* and is shown as the following [28]:

$$L_k = (0.7 + a\,(H - 1.75) + b\frac{(F_k - 1.79)\,H}{1.75})c \tag{14}$$

where L_k is the step length of k step; a, b and c are the tuning parameters; H is the height of user; F_k is the step frequency at k step. The pedometer is based on peak detection on the force of acceleration with the setting thresholds of peak value and time interval, as shown in Figure 4. Figure 5 shows the complete scheme of the contrast PDR system used in this study. The parameters of two KF and step length models were calibrated once based on data collected from one of the participants at the first experimental site (department building of NCKU) using a smartphone (HTC M7, HTC Corporation, New Taipei City, Taiwan). In addition, the rules of the proposed FDT system are trained once with the same data set.

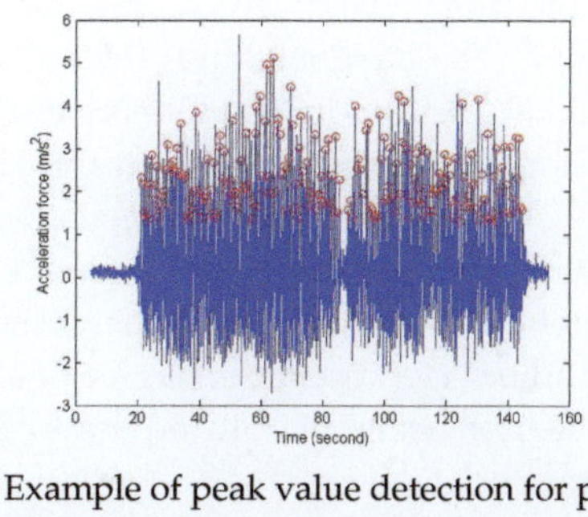

Figure 4. Example of peak value detection for pedometer.

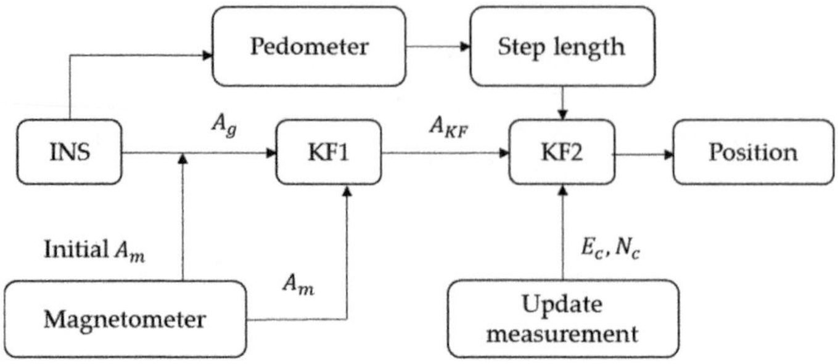

Figure 5. The contrast PDR system used in this study.

2.4. Fuzzy Decision Tree

The scheme for the proposed PDR system using a map-aided fuzzy decision tree is shown in Figure 6. The design of the pedometer is the same as in the contrast system described in Section 2.3. The map engine first loads the map data to construct a database with coordinates and their zone type. The zone type is one of the data attributes in the map database, which could be a corridor or plaza. The knowledge-based FDT has lots of fuzzy sets and rules inside. The fuzzy system can be easily shown in the following equation:

$$A = \{(x,\ \mu_A(x)) \mid x \in X\},\ \mu_A : X \to [0,1] \tag{15}$$

where x is the object element; $\mu_A(x)$ is the fuzzy membership function; A is a fuzzy set; X is a crisp set. When the $\mu_A(x)$ is closer to one, it means the element x is more representative of the set A. The fuzzy decision tree shown in the Figure 6 has three fuzzy systems: step length, zone type and moving direction. The following paragraphs will respectively describe these systems.

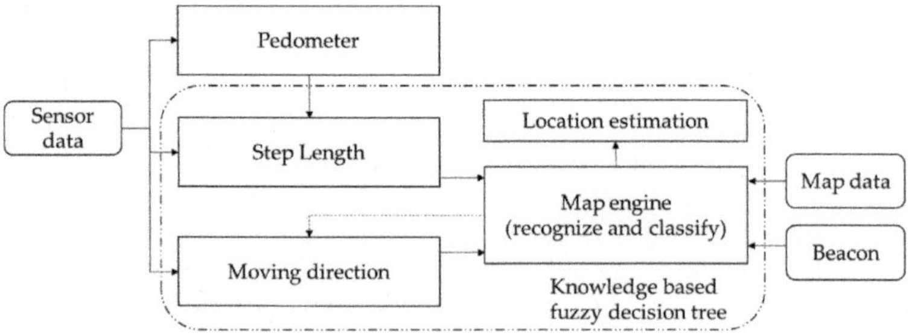

Figure 6. The architecture of the proposed system, with fuzzy decision tree.

The fuzzy system of step length can be seen in reference to the architecture proposed in [11], which is complete and detailed. Compared to this previous research, our design is simpler because we consider only one type of locomotion: walking. The membership function and fuzzy system of step length are shown in Figure 7. The membership functions of three fuzzy sets (step interval, user's height and sensor measurement) are used in the inference engine, which uses knowledge-based rules to determine three levels of step length (short, medium and long). In addition, three fuzzy sets and each sub level have different weights in the inference engine based upon the knowledge system. The membership degree of each step length level is then determined in an inference engine, and the results are used as the input of a later decision tree. Table 1 shows the example of rules in the inference engine of step length, where the numbers represent the degree of each fuzzy level. As shown in this table, if the step interval is short, the user's height is tall and the acceleration force is large; step length is then short and the corresponding output degree is set to 0.6. Another case is when the step interval is short, user's height is medium and acceleration force is large, then step length is medium and the corresponding output degree is set to 0.4.

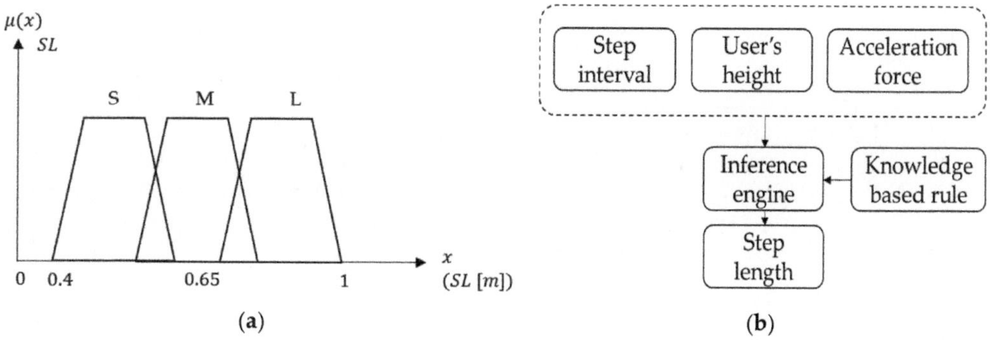

Figure 7. This figure shows fuzzy system of step length: (**a**) The membership function of step length; (**b**) The fuzzy system of step length with input of the three fuzzy sets.

Table 1. The example of IF-THEN rules for the fuzzy system of step length.

IF	Activation Degrees	THEN	Output Degrees
Step interval is short AND user's height is tall AND acceleration force is large	short 1, tall 0.8, large 1	Step length is short	0.6
Step interval is short AND user's height is medium AND acceleration force is large	short 1, medium 0.2, large 1	Step length is medium	0.4

The map engine recognizes key information from the input map, such as the direction of a corridor and where the zone of the plaza is. In general, the corridors are perpendicular to each other in a building; therefore, the type of input map can be divided into vertical or oblique. The rules in the FDT algorithm must be slightly modified after the type of map has been decided, after which the map engine generates some map point candidates with zone type attributes around the previous location based on the levels of step length.

The membership probability and class of next zone type is determined by the inference engine, which uses the membership degree and level of step length and individual minimum distance. The minimum distance is determined by the previous position and the location of each zone. The weight of minimum distance is larger than step length, but step length has influence when more than one zone is close to the user. Thus, the classes of next zone type and their probabilities can be used in subsequent decision trees. The membership function of minimum distance and the fuzzy system of the next zone type are shown in Figure 8. Table 2 shows the example of rules in the inference engine of next zone type, where the minimum distance corresponds to each zone type; zone type 1 is a kind of corridor, zone type 2 is another kind of corridor perpendicular to type 1, and type 3 represents the plaza. The example in Table 2 continuously uses the assumption in Table 1.

The fuzzy system of moving direction is based on the magnetic heading, gyro measurement and map knowledge, which includes the next possible zone, such as the east–west corridor, as shown in Figure 9. Map-derived headings cannot be used exclusively since a corridor may be wide enough to have a wide range of possible directions. The inference engine gives their different weights through expert knowledge and map information. After the inference engine, the moving directions are divided into four or eight directions according to different current zone types (corridor or plaza). The inference engine also decides the probabilities of each direction as determined by knowledge-based rules, and then uses these results as the input data for a later decision tree. Table 3 shows the example of rules in the inference engine of a moving direction that, in keeping with previous tables, suggests that the magnetic heading may have interference, but the result is corrected based upon other fuzzy sets.

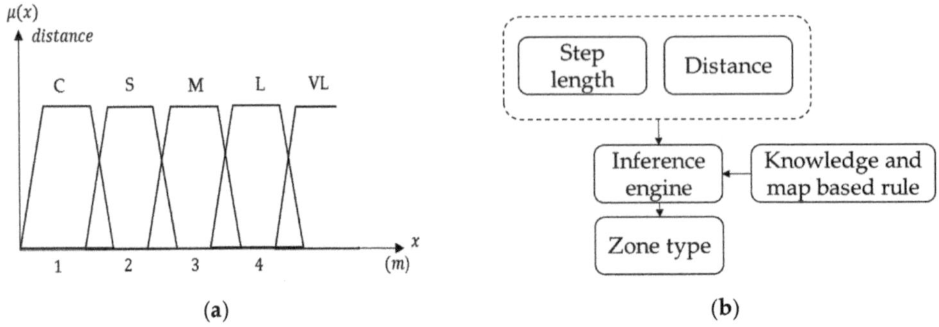

Figure 8. This figure shows the fuzzy system of the next zone type: (**a**) The membership function of minimum distance; (**b**) The fuzzy system of next zone type with the two inputs of fuzzy sets.

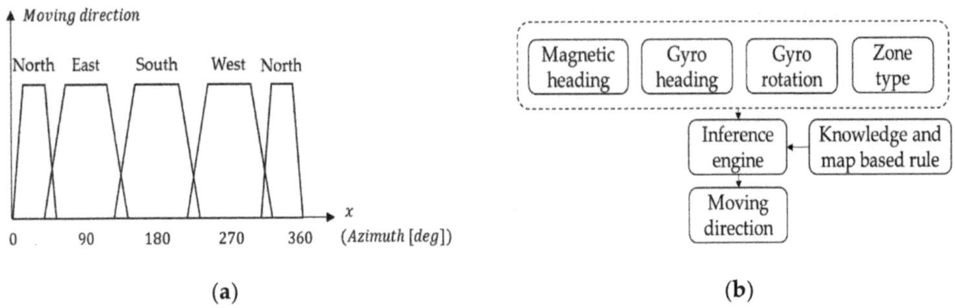

Figure 9. This figure shows the fuzzy system of the moving direction: (**a**) The membership function of moving direction; (**b**) The fuzzy system of moving direction with the input of four fuzzy sets.

Table 2. The example of IF-THEN rules for the fuzzy system of next zone type.

IF	Activation Degrees	Previous Decision	THEN	Output Degrees
Minimum distance of zone type-1 is close AND minimum distance of zone type-2 are close and short AND minimum distance of zone type-3 is long	close 1 (type-1), close 0.4 (type-2), short 0.6 (type-2), long 1 (type-3)	Step length is short	Next zone type is type-1	1 (type-1)
		Step length is medium	Next zone type are type-1 and type-2	0.7 (type-1) and 0.3 (type-2)

Table 3. The example of IF-THEN rules for the fuzzy system of moving direction.

IF	Activation Degrees	Previous Decision	THEN	Output Degrees
Magnetic heading is east and north AND gyro heading is east AND gyro rotation is medium	east 0.2 (Mag.), north 0.8 (Mag.), east 1 (Gyro), medium 1	Next zone type is type-1	Moving direction is east and north	east 0.9 and north 0.1
		Next zone type is type-2	Moving direction is east and north	east 0.7 and north 0.3

Instead of using a defuzzifier to obtain a crisp value for the set of the next zone type, step length and moving direction, the decision tree directly uses levels and degree (classes and probability) of the membership functions. If the three major components (next zone type, step length and moving direction) are collected, the decision tree calculates the final probability through individual branches. Each individual branch contains several nodes which decide the user's location by multiplying the probability (degree) of each node on a specific branch. In other words, it chooses one of the locations from the four or eight candidates according to the fuzzy sets and their probability, as determined by the decision tree. The tree architecture shown in Figure 10 is simplified for illustrative purposes, where A1, B1 and C1 are the fuzzy set of step length, their membership degree and levels; A2, B2 and C2 are the fuzzy set of the next zone type, their membership probability and classes; A3, B3 and C3 are the fuzzy set of the moving direction, their membership probability and classes; D4 is the example of final probability from one of the branches. The plaza zone type has eight directions and another has only four, as it is a simple perpendicular corridor. Therefore, we can choose one of the map point candidates as the next user's location based upon a direction which has maximum probability. Table 4 shows the example of the final decision tree which follows the previous tables after cutting off the branch of zero probability. The final decision is based upon the maximum probability of Branch 1.

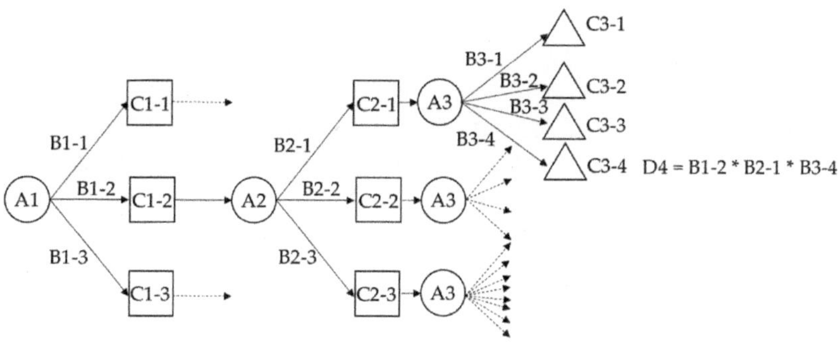

Figure 10. The fuzzy decision tree used in the proposed system with the knowledge-based rules and map information.

Table 4. One of the possible situation in FDT system.

	Step Length (%)	Next Zone Type (%)	Moving Direction (%)	Final Probability (%)
Branch 1	Short (0.6)	zone type 1 (1)	east (0.9)	0.54
Branch 2	Short (0.6)	zone type 1 (1)	north (0.1)	0.06
Branch 3	Medium (0.4)	zone type 1 (0.7)	east (0.9)	0.252
Branch 4	Medium (0.4)	zone type 1 (0.7)	north (0.1)	0.028
Branch 5	Medium (0.4)	zone type 2 (0.3)	east (0.7)	0.084
Branch 6	Medium (0.4)	zone type 2 (0.3)	north (0.3)	0.036
Final decision: Step length short, move to zone type 1 with *east* direction				
Choose the corresponding candidate point				

The sample data was collected in a departmental building in the NCKU campus by one person using smartphone HTC M7. This experiment was designed to verify and modify the architecture of FDT and related knowledge-based rules. The main architecture following the test and various refinement procedures has three groups of rules, and some rules to avoid possible exceptions during the procedure are added. Then, the defined FDT system is fixed and applied to other test fields using different smartphones with map recognition and corresponding modification (such as the position of beacon and definition of direction). In conclusion, the scheme of the proposed FDT system was implemented in the following steps:

1. Recognize the map in order to choose the corresponding algorithm and database
2. Count the step through peak detection
3. Generate the map candidates around the previous location
4. Determine the class and probability of step length, next neighbor zone type and moving direction with fuzzy logic

5. Choose the most likely user's location through the decision tree with the input of fuzzy membership value and knowledge-based rules.

2.5. Experimental Setting

The test smartphones are of different generations and include the HTC M7, SONY Z2 and SONY Z3 (Sony Corporation, Tokyo, Japan). Each smartphone has its own specification of inside sensors. Four participants, three males and one female, each held a smartphone in front of the torso and walked through the experimental routes. The characteristics such as height of each participant are different. The experimental fields include three different areas: the department building at NCKU, the office floor at the Institute for Information Industry (III) building and the hall of an enterprise. The area of the tested field at NCKU is about 1000 m^2, and includes classrooms and laboratories. The route begins at the end of a corridor and goes into a classroom and then back to the origin. The length of NCKU route is about 143 m. The area of the III tested field is about 2000 m^2, and contains offices with many computers and a circular walkway. The route starts at the elevator, which is a kind of POI, and goes around the office, and after some sharp turns and a circular walkway, goes back to the beginning. The length of the second experimental route is about 162 m. The third area is a hall at an enterprise field which is about 1600 m^2 at the ground floor. It is a typical company hall with many conference rooms and a coffee plaza, but there is no ceiling above the coffee plaza from the ground floor up to the second floor. In addition, there is an engine room on the third floor. The route starts at the elevator and proceeds along a corridor into a showroom and two plazas then goes back to the origin. The length of the third experimental route is about 175 m. Figure 11 shows the environments of III and enterprise. All the experimental fields are magnetic-hostile environments because of a lot of electronic equipment. Meanwhile, there are some spaces which are hard to determine the specific routes for with general map matching.

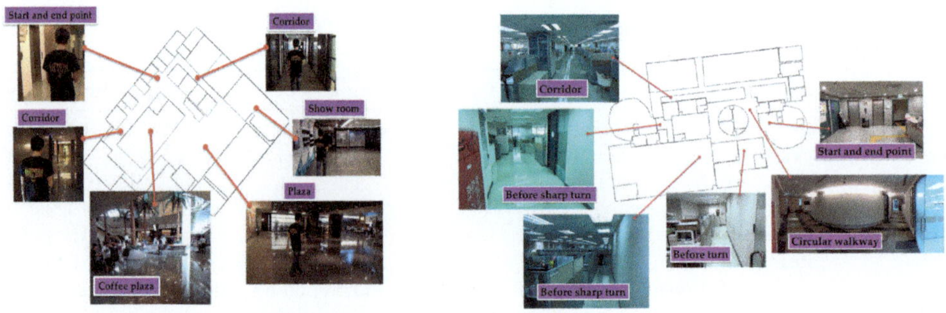

(a) (b)

Figure 11. This figure shows the experimental environments of III and enterprise: (**a**) The environment of enterprise; (**b**) The environment of III.

To precisely access the performance of the proposed FDT system, we have controlled the accuracy of the initial position by using proximity only. If the user starts the navigation in the plaza, the proposed system can still estimate the initial position using the bounding method. This method was not implemented in the test because its accuracy depends on the number of beacons, geometry of the distributed beacons and the degree of complication of the environment. For further information about initialization based on the bounding box method, refer to [29]. In order to show that the proposed FDT system can work in a harsh environment, frequent position updates and direction constraints from the proximity beacon were avoided. In the NCKU field test, only one proximity-type beacon was used to obtain the initial position. Other field tests have one beacon at the starting point for initialization and another one located near the beginning to correct the initial heading. The settings described in this paragraph are both used in the proposed FDT system and contrast PDR system.

3. Results and Discussion

The experimental results are presented in the following paragraphs. The developed Android software with proposed FDT algorithm was implemented to process the data in real time. Figures 12–14 show the trajectories at NCKU, III and the enterprise, respectively. The green lines shown in the figures in the first row represent the real experimental route. This route is also a survey by the IMMS (using navigation grade IMU) which is estimated by the EKF and RTS-smoother with NHC and frequent ZUPT constraint. Because these two systems were not implemented together (time synchronization), no further position difference is shown. The red lines shown in the figures represent the proposed FDT algorithm, and the light-blue lines represent the contrast PDR system. The black circles indicate the starting and ending points. The purple rectangles indicate the locations where beacons were used to trigger the update operation. Symbols A, B, C and D represent the four different people in the experiments. In all figures, the second row shows the results of HTC M7, the third row shows SONY Z2 and the fourth row shows SONY Z3.

Figure 12 shows the experimental trajectories at NCKU department building. The result in the third of the second row is user C using the HTC M7, which is the sample data for tuning parameters and establishing knowledge rules used in the contrast PDR and proposed FDT systems, respectively. In other words, all the parameters or rules for both systems are only tuned once with HTC M7 and user C at NCKU field, and there is no further training or tuning for other smartphones, users and fields. Therefore, the other tests are designed to verify the stability and adaptability for real-time application.

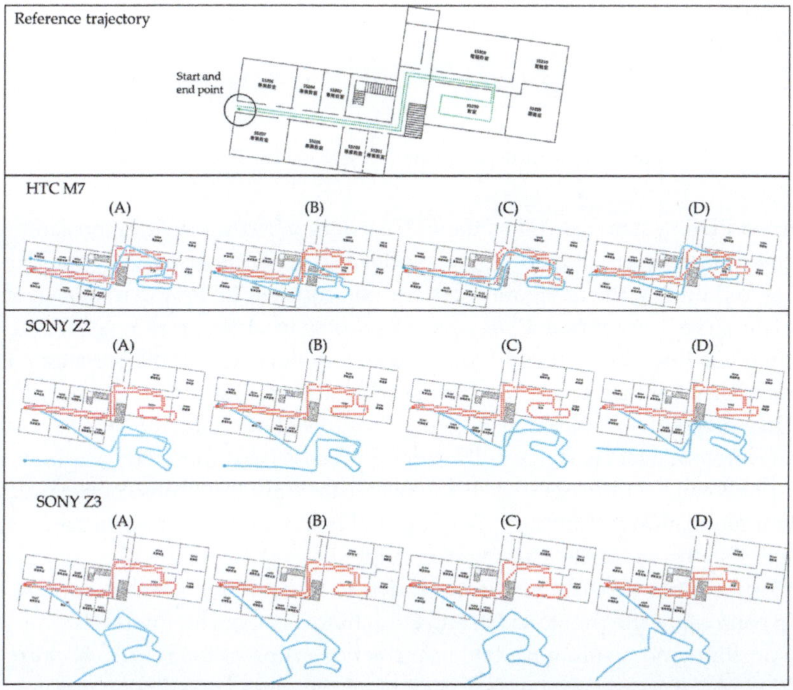

Figure 12. The experimental trajectories at *NCKU* department building.

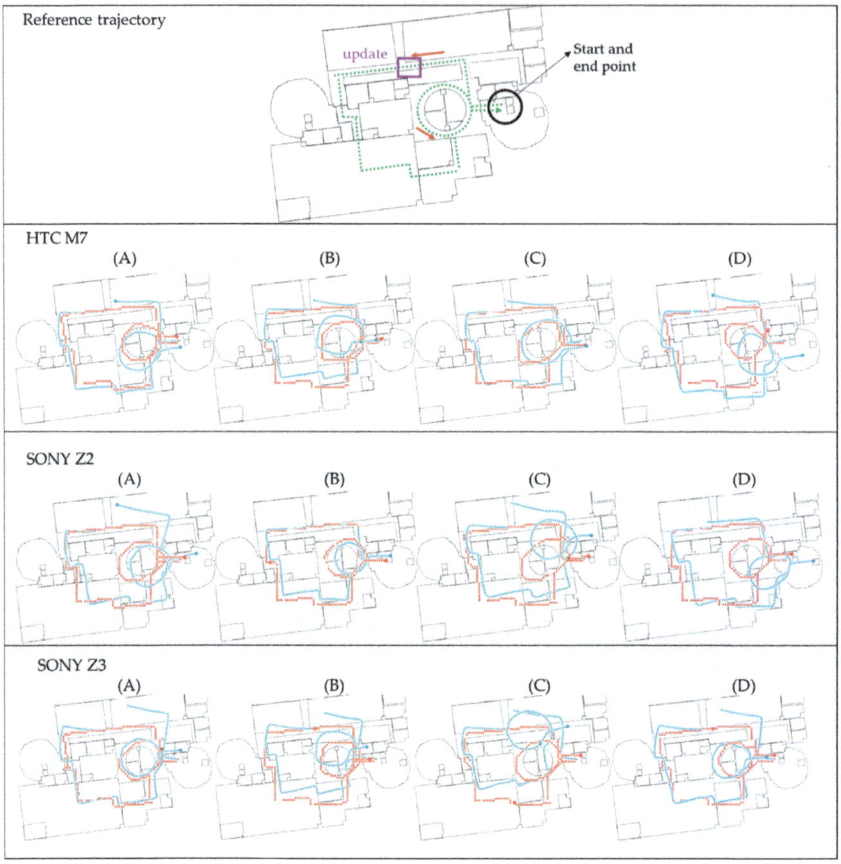

Figure 13. The experimental trajectories at *III* office floor.

The contrast PDR system as well as the FDT system performs well in the sample data because of the well-tuned parameters of step length model and KF in post-processing. Due to no update operation triggered during the navigation phrase, the contrast PDR system has no ability to correct the heading. Therefore, the contrast PDR performs worse in other cases at NCKU fields. Even if we correct the initial heading, the drift continues and clearly deteriorates the trajectory. It indicates that the parameters of contrast PDR in step length model and KF are not suitable to the different users and smartphones in a magnetic-hostile environment. Thus, the contrast PDR needs new calibration of the step length for different users and a new KF tuning process for different smartphones and magnetic fields in post-processing. Unfortunately, these procedures are time consuming and not suitable for real-time indoor navigation. In contrast, the proposed map-aided FDT system performs stably in real time at all the cases without further tuning requirement.

Figure 13 shows the experimental trajectories at III building which is an office with a lot of computers. In contrast to the previous NCKU field, there is a beacon (markedby a purple rectangle) to trigger the position and heading update near the beginning of the routes. Because of this beacon, the contrast PDR performs well at the beginning of all the cases, but still deteriorates over time. This phenomenon confirms that the parameters need individual calibration for each factor. However, the proposed FDT system still performs stably with good accuracy at III field. It is worth mentioning that the traditional topological map matching based on the polyline and nodes is hard to apply to this circular walkway, because the user may walk in a rectangular or a circular shape which cannot be matched with a fix route. The proposed system using the matching of eight directions can mitigate this problem.

Figure 14 shows the experimental trajectories at the enterprise building which has plazas, engine room and a high ceiling. There is also a beacon installed near the beginning to trigger the position and heading update. The contrast PDR still drifts over time because of a magnetic-hostile environment and inaccurate step length. The proposed system uses the matching of eight directions at the plazas without the specific route which is required by the traditional map matching. In addition, the RF-based positioning usually does not work in the high-ceilinged space or in spaces with serious electromagnetic interference caused by a large number of electronic devices. However, the results show the proposed FDT system still performs stably with good quality.

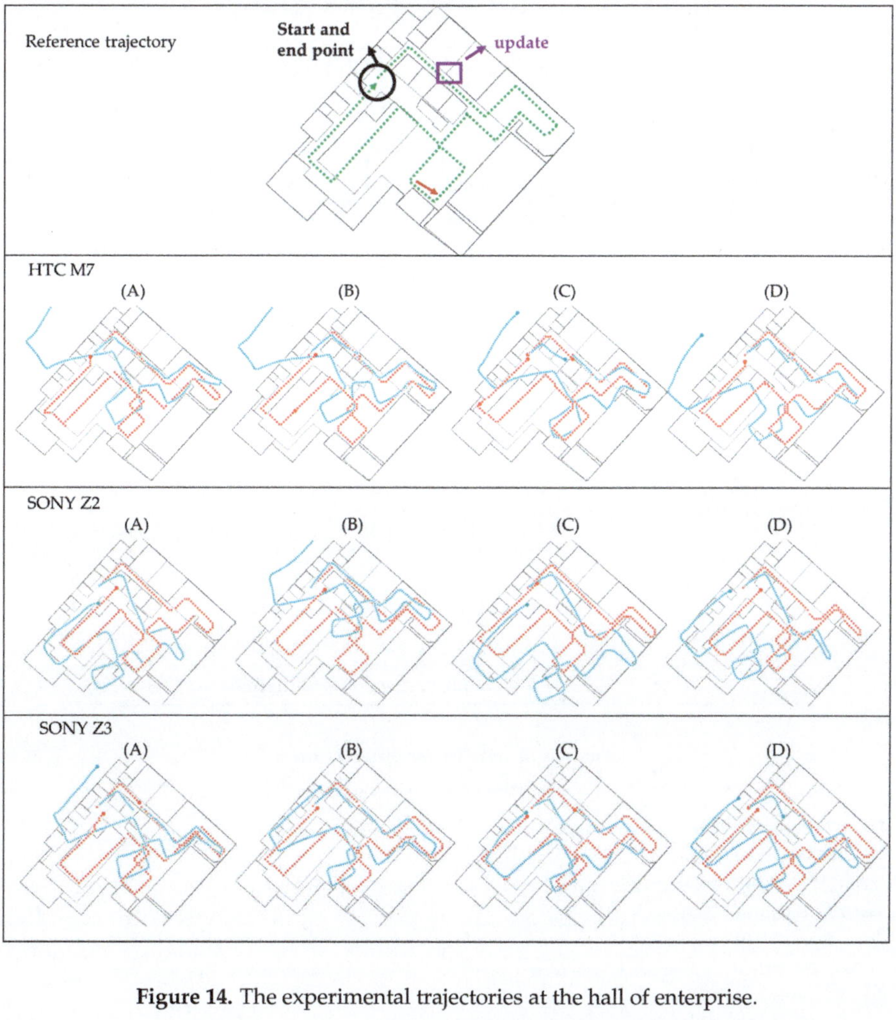

Figure 14. The experimental trajectories at the hall of enterprise.

Table 5 summarizes the accuracy of all the results. The accuracy index is the percentage of loop closure error and total distance. For example, the accuracy of 1% means every one hundred travelled there will have one meter error accumulated. As demonstrated in Figures 12–14 the proposed system has better performance than the contrast PDR system. Table 3 shows that the loop closure errors of the proposed FDT algorithm for all the experimental results and mean value is 1.3%. It is more accurate than the mean error of 8.21% in the contrast PDR. The standard deviation of the loop closure errors of the map-aided FDT system is less than 1%, which means it has high stability for different users, smartphones and fields.

Figure 15 shows the mean values in terms of different users in the same field and using the same smartphone based upon Table 5. In other words, this figure indicates the stability of different users in the same situation. For example, the mean error of 0.62% at the first column represents the mean error of FDT system calculated from all users that use Sony Z2 at the NCKU field. The performances of contrast PDR are worse at the NCKU field because there is no update beacon at the beginning for initial heading correction. However, using the HTC M7 is better in this field because the parameters have been tuned. The figure shows that the map-aided FDT system has better stability when the user is different, regardless of smartphone or field.

Table 5. The accuracy of FDT and contrast PDR system.

Field	User	SONY Z2		SONY Z3		HTC M7	
		PDR	FDT	PDR	FDT	PDR	FDT
NCKU	A	12.07	0.61	29.96	0.66	2.43	1.09
143 m	B	14.00	0.66	33.44	0.68	1.30	0.74
	C	12.45	0.60	32.90	0.64	0.91	0.64
	D	9.40	0.61	30.42	0.77	5.79	0.66
III	A	3.19	0.89	1.84	3.86	1.28	0.93
162 m	B	1.98	0.99	3.28	1.33	0.78	0.61
	C	4.48	0.86	7.12	1.19	1.14	1.16
	D	5.07	1.17	1.71	0.87	4.36	3.87
Enterprise	A	3.47	3.39	6.99	1.66	13.03	0.54
175 m	B	9.91	2.31	3.47	0.89	11.12	1.19
	C	3.02	1.75	1.48	1.94	7.98	0.82
	D	2.59	3.38	2.53	1.71	8.79	1.25
	Mean of FDT: 1.3, Standard deviation: 0.94						
:	Mean of contrast PDR: 8.21, Standard deviation: 9.26						
	Unit: loop closure error/travelling distance (%)						

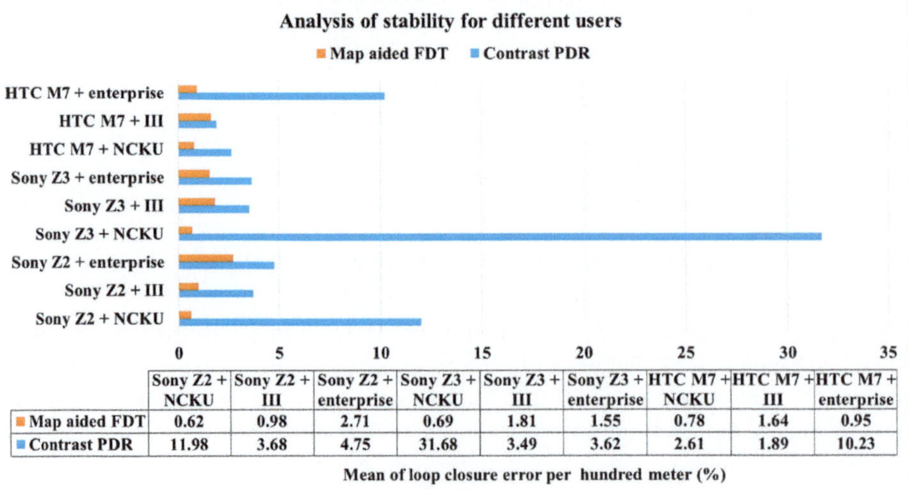

Figure 15. Analysis of stability for different users.

Figure 16 shows the mean values in terms of the same user using the same smartphone at different sites based on Table 5, where it implies adaptability to the place. In other words, this figure indicates the stability of algorithms in different environments in the same situation. For example, the mean error of 1.63% at the first column represents the mean error of the FDT system calculated from the user *A* using Sony Z2 in all fields. The figure shows that the map-aided FDT system has better stability when the field is different, regardless of user and smartphone. In addition, because the parameters of HTC M7 have been tuned with user C at NCKU, the group of HTC M7 performs better than other smartphones. It is obvious that the parameters of HTC M7 are not suitable for other smartphones which have different specifications and need individual tuning to reach optimum performance.

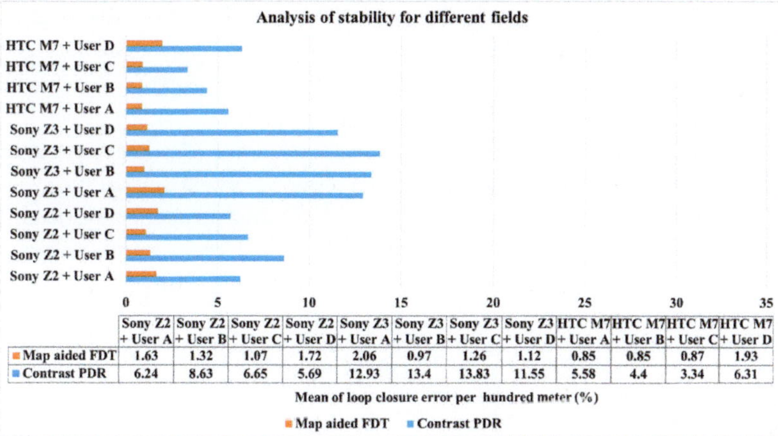

	Sony Z2 + User A	Sony Z2 + User B	Sony Z2 + User C	Sony Z2 + User D	Sony Z3 + User A	Sony Z3 + User B	Sony Z3 + User C	Sony Z3 + User D	HTC M7 + User A	HTC M7 + User B	HTC M7 + User C	HTC M7 + User D
Map aided FDT	1.63	1.32	1.07	1.72	2.06	0.97	1.26	1.12	0.85	0.85	0.87	1.93
Contrast PDR	6.24	8.63	6.65	5.69	12.93	13.4	13.83	11.55	5.58	4.4	3.34	6.31

Figure 16. Analysis of stability for different fields.

Figure 17 shows the analysis for each smartphone when used by the same user in the same field, where it gives some indication of the acceptance of algorithms by the different devices. For example, the mean error of 0.79% in the first column represents the mean error of the FDT system calculated from user *A* using three kinds of smartphones at NCKU. The NCKU cases of contrast PDR are worse because there is no initial heading correction, but the NCKU cases of FDT are better than in other fields. The group of PDR values in III has better performance than the group of PDR values in enterprise, indicating that the different characteristics between two fields need an individual tuning process. This figure shows that the map-aided FDT system has better stability even when the specification of sensors is new to this system.

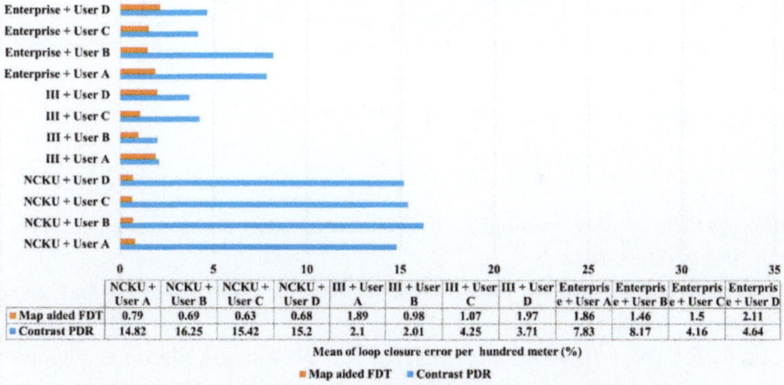

	NCKU + User A	NCKU + User B	NCKU + User C	NCKU + User D	III + User A	III + User B	III + User C	III + User D	Enterprise + User A	Enterprise + User B	Enterprise + User C	Enterprise + User D
Map aided FDT	0.79	0.69	0.63	0.68	1.89	0.98	1.07	1.97	1.86	1.46	1.5	2.11
Contrast PDR	14.82	16.25	15.42	15.2	2.1	2.01	4.25	3.71	7.83	8.17	4.16	4.64

Figure 17. Analysis of stability for different smartphones.

Ideally, the parameters and corresponding weights in KF of general PDR should be tuned for different fields and smartphones in order to compensate for the different levels of magnetic interference and differences within the sensor grade. This is because the accuracy of heading fusion depends on the accuracy of magnetometer and gyro as well as the parameters of the filter. The grade of accelerometers not only affects the pedometer but also indirectly influences the determination of step length. The model parameters of step length also depend on the characteristics of the user which needs calibration. According to these reasons, the contrast PDR system will perform poorly without individual pre-calibration and post-processing for each user, smartphone and field. However, the proposed FDT algorithm has better accuracy and stability in all cases without any further processing. It uses some simpler rules derived from one experiment and map information to mitigate the error from an inaccurate step length model and heading estimation. Therefore, the proposed map-aided FDT system provides stable average accuracy of about 1.3% based on 36 experiments in this study.

4. Conclusions

In order to simplify the necessary computations and time-consuming pre/post-processing needed in a practical indoor navigation algorithm, this research proposes an FDT algorithm which significantly improves the stability of pedestrian indoor navigation with an average loop closure error of about 1.3% as well as the standard deviation of about 0.94%. Even if some algorithms can achieve higher accuracy, they usually require a lot of time in pre-calibration and post-processing for individually tuning parameters such as the contrast PDR system in this study. The proposed system has better stability and adaptability without any further tuning for different users, places and smartphones. Furthermore, some systems achieve the same stability and accuracy based upon a complex algorithm, a large amount of calculation and various integrations for real-time indoor navigation. However, the proposed FDT system only uses the rules to integrate the map information and sensor data rather than topological or graphic matching and mathematical filter. The rule-based algorithm used has relatively low complexity because of the semantic logic conditions employed and many calculations reduced.

The map database and Bluetooth beacon are another important aiding source used in this study. Both the proposed FDT system and contrast PDR use the map database and Bluetooth beacon technique to enable both initialization and update measurements in real time. Although it is true that one can install more beacons to maintain navigational accuracy over a long time, the advantage of the FDT system is that it can provide acceptable accuracy even with only a small number of distributed beacons. Finally, the experiments performed for this research clearly demonstrate that the proposed FDT algorithm performs well for real-time pedestrian indoor navigation and also has high stability in different scenarios.

Acknowledgments: The authors would like to acknowledge the financial support provided by the Ministry of Science and Technology NSC (102-2221-E-006-137-MY3), all of the technical assistance from our supervisor, Kai-Wei Chiang, and field support from National Cheng Kung University, Institute for Information Industry.

Author Contributions: Kai-Wai Chiang conceived and designed the experiments; Jhen-Kai Liao and Guang-Je Tsai performed the experiments; Hsiu-Wen Chang and Jhen-Kai Liao analyzed the data and wrote the paper.

Conflicts of Interest: The authors declare no conflict of interest.

References

1. Mautz, R. Indoor positioning technologies. Habilitation Thesis, Eidgenössische Technische Hochschule Zürich, Zurich, Switzerland, 2012.
2. Harle, R. A survey of indoor inertial positioning systems for pedestrians. *IEEE Commun. Surveys Tutor.* **2013**, *15*, 1281–1293. [CrossRef]
3. Groves, P.D.; Pulford, G.W.; Littlefield, C.A.; Nash, D.L.J.; Mather, C.J. Inertial navigation versus pedestrian dead reckoning optimizing the integration. In Proceedings of the 20th International Technical Meeting of the Satellite Division of The Institute of Navigation, Fort Worth, TX, USA, 25–28 September 2007.

4. Lim, J.-M.; Yoo, S.-H.; Lee, K.-J.; Sung, T.-K. Integration of Pedestrian dr and Beacon-ap Based Location System for Indoor Navigation. Available online: http://ignss.org/LinkClick.aspx?fileticket=p%2BiDjFuLf2-o%3D&tabid=130&mid=527 (accessed on 24 December 2015).
5. Faragher, R.; Harle, R. An analysis of the accuracy of bluetooth low energy for indoor positioning applications. In Proceedings of the 27th International Technical Meeting of The Satellite Division of the Institute of Navigation (ION GNSS+ 2014), Tampa, FL, USA, 8–12 September 2014.
6. White, C.E.; Bernstein, D.; Kornhauser, A.L. Some map matching algorithms for personal navigation assistants. *Transp. Res. Part C Emerg. Technol.* **2000**, *8*, 91–108. [CrossRef]
7. Link, J.A.B.; Smith, P.; Viol, N.; Wehrle, K. Footpath: Accurate map-based indoor navigation using smartphones. In Proceedings of the International Conference on Indoor Positioning and Indoor Navigation (IPIN), Guimaraes, Portugal, 21–23 September 2011.
8. Fuller, R. *Fuzzy Reasoning and Fuzzy Optimization*; Turku Centre for Computer Science: Turku, Finland, 1998.
9. Trawiński, K.; Alonso, J.M.; Hernández, N. A multiclassifier approach for topology-based WiFi indoor localization. *Soft Comput.* **2013**, *17*, 1817–1831. [CrossRef]
10. Lai, Y.C.; Chang, C.C.; Tsai, C.M.; Lin, S.Y.; Huang, S.C. Development of a pedestrian indoor navigation system based on multi-sensor fusion and fuzzy logic estimation algorithms. Available online: http://www.int-arch-photogramm-remote-sens-spatial-inf-sci.net/XL-4-W5/81/2015/isprsarchives-XL-4-W5-81-2015.pdf (accessed on 24 Decmeber 2015).
11. Moafipoor, S.; Grejner-brzezinska, D.A.; Toth, C.K. A fuzzy dead reckoning algorithm for a personal navigator. *Navigation* **2008**, *55*, 241–254. [CrossRef]
12. Haibin, S.; Jiansheng, T.; Chaozhen, H. A integrated map matching algorithm based on fuzzy theory for vehicle navigation system. *Int. Conf. Comput. Intell. Secur.* **2006**, *1*, 916–919.
13. Zhang, Y.; Gao, Y. A fuzzy logic map matching algorithm. *Fifth Int. Conf. Fuzzy Syst. Knowl. Discov.* **2008**, *3*, 132–136.
14. Wu, Q.; Gu, X.; Luo, J.; Zhang, P.; Fang, X. A vehicle map-matching algorithm based on measure fuzzy sorting. *J. Comput.* **2014**, *9*, 1058–1065. [CrossRef]
15. Ren, M.; Karimi, H.A. A fuzzy logic map matching for wheelchair navigation. *GPS Solut.* **2011**, *16*, 273–282. [CrossRef]
16. Syed, S.; Cannon, M.E. Fuzzy logic based-map matching algorithm for vehicle navigation system in urban canyons. In Proceedings of the 2004 National Technical Meeting of the Institute of Navigation, San Diego, CA, USA, 26–28 January 2004.
17. Yim, J. Introducing a decision tree-based indoor positioning technique. *Expert Syst. Appl.* **2008**, *34*, 1296–1302. [CrossRef]
18. Sanchez-Rodriguez, D.; Hernandez-Morera, P.; Quinteiro, J.M.; Alonso-Gonzalez, I. A low complexity system based on multiple weighted decision trees for indoor localization. *Sensors* **2015**, *15*, 14809–14829. [CrossRef] [PubMed]
19. Renaudin, V.; Susi, M.; Lachapelle, G. Step length estimation using handheld inertial sensors. *Sensors* **2012**, *12*, 8507–8525. [CrossRef] [PubMed]
20. Umano, M.; Okamoto, H.; Hatono, I.; Tamura, H.; Kawachi, h.; Umedzu, S.; Kinoshita, J. Fuzzy decision trees by fuzzy ID3 algorithm and its application to diagnosis systems. In Proceedings of the Third IEEE Conference on Fuzzy Systems, IEEE World Congress on Computational Intelligence, Orlando, FL, USA, 26–29 June 1994.
21. Yuan, Y.; Shaw, M.J. Induction of fuzzy decision trees. *Fuzzy Sets Syst.* **1995**, *69*, 125–139. [CrossRef]
22. Mitra, S.; Konwar, K.M.; Pal, S.K. Fuzzy decision tree, linguistic rules and fuzzy knowledge-based network: Generation and evaluation. *IEEE Trans. Syst. Man Cybern. Part C Appl. Rev.* **2002**, *32*, 328–339. [CrossRef]
23. Olaru, C.; Wehenkel, L. A complete fuzzy decision tree technique. *Fuzzy Sets Syst.* **2003**, *138*, 221–254. [CrossRef]
24. Chiang, K.W.; Duong, T.T.; Liao, J.K. The performance analysis of a real-time integrated ins/gps vehicle navigation system with abnormal gps measurement elimination. *Sensors* **2013**, *13*, 10599–10622. [CrossRef] [PubMed]
25. Trimble. Timms Indoor Mapping. Available online: http://www.applanix.com/solutions/land/timms.html (accessed on 25 December 2015).

26. Chawathe, S.S. Beacon placement for indoor localization using bluetooth. In Proceedings of the 11th International IEEE Conference on Intelligent Transportation Systems, Beijing, China, 12–15 October 2008.
27. Boukerche, A.; Oliveira, H.A.B.; Nakamura, E.F.; Loureiro, A.A.F. Localization systems for wireless sensor networks. *Wirel. Commun.* **2007**, *14*, 6–12. [CrossRef]
28. Chen, R.; Pei, L.; Chen, Y. A smart phone based pdr solution for indoor navigation. In Proceedings of the 24th International Technical Meeting of the Satellite Division of the Institute of Navigation, Portland, OR, USA, 20–23 September 2011.
29. Zhou, Z.-M. The performance analysis of seamless navigation systems using smartphones. Master Thesis, National Cheng Kung University, Tainan, Taiwan, 2015.

6

A Novel Method of Multi-Information Acquisition for Electromagnetic Flow Meters

Wenhua Cui *, Bin Li *, Jie Chen and Xinwei Li

Academic Editor: Andreas Hütten

School of Mechatronic Engineering and Automation, Shanghai University, Shanghai 200072, China; jane.chen@shu.edu.cn (J.C.); LiXinwei@shu.edu.cn (X.L.)
* Correspondence: cuiwhlib@gmail.com (W.C.); sulibin@shu.edu.cn (B.L.)

Abstract: In this paper, a novel method is proposed for multi-information acquisition from the electromagnetic flow meter, using magnetic excitation to measure the fluid velocity and electrochemistry impedance spectroscopy (EIS) for both the fluid quality and the contamination level of the transducer. The impedance spectra of the transducer are measured with an additional electrical stimulus in series with the electrode measurement loop. The series connection mode instead of the parallel one improves the signal-to-noise ratio (SNR) of the fluid velocity measurement and offers a wide range of impedance measurements by using a sample capacitance. In addition, a multi-frequency synchronous excitation source is synthesized based on the method of dual-base power sequences for fast EIS measurement. The conductivity measurements in the range of 1.7 μS/cm–2 mS/cm showed a relatively high accuracy with a measurement error of 5%, and the electrode adhesion detection on both with coating and no coating showed the ability of the qualitative determination of the electrode adhesion, which validated the feasibility of the multi-information acquisition method for the electromagnetic flow meter (EMFM).

Keywords: electromagnetic flow meter; multi-information; electrical excitation; impedance spectroscopy; multi-frequency synchronous signal; photovoltaic cell

1. Introduction

With the development of industry, the traditional electromagnetic flow meter (EMFM) only for fluid-velocity measurement could not meet the growing demand for monitoring both the fluid quality (e.g., fluid conductivity) and the contamination level of the transducer (e.g., electrode adhesion, empty pipe, *etc.*) [1] in real time. Generally, the fluid information is integrated from multiple sensors [2,3], and the contamination level of the transducer is inspected with much manpower, which would result in poor real-time performance, low accuracy and high maintenance cost. Due to these facts, a high-efficiency method, not at the cost of multiple sensors and much manpower, is desirable for the EMFM to obtain multiple types of information on both the fluid and the transducer [4–6].

A promising tool is to measure the wetted electrode impedance of the EMFM with an additional electrical stimulus. This method takes full advantage of the sensing electrodes in direct contact with the conductive fluid, and the internal information of the transducer could be obtained *in situ* without installation of other sensors. Some papers and patents have been published on two aspects, including the electrical excitation method and the parameter extraction theory [7], in the recent decade. The works in [8,9] introduce an approach to monitor the change of the fluid conductivity and the electrode adhesion with impedance measurement at only one frequency. However, the transducer of the EMFM is a complicated system, whose impedance is influenced by many factors, such as fluid, electrode adhesion, temperature, *etc.*, and changes with the frequency. The measured single-frequency impedance could only be used for a rough evaluation. The work in [5] describes an exploration of the process diagnosis in the EMFM with the electrochemistry impedance spectroscopy (EIS) technique. However, it gives out only a preliminary result of 20% error for the measurement of medium conductivity in the range of about 25 μS/cm–4 mS/cm. The work in [10] presents an idea to detect the conductivity and electrode adhesion with multi-frequency impedance spectroscopy measurement. In all of the above methods, it should be noted that the additional electrical excitation source, is in parallel connection with the electrode measurement loop (called "parallel electrical excitation mode" in this paper) [11,12]. Because the electrical excitation source has the same ground with the instrumentation amplifier (INA), the equivalent input impedance of the INA would be reduced, leading to a further attenuation of the weak fluid-velocity signal and a deterioration of the SNR. Some product guides have stated to require the fluid conductivity to be higher than 10 μS/cm due to the imported electrical excitation source [13].

On the basis of our previous work [14], this paper introduces a novel method for the EMFM to acquire more valuable information with an additional electrical excitation. The electrical excitation is in series with the electrode measurement loop of the EMFM (called "serial electrical excitation mode" in this paper), overcoming the drawback of fluid signal attenuation in the parallel electrical excitation mode; a sample capacitance, identical to an open circuit for the sufficient low frequency during magnetic excitation, ensures a wide impedance measurement range in the period of electrical excitation; a multi-frequency synchronous (MFS) excitation source based on a photovoltaic (PV) cell [15], synthesized with the dual-base power sequences (DBPS) method proposed by the author in a previous paper [16], is employed for a fast impedance spectroscopy measurement. Furthermore, the electrical excitation module is electrically isolated from the electrode measurement loop by using a photoelectric-coupling control method.

The remainder of this paper is organized as follows. In Section 2, an equivalent model of the transducer is presented at first. The fluid-velocity signal loss under different modes of electrical excitation is then discussed. After that, a dual-excitation EMFM based on a PV cell is introduced, and the measurement principles of both the fluid-velocity and the electrode impedance spectroscopy are introduced. In Section 3, a brief description is given of the lab apparatus for the EMFM to obtain multiple information based on the data acquisition (DAQ) module and LabVIEW program. Additionally, the multi-frequency synchronous excitation signal based on the DBPS method is also introduced. In Section 4, two experiments, including medium conductivity identification and electrode adhesion detection, were carried out, and the experimental results are discussed.

2. Detecting Mechanism

2.1. Model of the Transducer

In the conventional EMFM, a pair of sensing electrodes are directly immersed in the conductive fluid, which would result in the formation of electrode-electrolyte interfaces. Since the conductive fluid moves in a direction perpendicular to the magnetic field with some velocity, induced voltage would be generated between the electrodes. With the effect of the Lorentz force, charged ions in the solution are attracted in the electrode-electrolyte interface [17–19]. This phenomenon is usually called the electrical double layer (EDL), the impedance of which could be described by a constant phase element (CPE) as follows:

$$Z_{CPE} = \frac{1}{Q(s)^n} = \frac{1}{Q(j\omega)^n} \tag{1}$$

where the exponent n is between 0.5 and 1.0, with 0.5 corresponding to the case of a Warburg impedance, a charge diffusion dominated element, and 1.0 to a pure capacitance element.

Some contributions have been made on the model of the transducer [5,7]. An equivalent circuit that models the transducer full of fluid is illustrated in Figure 1. The impedance (denoted "Z_x") between the electrode (s_1 or s_2) and the ground comprises a parallel connection of a capacitor C_d

representing the distributed capacitance between the electrode and the ground, a resistance R_m representing the solution resistance and a constant phase element (CPE) Z_{cpe}, which describes the behavior of the EDL, wherein the latter are connected in series with each other. Then, the impedance Z_x could be represented as:

$$Z_x(R_m, Q, n, C_d, \omega) = \frac{(Qs^n)^{-1} + R_m}{1 + ((Qs^n)^{-1} + R_m)C_d s} \quad (2)$$

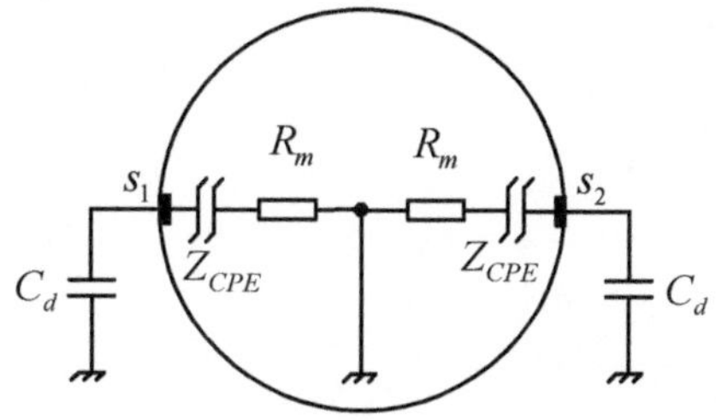

Figure 1. Equivalent circuit model of the transducer full of fluid.

Consider some values of R_m in the range of 0.25 up to 250 kΩ, which corresponds to the solution conductivity from 2000 down to 2 μS/cm, $Q = 10^{-5}(S \cdot cm^{-2} \cdot s^{-n})$, n = 0.8, C_d = 100 pF representing a short signal line. According to Equation (2), the Nyquist plot of the impedance Z_x over a frequency range from 2 Hz–10 kHz (restricted from the limited practical resources) is shown in Figure 2. In the low frequency range, the impedance Z_x appears as a line of a certain slope, mainly dependent on the EDL impedance Z_{cpe}. Yet, in the intermediate frequency range, Z_x is mostly determined by the solution resistance R_m, especially as the reactance of Z_x is close to zero. In the relatively high frequency range, Z_x, contributed mainly by the capacitance C_d and the solution resistance R_m, appears as a semicircle at the condition of R_m large enough.

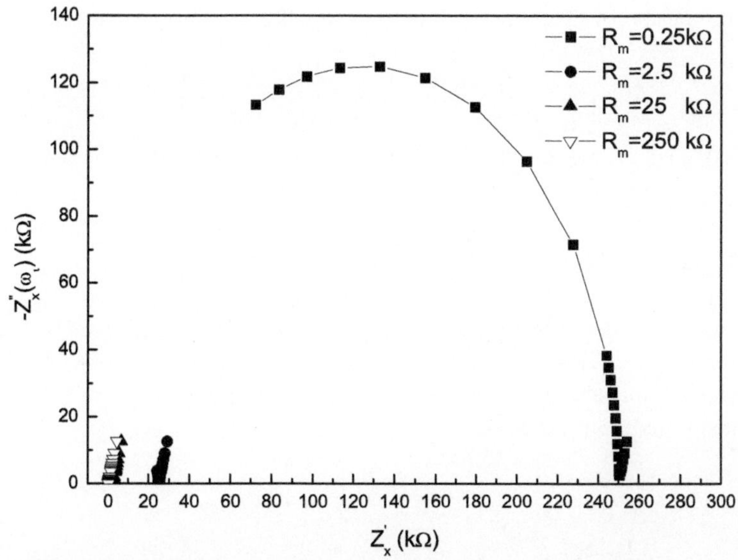

Figure 2. Nyquist plot of the impedance Z_x for known parameters.

Thus, the impedance Z_x contains the properties of the EMFM. By using the electrical impedance spectroscopy technique, the related valuable parameters could be extracted from the impedance spectra of the transducer.

2.2. Fluid-Velocity Signal Loss under Different Modes of Electrical Excitation

Without electrical stimulus, the electrode measurement loop of the EMFM is illustrated schematically in Figure 3a. It consists of flow-induced voltage E_m, electrode impedance Z_x and an INA, whose input impedance is Z_i and amplification coefficient is K_i. Deduced from Ohm's law, the differential voltage of the INA's input signal is given by:

$$V_{in-n} = \frac{Z_i}{Z_x + Z_i} \times E_m \tag{3}$$

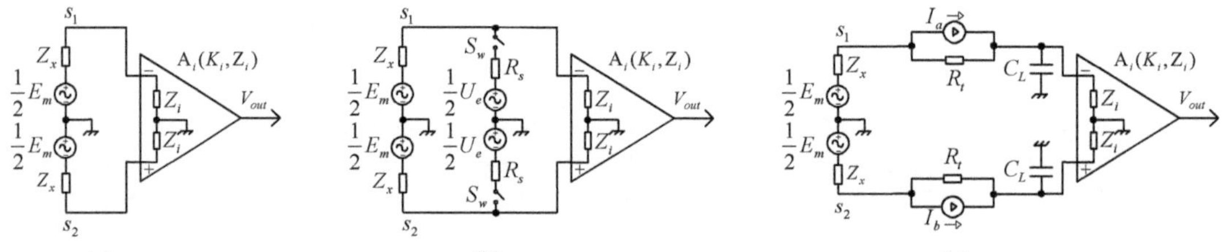

Figure 3. Principal diagram of the electrode measurement loop of the electromagnetic flow meter (EMFM). (**a**) No electrical excitation; (**b**) parallel electrical excitation; mode (**c**) serial electrical excitation mode.

To acquire the impedance spectrum of Z_x, there must be another known stimulus fed into the transducer via the sensing electrodes. As shown in Figure 3b, a common method is to import the electrical excitation in parallel connection mode, wherein the electrical excitation module, consisting of an excitation voltage source and a sample resistance R_s, is in parallel connection with the input of the INA via an analog switch S_w. Supposing the internal impedance of the voltage source U_e is zero, the impedance of the excitation module could be written as $Z_l = R_s + Z_w$, with Z_w denoting the switch impedance. Since magnetic excitation alternates with electrical excitation in the time domain in this research, U_e=0 in the period of magnetic excitation. Then, the differential voltage of the INA's input signal V_{in-p} is:

$$V_{in-p} = \frac{Z_i}{Z_x + Z_i} \times E_m \times (1 + \frac{Z_x}{(1 + Z_x/Z_i)Z_l})^{-1} \tag{4}$$

From Equations (3) and (4), the fluid-velocity signal loss percentage in the parallel electrical excitation mode could be calculated as:

$$p_{loss}(\%) = \frac{V_{in-n} - V_{in-p}}{V_{in-n}} = \frac{Z_x}{Z_x + (1 + Z_x/Z_i)Z_l} \times 100(\%) \tag{5}$$

Generally, Z_i is of the order of magnitude of 10^9 Ω, and that of Z_x would not be larger than 10^6 Ω, even at the condition of serious electrode adhesion. Moreover, Z_l is equivalent to Z_w during magnetic excitation, due to the fact that Z_w under off-state is of the order of magnitude of 10^7 Ω to 10^8 Ω, much larger than R_s. Thus, Equation (5) becomes:

$$p_{loss}(\%) = \frac{Z_x}{Z_x + Z_l} \times 100\% \tag{6}$$

Consider the impedance Z_x in cases such as low medium conductivity or serious electrode contamination. The related fluid-velocity signal loss percentages are summarized in Table 1. Supposing the value of Z_x=0.1 MΩ corresponding to the medium conductivity of 10 µS/cm, p_{loss} is calculated for some values of Z_x in the range from 0.1–10 MΩ at the conditions of Z_l = 10 MΩ, 30 MΩ, 100 MΩ. If Z_l = 10 MΩ, p_{loss}= 0.99% at Z_x = 100 kΩ; p_{loss} = 9.08% at Z_x = 1 MΩ; p_{loss} = 49.75% at Z_x = 10 MΩ. If Z_l = 100 MΩ, p_{loss} = 0.1% at Z_x = 100 kΩ; p_{loss} = 0.99% at

$Z_x = 1$ MΩ; $p_{loss} = 9\%$ at $Z_x = 10$ MΩ. In addition, to obtain a wide measurement range, several sample resistors with resistances of different orders of magnitude are required to be connected in parallel to match the measured impedance Z_x, making the fluid signal loss more serious.

Table 1. Fluid-velocity signal loss in the parallel electrical excitation mode.

Z_x (MΩ)	p_{loss}		
	$Z_l = 10$ MΩ	$Z_l = 30$ MΩ	$Z_l = 100$ MΩ
0.1	0.99%	0.33%	0.1%
1	9.08%	3.22%	0.99%
10	49.75%	24.81%	9%

The electrode measurement loop of the EMFM in the serial electrical excitation mode proposed in this paper, is illustrated schematically in Figure 3c. The electrical excitation module, consisting of a current source I_a (and I_b) and a resistance R_t, serving as a current-to-voltage converter (R_t=3.6 kΩ), is in series between the sensing electrode and the input of the INA. Moreover, there is a pair of sample capacitances C_l connected to the inputs of the INA. During magnetic excitation, the excitation source with zero current ($I_a = 0$ and $I_b = 0$) is identical to an ideal open circuit. On the basis of transient circuit analysis, the differential voltage of the INA's input signal V_{in-s} is:

$$V_{in-s} = \frac{Z_i}{Z_x + R_t + Z_i} \times E_m \times (1 - e^{-t/\tau}) \quad (7)$$

where $\tau = (Z_x + R_t) \times C_l$. Having $R_t \ll Z_i$, so Equation (7) becomes:

$$V_{in-s} = \frac{Z_i}{Z_x + Z_i} \times E_m \times (1 - e^{-t/\tau}) \quad (8)$$

By Equation (8), provided that the time interval between the magnetic excitation starting and the fluid-velocity signal sampling is larger than 5τ, the fluid-velocity signal loss percentage would be smaller than the value of $e^{-5} = 0.67\%$. Table 2 illustrates the related time intervals (5τ) to some values of Z_x in the range from 0.1–10 MΩ at the conditions of $C_l = 1$ nF, 10 nF, 100 nF. For most applications, such as medium conductivity measurement, electrode adhesion evaluation and empty pipe detection, the capacitance C_l would be selected with a value equal or smaller than 10 nF. If $C_l = 10$ nF, $5\tau = 50$ ms at $Z_x = 1$ MΩ; $5\tau = 500$ ms at $Z_x = 10$ MΩ. The above time intervals are feasible for the reason that the magnetic excitation cycle of the EMFM is usually 16 or 32 multiples of the power frequency cycle (20 ms in a 50-Hz system). Nevertheless, for special applications, such as the property detection of the electrode adhesion, to obtain the parameters of the EDL, the capacitance of C_l is suggested to be larger than 10 nF. At this time, even when the electrode is seriously contaminated (not fully insulated), the signal loss still could be small.

Table 2. Suitable time intervals in the serial electrical excitation mode.

Z_x (MΩ)	5τ (ms)		
	$C_l = 1$ nF	$C_l = 10$ nF	$C_l = 100$ nF
0.1	0.52	5.2	52
1	5	50	500
10	50	500	5000

Therefore, if the condition $t \geq 5\tau$ (t denotes the time interval) was fulfilled, the conventional flow-rate measurement equation [20] would be also available in the serial electrical excitation mode, given by:

$$V_O = K_i \times \frac{Z_i}{Z_x + Z_i} \times E_m = B \times D \times v \quad (9)$$

where B is the magnetic flux density, D is the diameter of the tube in meters and v is the flow velocity.

2.3. EIS Measurement Principle in Serial Electrical Excitation Mode

The schematic diagram of a dual-excitation EMFM based on the PV cell is shown in Figure 4. It consists of a pair of sensing electrodes (s_1 and s_2), a pair of specially-designed PV converters, a pair of sample capacitances C_l and an INA. Each PV converter consists of a PV cell and a resistor R_t (a current-to-voltage converter) in parallel connection. During electrical excitation, E_m = 0. The PV cell (PVC_a or PVC_b), activated by an adjacent LED(D_a or D_b), which is driven by a controllable AC current (S_a or S_b), generates the photovoltaic current (I_a or I_b) based on the photovoltaic effect. The current signal for electrical excitation could be expressed mathematically as:

$$\dot{I}_I = \dot{I}_a + \dot{I}_b \quad (10)$$

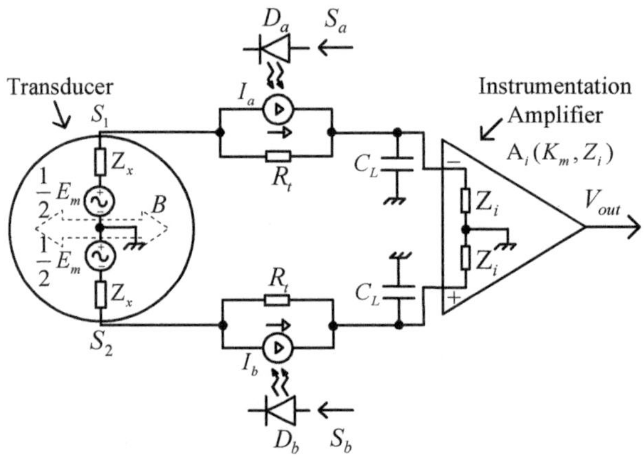

Figure 4. Schematic diagram of the dual-excitation EMFM based on a PV cell.

With the resistor R_t converting current to voltage, the terminal voltages of the two PV cells are E_a and E_b, respectively. Then, the voltage signal for electrical excitation becomes:

$$\dot{V}_I = \dot{E}_a + \dot{E}_b \quad (11)$$

Thus, the differential voltage of the INA's input signal could be calculated as:

$$\dot{V}_O = \frac{Z_{C_l}}{Z_x + Z_{C_l}} \times \dot{V}_I \quad (12)$$

where Z_{C_l} is the impedance of the sample capacitance C_l. Since the signal I_I generated for electrical excitation is a periodic odd signal, according to the Fourier series theory, V_I could be represented as an infinite summation of sinusoidal harmonics.

$$\dot{V}_I = \sum_{i=1}^{\infty} A_i sin(\omega_i t + \alpha_i) \quad (13)$$

where A_i and α_i are the magnitude and phase of the i-th harmonic, respectively.

At the same time, the resulting signal V_O also could be expressed as an infinite summation of sinusoidal harmonics.

$$\dot{V_O} = \sum_{i=1}^{\infty} B_i sin(\omega_i t + \beta_i) \quad (14)$$

where B_i and β_i are the magnitude and phase of the i-th harmonic, respectively. Therefore, substituting Equations (13) and (14) into Equation (12), solving for Z_x yields:

$$Z_x(\omega_i) = Z_x(\omega_i)' - jZ_x(\omega_i)'' = \frac{A_i}{B_i} \times \frac{sin\theta_i}{C_l\omega_i} - j(\frac{A_i}{B_i} \times \frac{cos\theta_i}{C_l\omega_i} - \frac{1}{C_l\omega_i}) \quad (15)$$

where $Z_x(\omega_i)$ is the impedance Z_x at the frequency f_i, $\theta_i = \alpha_i - \beta_i$. Then, from Equation (15), the amplitude and phase at the frequency f_i could be written as:

$$\begin{cases} |Z_x(\omega_i)| = \sqrt{(Z_x(\omega_i)')^2 + (Z_x(\omega_i)'')^2} \\ \angle Z_x(\omega_i) = -arctan(Z_x(\omega_i)''/Z_x(\omega_i)') \end{cases} \quad (16)$$

3. System Realization

After being amplified by the INA, both the fluid-velocity signal and the impedance signal will be processed with their respective circuits, since the signal amplitudes are not on the same order of magnitude. Figure 5 shows the block diagram of the EIS measurement system for the EMFM. It consists of the electrode measurement loop, a DAQ module and a LabVIEW program on a PC (personal computer). The signal processing for EIS measurement consists of two main parts: one is to generate a MFS excitation signal I_I; the other is to sample the resulting signals E_a, E_b and V_O. The three response signals, output from the respective amplifiers (INA_0, INA_1 and INA_2), are sampled simultaneously by the analog-to-digital converter located in the DAQ module. Next, the sampled signals would be sent to the PC via the USB cable and then processed by the LabVIEW program with the FFT technique. Finally, the impedance $Z_x(\omega_i)$ would be calculated according to Equation (15).

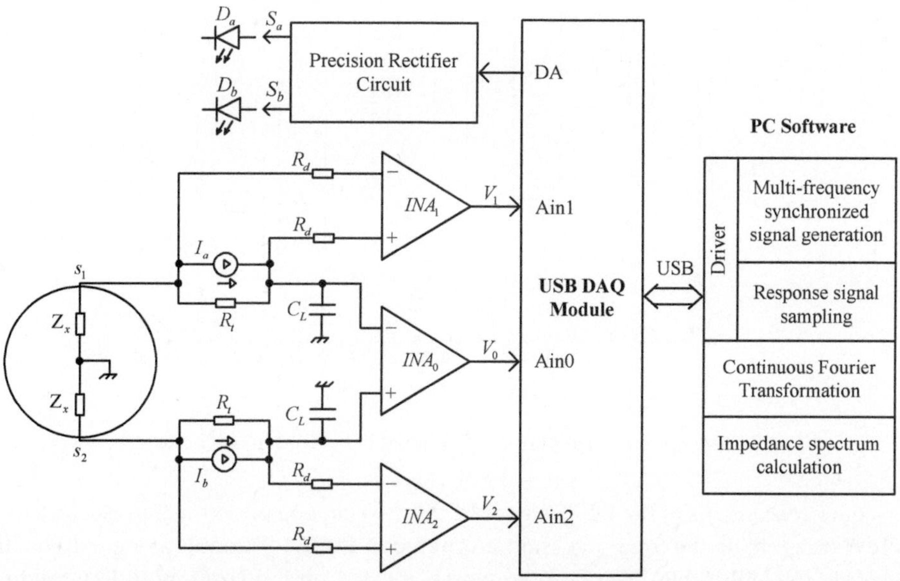

Figure 5. Waveform of the dual-base power sequence (DBPS) signal F(2, 3, ω_0, t) in one period.

For a fast impedance spectroscopy measurement, the DBPS method [16] is proposed for the synthesis purpose of the MFS signal. In this method, to make an M × N-frequency DBPS signal F(M, N, ω_0, t), M groups, each with N odd square waves, are applied simultaneously with a matrix of weight coefficients. The frequencies of all members of a group increase by powers of three, whilst the frequencies of the n-th ($0 \leq n \leq N-1$) member of each group could be constructed as a geometric sequence with a common ratio of two. The synthesized signal offers advantages, such as a wide frequency spectrum, a flat amplitude spectrum, simultaneous phase, high energy efficiency, *etc.* For simplicity, let M = 2, N = 3; the six-frequency signal F(2, 3, ω_0, t), containing primary harmonics at frequencies of $f_0, 2f_0, 3f_0, 6f_0, 9f_0$ and $18f_0$, could be expressed mathematically as:

$$F(2,3,\omega_0,t) = E \times \sum_{m=0}^{1} [2Sgn(sin(2^m \omega_0 t)) + 2\sum_{n=1}^{2} Sgn(sin(2^m 3^n \omega_0 t))] \tag{17}$$

where E is the amplitude, ω_0 is the fundamental angular frequency and Sgn(x) is the signum function.

$$Sgn(x) = \begin{cases} 1 & for\, x > 0 \\ 0 & for\, x = 0 \\ -1 & for\, x < 0 \end{cases} \tag{18}$$

Provided E = 1, six sequence functions from 3Sgn(sin(ω_0t)) to 2Sgn(sin(18ω_0t)) are applied to compose F(2, 3, ω_0, t) according to Equation (17). Since the highest frequency of the sequence functions is $18f_0$, the period T_0 of F(2, 3, ω_0, t) could be divided into 36 units; the resultant vector could be written as:

$$\begin{aligned} F(2,3,\omega_0,t) = [&14, 10, 10, 2, 10, 6, 6, 2, 10, -4, -4, -8, 8, 4, 4, -4, 4, 0, 0, -4, 4, \\ &-4, -4, -8, 8, 4, 4, -10, -2, -6, -6, -10, -2, -10, -10, -14] \end{aligned} \tag{19}$$

The time-domain waveform of F(2, 3, ω_0, t) is shown in Figure 6.

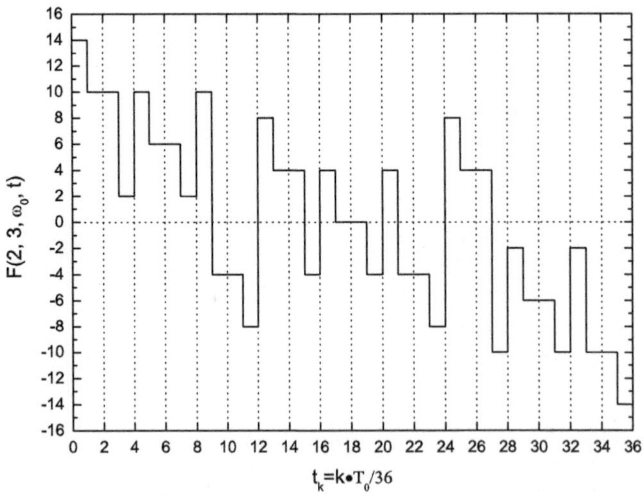

Figure 6. Waveform of the DBPS signal F(2,3,ω_0,t) in one period.

The sequence elements of the DBPS signal F(2, 3, ω_0, t) in one period would be constructed with the LabVIEW program on the computer and transmitted to the DAQ module along with the frequency $f_{clk} = 36f_0$ via the USB cable. The D/A converter located on the DAQ module would circularly produce every element with the frequency f_{clk} and output via the terminal DA (output port of the DAQ module depicted in Figure 5). Thus, a voltage signal of F(2, 3, ω_0, t) is generated and then would be split into positive and negative (S_a and S_b) via a precision rectifier circuit, to drive the LEDs of both Da and Db, respectively. Based on the photovoltaic effect, the PV cells would produce proportional induced currents I_a and I_b, which exactly constitutes the excitation current I_I. The frequency of the

DBPS signal for the EIS measurement could be adjusted via the frequency f_{clk}; e.g., if f_{clk} = 360 Hz was set, the primary harmonic frequencies of the DBPS signal would be in the range of 10–180 Hz; If f_{clk} = 36 kHz was set, they would be in the range of 1–18 kHz.

4. Experimental Results and Discussion

To validate the method of multi-information acquirement for the EMFM, two experiments, including medium conductivity identification and electrode adhesion detection, were carried out with stationary fluids. The EMFM used for experiments had a nominal diameter of 100 mm, and the signal line was 1 m long.

4.1. Medium Conductivity Identification

The conductivity identification experiments covered a conductivity range from 1.7 μS/cm–2.1 mS/cm. The electrolyte solutions for experiments were mixed with different portions of deionized water (conductivity of about 0.1 μS/cm, a resistance of about 10 MΩ) and KCl, at a temperature of about 25 °C. Additionally, the reference conductivity values were provided by a commercial conductivity meter.

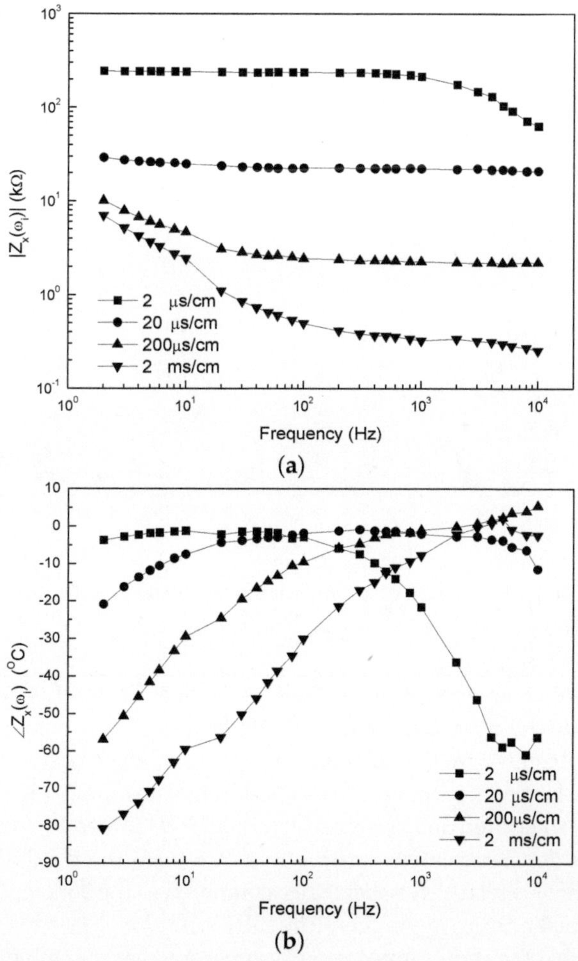

Figure 7. Bode plot of the measured impedance spectrum of the transducer for different solution conductivities. (**a**) Absolute value of impedance *vs.* frequency; (**b**) Phase of impedance *vs.* frequency.

The measured impedance spectrum of the transducer for different solution conductivities is illustrated in Figure 7 (Bode plot) and Figure 8 (Nyquist plot). The Nyquist plot is in agreement with the one shown in Figure 2, so that the model of the transducer introduced in Section 2.1 is feasible. In the low-frequency range of 2–100 Hz, the absolute value of impedance $|Z_x(\omega_i)|$ for the conductivities, such as 200 μS/cm and 2 mS/cm, have an obvious down-trend with the increase of the frequency in Figure 7a, which is shown as an oblique line in the Nyquist plot. The dominant effect of the linear portion comes from the EDL, and additionally, it would occur in a higher frequency range as the solution conductivity increases. In the intermediate-frequency range of about 0.1–1 kHz, the spectrum of Z_x keeps flat for the conductivities in the range of 2 μS/cm–2 mS/cm in Figure 7a, and the related $-Z''_x(\omega_i)$ is close to zero in Nyquist plot. This is because Z_x is mainly determined by the solution resistance at this moment. Especially, when $\angle Z_x(\omega_i)$ is zero, $|Z_x(\omega_i)|$ equals the solution resistance. In the high-frequency range (>1 kHz), $|Z_x(\omega_i)|$ for the conductivity of 2 μS/cm decreases with increasing the frequency, which is shown as a semicircular arc in the Nyquist plot. According to the EIS theory, the arc portion is mainly caused by a resistor parallel with a capacitance. In our cases, this is caused by the solution resistance and the distributed capacitance of the signal line, because the effect of the dielectric constant of the fluid would become visible only at much higher frequencies. Whilst for the solution conductivity equal or higher than 200 μS/cm, the arc portion disappears because the solution resistance is so small that the effect of the distributed capacitance is not obvious in the considered frequency region.

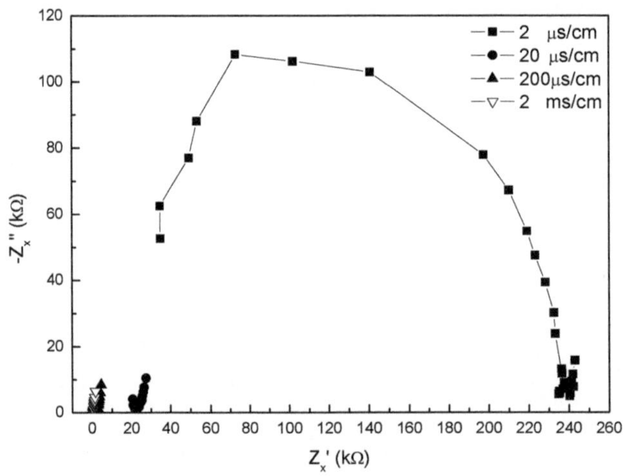

Figure 8. Nyquist plot of the measured impedance spectrum of the transducer for different solution conductivities.

According to the above analysis, the solution resistance R_m nearly equals the absolute value of impedance $|Z_x(\omega_i)|$ at the condition of the related phase $\angle Z_x(\omega_i)$ close to zero. The medium conductivity could be calculated by the formula $\sigma = k/(2R_m)$, where k is a constant associated with the electrode area and electrode spacing. The experimental results (in Figure 9) showed a linear relationship between the measured and reference conductivity. Limited by the measurement accuracy of the current system, the measurement error is within 5% for the conductivities in the range of 1.7 μS/cm–2 mS/cm, which is relatively satisfactory compared to the 20% for the conductivities in the range of about 25 μS/cm–4 mS/cm in the literature [5].

It should be stated that the above impedance spectroscopy was measured by using the traditional frequency-sweep approach. Nevertheless, in practical applications, the effective frequencies for conductivity identification mainly concentrate in the range of 0.1–1 kHz. Thus, the solution conductivity could be measured with one-time electrical excitation by using a 2 × 3-frequency DBS signal, which contains six frequencies in the range of 0.1–1.8 kHz.

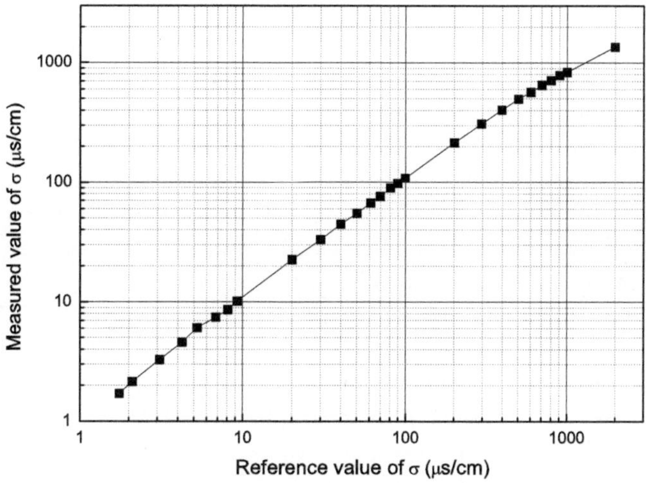

Figure 9. Measured value of the medium conductivity *vs.* the reference value.

4.2. Electrode Adhesion Detection

It is a little difficult to build a practical environment for the experiment of electrode adhesion detection. In this research, the experiment was carried out with four cases of soft cream, hard cream, chewing gum and MoS2. In each case, the electrode surface was cleaned at first and then fully covered with dirt. After that, the pipe was filled with tap water. In addition, for the purpose of comparative analysis, the impedance spectroscopy of the transducer full of tap water was also tested at the condition of no electrode fouled. The measured impedance data for both medium conductivity identification and electrode adhesion detection are analyzed using the Zsimpwin software. The parameters, such as R_m (the solution resistance), Q and n (the parameters to describe the EDL) for different cases, obtained based on the model of the transducer defined by Equation (2), are listed in Table 3.

Table 3. Three parameters R_m, Q and n for different cases. Case 1: no electrode fouled; Case 2: both electrodes fouled.

Conditions		R_m k(Ω)	Q $(S \cdot cm^{-2} \cdot s^{-n})$	n
Case 1	200 μS/cm	2.4	14.7×10^{-6}	0.8
	400 μS/cm	1.24	15.7×10^{-6}	0.81
	600 μS/cm	0.85	16×10^{-6}	0.81
	1 mS/cm	0.5	16.7×10^{-6}	0.75
	tap water	1.518	15×10^{-6}	0.78
Case 2	soft cream	0.946	26×10^{-6}	0.86
	hard cream	2.57	21×10^{-6}	0.8
	chewing gum	18	12×10^{-6}	0.63
	MoS2	0.007	5×10^{-6}	0.23

Obviously, for the case of no electrode fouled (Case 1), the parameter R_m changes with conductivity, whereas the parameters Q and n, which denote the properties of the EDL, almost keep constant, such as Q in the range of about 14.7×10^{-6}–16.7×10^{-6} $(S \cdot cm^{-2} \cdot s^{-n})$, n of about 0.8. Yet, when the electrodes are fouled, all three parameters R_m, Q and n, change with the coating. Compared to the reference values of tap water ($Q = 15 \times 10^{-6}$, $n = 0.78$), Q increases to 21×10^{-6}, 26×10^{-6}, when the electrodes are coated with adhesive materials of relatively high conductivity, such as soft cream, hard cream. Additionally, when the adhesive conductivity is a bit low, such as gum, MoS2, the related Q decreases to 12×10^{-6}, 5×10^{-6}, and the related n decreases to 0.63, 0.23.

Thus, the parameters Q and n could be used to determine whether electrode contamination occurs and to evaluate the properties of the adhesive material qualitatively; e.g., when both Q and n change, the electrode of the transducer should have been contaminated. Furthermore, if Q and n get smaller, the adhesive conductivity would be relatively low, and the related electrode impedance Z_x at the magnetic excitation frequency might be large enough to affect the fluid velocity signal measurement. If Q gets larger, the adhesive conductivity would be relatively high, and the related electrode impedance Z_x would not be large enough to affect the fluid signal measurement. The relationship between the impedance Z_x and the fluid-velocity signal, expressed as Equation (9), could be used to compensate the fluid signal loss caused by the change of Z_x, especially for the cases of the electrode seriously contaminated or the fluid with low conductivity.

It should be stated that, for the reason that the practical contamination is very much complicated, the above results obtained based on only one simple model are rough. This paper is only a primary work; bigger or better studies need to be carried out to investigate the electrode adhesion detection.

5. Conclusions

This research work mainly contributes a multi-information acquisition method for the EMFM. A dual-excitation electromagnetic flow meter based on PV cells was designed and implemented. With fluid velocity measured with magnetic excitation, more valuable parameters could be obtained based on the EIS measurement of the transducer, by applying an additional electrical excitation in series connection with the original electrode measurement loop and a sample capacitance. The flow-rate signal loss would be lower than 0.67%, so long as a time interval of 5τ is taken into account between the magnetic excitation starting and the fluid-velocity signal sampling. The multi-frequency synchronous excitation signal based on the DBPS method has features such as a wide frequency spectrum, flat amplitude spectrum, high efficiency, *etc.* Conductivity identification experiments were carried out for the solution conductivities in the range of 1.7 μS/cm–2 mS/cm, with measurement error within 5%. Based on the proposed transducer model, the properties of the electrode coating could be judged roughly.

Further research should focus on the electrode adhesion detection. Moreover, both the software and the hardware of this system are required to be improved for higher accuracy.

Acknowledgments: This research was supported by the Fund for the Doctoral Program of Higher Education of China (20123108110011). The authors also gratefully acknowledge the work of the reviewers.

Author Contributions: Wenhua Cui performed the experiments and analyzed the data. All together planned the experiments, evaluated the results and wrote the paper.

Conflicts of Interest: The authors declare no conflict of interest.

References

1. Schrag, D.; Hencken, K.; Andenna, A. From Flowmeter to Advanced Process Analyser. In Proceedings of the International Conference on SENSOR + TEST'11, Nurnberg, Germany, 6–9 June 2011; pp. 67–72.
2. Aisopou, A.; Stoianov, I.; Graham, N.J.D. In-pipe water quality monitoring in water supply systems under steady and unsteady state flow conditions: A quantitative assessment. *Water Res.* **2012**, *46*, 235–246.
3. Lambrou, T.P.; Anastasiou, C.C.; Panayiotou, C.G.; Polycarpou, M.M. A low-cost sensor network for real-time monitoring and contamination detection in drinking water distribution systems. *IEEE Sens. J.* **2014**, *14*, 2765–2772.
4. Hencken, K.; Schrag, D.; Grothey, H. Statistical Signal Processing for the Detection of Gas Bubbles in a Magnetic Flowmeter. In Proceedings of the IEEE Sensors, Lecce, Italy, 26–29 October 2008; pp. 1410–1413.
5. Hencken, K.; Schrag, D.; Grothey, H. Impedance spectroscopy for diagnostics of magnetic flowmeter. In Proceedings of the IEEE Sensors, Lecce, Italy, 26–29 October 2008; pp. 1112–1115.
6. Cao, J.L.; Li, B. Study on methods of empty pipe detection for electromagnetic flowmeter. *Chin. J. Sci. Instrum.* **2006**, *27*, 643–647.

7. Maalouf, A.I. A validated model for the electromagnetic flowmeter's measuring cell: Case of having an electrolytic conductor flowing through. *IEEE Sens. J.* **2006**, *6*, 623–630.
8. Ishikawa, I.; Oota, H. Electromagnetic Flowmeter. U.S. Patent 6,804,613, 12 October 2004.
9. Schrag, D.; Grothey, H.; Hencken, K.; Naegele, M. Method and Device for Operating a Flow Meter. U.S. Patent 7,546,212, 9 June 2009.
10. Rufer, H.; Drahm, W.; Schmalzried, F. Method for Predictive Maintenance and/or Method for Determining Electrical Conductivity in a Magneto-inductive flow-Measuring Device. U.S. Patent 8,046,194, 25 October 2011.
11. Cao, J.L. Study on Muti-parameter Measurement Electromagnetic Flowmeter and its Realization Techniques. Ph.D. Thesis, Shanghai University, Shanghai, China, 2007.
12. Shen, T.F.; Li, B.; Cao, J.L. Parallel Type Electromagnetic Flowmeter with Dual Excitations. CN Patent 100,491,928, 27 May 2009.
13. Yokogawa Electric Corporation. User's Manual-AXF Magnetic Flowmeter Integral Flowmeter [Software Edition]. Available online: https://www.yokogawa.com/fld/pdf/AXF/IM01E20C02-01E.pdf (accessed on 23 December 2015).
14. Cui, W.H.; Li, B. A new Approach for the Parameter Measurement of Fluid Conductivity in an Electromagnetic Flow Meter. In Proceedings of the 16th International Flow Measurement Conference, Paris, France, 24–26 September 2013; pp. 274–277.
15. Kitai, A. *Principles of Solar Cells, LEDs and Diodes: The Role of the PN Junction*; John Wiley & Sons: Hoboken, NJ, USA, 2011.
16. Cui, W.H.; Li, B.; Li, X.W. A New Synthesis Method of Uniform-amplitude Multi-frequency Synchronous Excitation Signals Based on Dual-base Power Sequences. *Math. Probl. Eng.* under review.
17. Chang, B.Y.; Park, S.M. Electrochemical impedance spectroscopy. *Annu. Rev. Anal. Chem.* **2010**, *3*, 207–229.
18. Macdonald, J.R.; Barsoukov, E. *Impedance Spectroscopy: Theory, Experiment, and Applications*; Wiley-Interscience: New York, NY, USA, 2005.
19. Bard, A.J.; Faulkner, L.R. *Electrochemical Methods: Fundamentals and Applications*, 2nd ed.; John Wiley & Sons Inc.: Hoboken, NJ, USA, 2001.
20. Shercliff, J. *The Theory of Electromagnetic Flow-Measurement*; Cambridge University Press: London, UK, 1962.

7

Moving Object Detection on a Vehicle Mounted Back-Up Camera

Dong-Sun Kim * and Jinsan Kwon

Academic Editor: Felipe Jimenez

Embedded Software Convergence Research Center, Korea Electronics Technology Institute, Saenari-ro 25, Bundang-gu, Seongnam-si, Gyeonggi-do 13509, Korea; jinsan.kwon@keti.re.kr
* Correspondence: dskim@keti.re.kr

Abstract: In the detection of moving objects from vision sources one usually assumes that the scene has been captured by stationary cameras. In case of backing up a vehicle, however, the camera mounted on the vehicle moves according to the vehicle's movement, resulting in ego-motions on the background. This results in mixed motion in the scene, and makes it difficult to distinguish between the target objects and background motions. Without further treatments on the mixed motion, traditional fixed-viewpoint object detection methods will lead to many false-positive detection results. In this paper, we suggest a procedure to be used with the traditional moving object detection methods relaxing the stationary cameras restriction, by introducing additional steps before and after the detection. We also decribe the implementation as a FPGA platform along with the algorithm. The target application of this suggestion is use with a road vehicle's rear-view camera systems.

Keywords: moving object detection; mixture of Gaussians; pyramidal Lucas-Kanade optical flow; backup collision intervention

---

1. Introduction

Improvements to the function, reliability, and manufacturing process of various sensors making them small, sensitive, and yet strong enough, have made it possible to use those sensors in automobiles [1–4]. One of its major applications is Backup Collision Intervention (BCI), which informs drivers of obstacles behind the vehicle by giving visual, aural, or tactile feedback. So far this technology has utilized laser, ultrasonic, microwave radar, and vision sensors to measure the distance from the vehicle to obstacles. Although they differs by specific application, those on-board sensors work as the main components for the BCI function independently or in conjunction with others. Drivers of vehicles with extended bodies or carrying oversized loads may have difficulty in seeing behind them directly through the rear-view mirror. As a part of BCI components, on-board monitors connected to a camera mounted on the back of the vehicle provide a visual aid to its user, helping the driver avoid obstacles behind while backing-up. In addition to covering the blind spots, further processing of the video feeds may provide more features such as moving object detection. The processed information then can be integrated back to the original video feed.

Although many other sensors detect stationary obstacles well based on the distance between objects and the vehicle, we focus on moving objects that are beyond the active sensor's range that might still require the user's attention while backing up. As a practical approach, we do not assume the obstacles to be of any specific type, therefore, in our suggested method the target objects can range from pedestrians to other vehicles. If available, the detection results may also further processed in combination with the results from other sensors to provide on the screen information in integrated form such as distance, speed, and direction.

Meanwhile, vision-based moving object detection is a major theme in the computer vision arena, and vast number of suggestions has been made. However, many of them assume its viewpoint to be stationary [5–8], because the movement introduced to its vision source is also reflected on the background of the taken scene. Without proper distinction, the mixed motion between the background and foreground object is hard to separate out. Since our target application is a vision-based BCI system mounted on the back of a vehicle, movement of the viewpoint is inevitable. Therefore, in our suggestion, we first compensate for the background movement, called ego-motion, for later processing using the traditional object detection methods. This is possible because of the unique characteristics of the slow-moving back-up movement, which introduces mild motions across the scene. The uniform vectors found all over the frame then be extracted, and based on those vectors, a rough compensation of the background will be made. The actual object detection routine is then performed using the difference between a frame and the compensated background frame at the time. This paper first examines the previous methods for object detection. Details about the suggested procedure then follow, and we discuss about the implementation of the suggestion as a hardware platform.

2. Background Works

In computer vision, finding target objects without additional information from other sensing sources is a challenging job [9]. Despite its ill-posed characteristics, various object detection algorithms having been introduced for vision-based applications such as surveillance cameras, robotics, intelligent systems, and smart devices. Yilmaz *et al.* classified those algorithms into four types according to the completeness of the detected target objects: feature point detectors, image segmentation, supervised learning, and background subtraction [10].

Among other methods, we focus on background subtraction methods to extract the moving objects from image frames. Supervised learning methods, on the other hand, use object classifiers in the form of decision trees or networks [11–13]. To build those decision-making networks, analysis of the every type of target object is required. However, due to the fact that any kind of object can obstruct the vehicle's course, it is impossible to prepare a tree for all types of objects, including unknown ones. Therefore learning-based object detection should not be used in a vehicle backing-up scenario.

In this section, we introduce background subtraction methods first, followed by optical flow which widely finds motion vectors between two consecutive frames. In fact, those vectors are key components for deriving the movement of vision source itself. The optical flow will also be used later in detecting, distinguishing, and finding objects which are moving in the source frames.

2.1. Background Subtraction

Background subtraction is a method of separating moving foreground objects from stationary background images. In a video sequence taken from a traffic monitoring camera, for example, vehicles can be separated out from stationary objects such as road markings and traffic signs which are found on the same location over frames. This separation is necessary to transform the video data into traffic information since the moving vehicles are of more interest than the road itself in most cases. To separate the foreground and background objects, building the information about background objects is essential. For this purpose, the background image is kept in various forms during the subtraction process. The actual form differs among methods though, keeping the background information as accurate as possible is critical. Among the background subtraction methods, one common background extraction method is to hold pixels relatively consistent across two or more consequent images, while rejecting other pixels whose value rapidly changes.

Frame differences [5] were one of earliest suggestions of a background subtraction process. This generates a background image by simple frame differences between two consequent images, making the method depend solely on the previous frame. Due to its simplicity, this method has modest computational loads and the background image can be generated with only two frames. The simplicity also makes the background references highly adaptive, resulting in very fast updating of background changes. The major drawback is however that moving foreground objects may easily become part of the background if the object is stopped for more than one frame. The interior pixels of an object also may not be distinguished correctly if the inside has a uniformly distributed intensity value, leaving no difference between the two frames.

Instead of simple subtraction, approximated median [6] keeps an accumulation of approximated pixel values, continuously updating the background image in the result. During the approximation, the generated background converges to a median value of all frames, therefore, background objects which are stationary across several frames may be better imprinted than foreground moving objects which have less opportunity to converge. Although the number of frames used in this convergence process depends on the pixel values of the objects, it is generally much longer than the simple frame difference. Temporarily stopped foreground objects or ones with a plain interior can now be better filtered than with simple frame differences, unless the object becomes part of the background by appearing in the same position longer than the originally occluded background.

Gaussian mixture models [7,8], on the other hand, process the input frames in different ways. Unlike the abovementioned methods, background components are accumulated as terms of Gaussian distribution functions, and the terms do not directly depict a background image at a specific moment. Rather, the Gaussian distribution functions decide whether a pixel from the input image is foreground or background. This statistical determinant is effective for minor changes in the background, such as waving leaves, moving clouds, or raindrops. Furthermore, using mixtures of such functions makes the method multimodal. That is, in the moving clouds example, both clouds and occluded sky should be considered as background. Thus, the distribution must be at least bimodal to classify the clouds and sky as background at the same time.

Recent suggestions on the background subtraction, on the other hand, use sophisticated techniques in combination with those basic ones. Several methods use advanced statistic models to extract backgrounds from moving viewpoints [14–17], while others suggests the use of neural networks or outlier detection models [18,19]. These subtraction models are further categorized into 17 groups according to their main characteristics [20].

2.2. Optical Flow

Rather than background subtraction, another strategy to find moving objects from consequent video frames is by using the optical flow of the frames. According to its definition, optical flow is a motion of apparent movement of brightness patterns in an image, making it sensitive to light sources [21]. Finding the movement can be described as finding the displacement δ which minimizes ε in the following equation:

$$\varepsilon(\delta_x, \delta_y) = \sum_{x=u_x-w_x}^{u_x+w_x} \sum_{y=u_y-w_y}^{u_y+w_y} \left(I_1(x,y) - I_2(x+\delta_x, y+\delta_y)\right) \quad (1)$$

with a given point x, y in an image I. Searching for such a value is performed within a small window w. The window should be introduced because it becomes an aperture problem without the window. An adequate window size provides hints on the direction of vectors from searching through neighbouring points. To find displacements between two consequent frames, various optical flow estimation techniques are introduced [21–24]. The derived displacements, or flow vectors, contain both the directions and magnitude of pixels, and these vectors are grouped together to form objects. Besides optical flow methods, further classification processes are needed to separate the vector group, in turn which represents the movement of an object, into foreground or background. For this purpose, various statistical grouping algorithms, such as RANSAC and K-means clustering are used [25,26].

Both background subtraction and optical flow methods have been widely used for applications including intrusion or motion detection from video stream with stationary cameras. Recent studies showed that the use of grouping algorithms with optical flow also alleviates the limitation of using stationary cameras for some level by further classification of feature points into two different motion groups [25,26]. Rather using classification methods, in our suggestion, we combine a background subtraction and an optical flow process to eliminate the stationary restriction and derive backgrounds at the same time. The key benefits of using this combinational approach are that the background motion, known as ego-motion, can be cancelled by its optical flow vectors, and the each detection result from optical flow and background subtraction can be cross-checked later.

3. Moving Object Detection

In this section, we present the proposed procedure for moving object detection. To achieve better detection and less false positives than previous standalone approaches with moving view sources, our proposed scheme uses both background subtraction and optical flow methods simultaneously. As shown in the Figure 1, this combinational scheme is composed with the following two step approach: background motion compensation, and object detection. The actual detection of moving objects is done in the object detection part by using the optical flow between a frame and the derived background at the moment. In our suggestion, a background frame is defined by the difference between two video frames, thus, object detection can be started with a sequence having at least two frames.

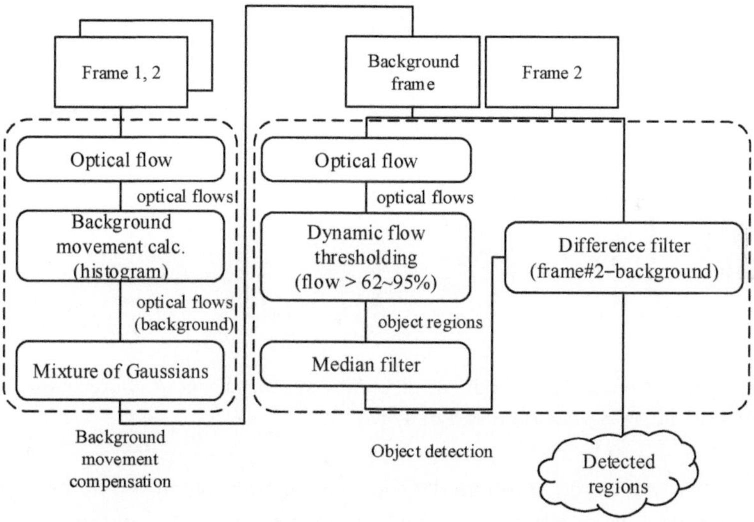

Figure 1. Overall detection procedure of the proposed scheme.

3.1. Background Motion Compensation

As the camera on the vehicle moves freely, the image sequences obtained from the camera contain motions of both background and target objects. Therefore, when a new frame arrives, the ego-motion of background at that time needs to be eliminated first. In our scheme, the ego-motion is induced by optical flow estimation between the newly captured frame and previously derived background components. The background components, on the other hand, are derived by Gaussian mixture models. Since we have movements in the background, the multimodal distribution functions discussed previously are needed for distinguishing the moving background objects. In this approach, background frames are stored as Gaussian mixture function terms, and if the ego-motion vector is induced, the entire background terms can be shifted back along the vector, resulting in compensated background terms.

To derive the motion vector disparities of each pixel, we use an optical flow method which is based on the pyramidal Lucas-Kanade optical flow and warp theory [23,27]. In this method, the overall optical flow is derived through several scales. The input images are scaled down into specified sizes for each level, similar to the pyramid structure. Optical flow results on each level are then scaled back and summed to form the overall disparities for all pixels between the two frames like one in the Figure 2. After the disparity between previous background and current frame has been derived, it is possible to extract the ego-motion from the entire pixel flow. This normalization process incorporates the generation of histogram statistics on all displacements and selection of the most frequent flow as the representative background movement. In our suggestion, the size of each histogram section is adjusted so there exist 100 sections. The estimated ego-motion vector is composed with the average values of the elected sections on each axis, such as −0.75 and −0.02 in Figure 3.

Figure 2. Optical flow between two consecutive frames. In this result, the vectors representing background motions stay consistent while moving object region shows major changes.

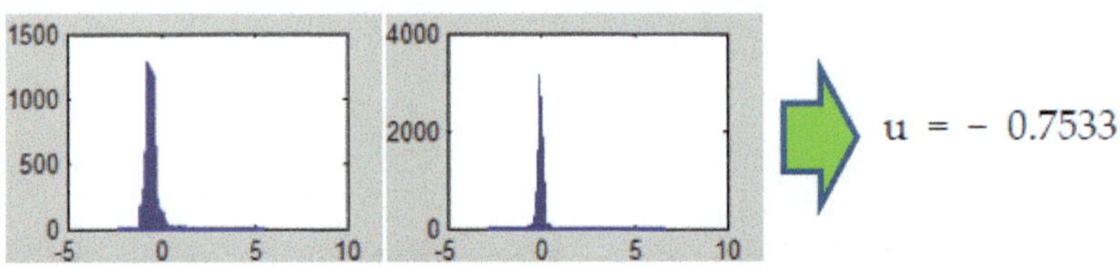

Figure 3. Histograms on the optical flow results. We assume the average value of the most frequent section found on the histogram as the ego-motion.

Before compensating the ego-motion, the background component should be built first. There are several methods to do this, as we discussed in the previous section, but we use Gaussian mixture models for their better adaptation to partial movements of the background itself. In this model, pixels of the background frame are stored as mean terms for the Gaussian functions. The mean and deviation terms are affected by upcoming frames with predefined update factors, therefore, the mean term can be changed over frames. At the very beginning of background subtraction, the mean term is initialized by the first frame directly, assuming it is a background image.

With the estimated vector and the subtracted background components, now it is possible to compensate the background motion. This action is done by a simple shift of the mean and deviation terms of the Gaussian Mixture functions along with the estimated vector, but being aware that the estimated movement is usually less than one, so the shift action should be handled as a convolution between the terms and the motion vector. The mean term after the shift action now can be seen as the compensated background.

3.2. Object Detection

Although it may seem possible to extract target objects using the frame differences between the compensated background and the current frame, the simple comparison between those frames contains both false-positives and false-negatives. This leads the detection results to have way too much noise and yet even less detection tendency than the results obtained sing optical flow methods alone.

To solve this problem, we use optical flow between the background and the current frame to deduce the motion vectors of all pixels. At this point, the resulting vectors which have different magnitude and direction than the background can be grouped into moving regions, and the regions can be reported as the moving objects.

To group the regions from scattered motion vectors, a threshold magnitude value should be presented. In our suggestion, we added a routine that chooses the threshold which can bisect and holds at most 5%–38% of overall vectors in a frame. This is because the optical flow between a background and a frame with relatively big ego-motion tends to yield many more false-positives. In this case, the threshold magnitude needs to be big enough to be less sensitive to both the background and foreground movement, resulting in reduced false-positives. The background motion magnitude determinant has to be chosen based on its application, so that the routine can choose the percentage proportionally according to the determinants.

Before reporting the final detection results, it is also possible to further apply a simple filtering process which can reduce false-positives. Since we derived the detection results solely from the optical flow method, the aforementioned frame difference between the background and the current frame is still valid for cross-checking with the results. In this usage, the subset of optical flow results, shown as red and green rectangles in the left picture of Figure 4, can be reported as moving objects which also have been confirmed as moving objects on the frame difference results. Other filtering methods including a median filter on the results are also possible.

Figure 4. Derived frame difference (**right**) for false-positive filtering. The difference is derived between a frame (**left**) and its background (**middle**) with adequate threshold.

4. Experimental Results

To evaluate our suggestion and compare the results with other methods, we present two experimental results in this section. In most cases, evaluations on optical flow based methods are usually focused on the terms of precision and recall, relative distance, angular error, and computation time. Since the accuracy is what matters in our target application, we mainly state the precision and recall terms. Precision and recall are defined as follows:

$$\text{Precision} = \frac{\text{True positive}}{\text{True positive} + \text{False positive}} \tag{2}$$

$$\text{Recall} = \frac{\text{True positive}}{\text{True positive} + \text{False negative}} \tag{3}$$

where true positive is the region first detected by the algorithm that turned out to be actually moving. False positive is the region reported by the algorithm as moving when it actually only contains background. False negative is the region which should be reported as a moving object but not detected by the algorithm. Thus, the precision term reflects the accuracy of overall detected results while recall depicts the sensitivity of the algorithm.

The first sample sequence in Figure 5 consists of 200 consecutive frames capturing three vehicles moving from right to left [28]. The camera also moves right to left along with the first vehicle, leaving ego-motion on its background pixels. In this sample, two rank-constraint models [18,26] result precision of 0.83, 0.95 and recall of 0.99, 0.92 are used, respectively, while the suggested procedure gives precision of 0.98 and recall of 0.78 (see in Table 1). Since the scenario of the target application focuses on rather closer obstacles, better precision on those can draw the immediate attention of the drivers by reducing false-alarms.

Figure 5. Car example with 200 consecutive frame sets. The rank-constraint models give better results in terms of recall (**left**), however, the precision of overall results is better with the suggested method (**right**) due to the additional filters.

Table 1. Precision and recall comparison

Algorithm	Precision	Recall
Rank-constraint 1 [26]	0.83	0.99
Rank-constraint 2 [18]	0.95	0.92
Suggested method	0.98	0.78

To evaluate the proposed scheme in everyday use, we implemented the scheme on a FPGA platform [29,30] and mounted it on a test vehicle. The platform consists of a camera with resolution of 800 × 480 pixels and a HDMI recorder for further analysis (Figure 6).

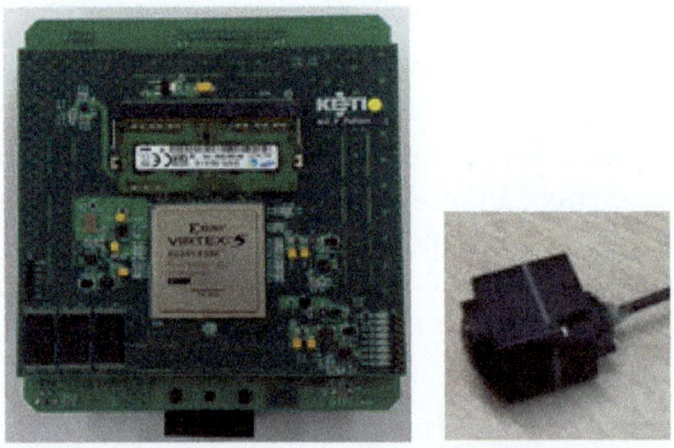

(a) (b)

Figure 6. The FPGA platform with camera. The (**a**) platform takes LVDS signal from (**b**) camera and outputs as HDMI signal. External HDMI recorder was needed for further analysis as the platform processes the input frames on-the-fly.

For the test, it took 19 videos 150 frames each, at seven frames per second, having both ego-motion towards the rear side and the target object's movement. Figure 7 and Table 2 show parts of the sequences with false-positive and negative analysis results on all 19 test scenarios.

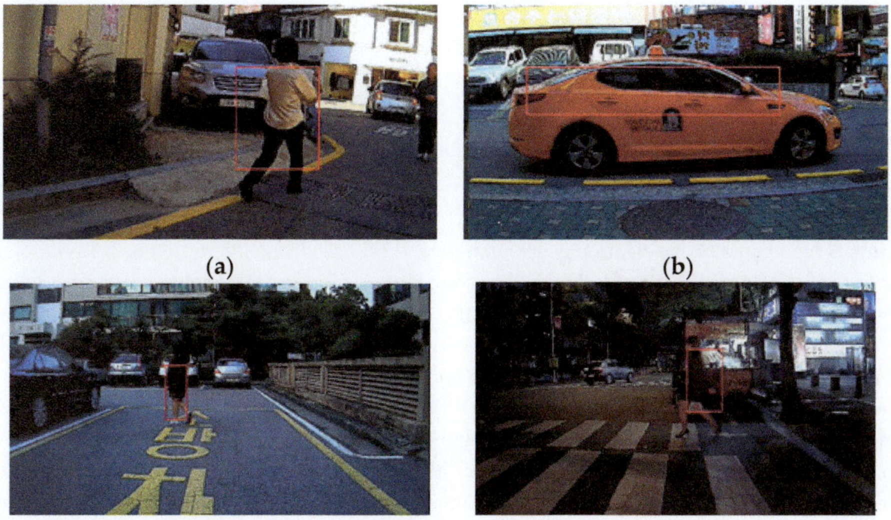

(c) (d)

Figure 7. Part of detection results from the test vehicle. Tests were conducted in several places and under various conditions such as (**b**,**c**) parking lots and (**a**,**d**), streets ranging from (**a**,**c**) daytime to (**d**) dusk. All 19 videos including the above four were taken while the vehicle was in backwards motion.

Table 2. False positive and negative results on the all 19 videos (150 frames each).

Scenarios					
False-Positives	0	2	2	0	3
False-Negatives	0	0	1	4	0
Scenarios					
False-Positives	1	0	2	0	1
False-Negatives	1	0	0	0	1
Scenarios					
False-Positives	0	0	0	1	0
False-Negatives	0	1	2	0	0
Scenarios					
False-Positives	0	0	0	0	
False-Negatives	0	1	0	0	

5. Conclusions

In this paper, we discussed two types of methods for moving object detection and their constraints. We suggested a combinational method and false-positive filters to relieve the stationary viewpoint constraint. Experimental results show the overall improvement in the term of precision, which is an important feature for use with BCI applications.

Acknowledgments: This work was supported by the IT R&D program of MKE, Korea (10041125, High Resolution SoC Development for Automotive SXGA Camera System).

Author Contributions: D.S. Kim conceived and designed the experiments; D.S. Kim and J.S. Kwon wrote the paper.

Conflicts of Interest: The authors declare no conflict of interest.

References

1. Fleming, B. New automotive electronics technologies. *IEEE Veh. Technol. Mag.* **2008**, *3*, 10–12.
2. Jiménez-Pinto, J.; Torres-Torriti, M. Optical flow and driver's kinematics analysis for state of alert sensing. *Sensors* **2013**, *13*, 4225–4257. [CrossRef] [PubMed]
3. Llamazares, A.; Ivan, V.; Molinos, E.; Ocaña, M.; Vijayakumar, S. Dynamic obstacle avoidance using bayesian occupancy filter and approximate inference. *Sensors* **2013**, *13*, 2929–2944. [CrossRef] [PubMed]
4. Doğan, S.; Temiz, M.S.; Külür, S. Real time speed estimation of moving vehicles from side view images from an uncalibrated video camera. *Sensors* **2010**, *10*, 4805–4824. [CrossRef] [PubMed]
5. Zhan, C.; Duan, X.; Xu, S.; Song, Z.; Luo, M. An improved moving object detection algorithm based on frame difference and edge detection. In Proceedings of the Fourth International Conference on Image and Graphics, Chengdu, China, 22–24 August 2007.
6. McFarlane, N.J.B.; Schofield, C.P. Segmentation and tracking of piglets in images. *Mach. Vis. Appl.* **1995**, *8*, 187–193. [CrossRef]
7. Friedman, N.; Russell, S. Image segmentation in video sequences: a probabilistic approach. In Proceedings of the 13th Conference on Uncertainty in Artificial Intelligence, Providence, RI, USA, 1–3 August 1997.
8. Chris, S.; Grimson, W.E.L. Adaptive background mixture models for real-time tracking. In Proceedings of the IEEE Computer Society Conference on Computer Vision and Pattern Recognition, Fort Collins, CO, USA, 23–25 June 1999.
9. Poggio, T.; Torre, V.; Koch, C. Computational vision and regularization theory. *Nature* **1985**, *317*, 314–319. [CrossRef] [PubMed]
10. Yilmaz, A.; Javed, O.; Shah, M. Object tracking: A survey. *J. ACM Comput. Surv.* **2006**, *38*, 1–45. [CrossRef]
11. Dalal, N.; Triggs, B. Histograms of oriented gradients for human detection. In Proceedings of the IEEE Computer Society Conference on Computer Vision and Pattern Recognition, San Diego, CA, USA, 25 June 2005.
12. Viola, P.; Jones, M. Rapid object detection using a boosted cascade of simple features. In Proceedings of the IEEE Computer Society Confercence on Computer Vision and Pattern Recognition, Kauai, HI, USA, 8–14 December 2001.
13. Lienhart, R.; Kuranov, A.; Pisarevsky, V. Empirical Analysis Of Detection Cascades of Boosted Classifiers for Rapid Object Detection. In *Pattern Recognition—25th DAGM Symposium, Magdeburg, Germany, September 10–12, 2003*; Springer Berlin Heidelberg: Berlin, Germay, 2003.
14. Huang, S.C.; Chen, B.H. Highly accurate moving object detection in variable bit rate video-based traffic monitoring systems. *Neural Netw. Learn.* **2013**, *24*, 1920–1931. [CrossRef] [PubMed]
15. Huang, S.C.; Chen, B.H. Automatic moving object extraction through a real-world variable-bandwidth network for traffic monitoring systems. *IEEE Trans. Ind. Electron.* **2013**, *61*, 2099–2112. [CrossRef]
16. Cheng, F.C.; Huang, S.C.; Chen, B.H. A hybrid background subtraction method with background and foreground candidates detection. *ACM Trans. Intell. Syst. Technol.* **2015**, *7*, 1–14. [CrossRef]
17. Guo, J.M.; Hsia, C.H.; Liu, Y.F.; Shih, M.H.; Chang, C.H.; Wu, J.Y. Fast background subtraction based on a multilayer codebook model for moving object detection. *IEEE Trans. Circuits Syst. Video Technol.* **2013**, *23*, 1809–1821. [CrossRef]
18. Zhou, X.; Yang, C.; Yu, W. Moving object detection by detecting contiguous outliers in the low-rank representation. *IEEE Trans. Pattern Anal. Mach. Intell.* **2012**, *35*, 597–610. [CrossRef] [PubMed]
19. Huang, S.C.; Do, B.H. Radial basis function based neural network for motion detection in dynamic scenes. *IEEE Trans. Cybern.* **2013**, *44*, 114–125. [CrossRef] [PubMed]
20. Bouwmans, T. Traditional and recent approaches in background modeling for foreground detection: An overview. *Comput. Sci. Rev.* **2014**, *11–12*, 31–66. [CrossRef]

21. Sharmin, N.; Brad, R. Optimal filter estimation for Lucas-Kanade optical flow. *Sensors* **2012**, *12*, 12694–12709. [CrossRef]
22. Black, B.J.; Anandan, P. Robust dynamic motion estimation over time. In Proceedings of the IEEE Computer Society Conference on Computer Vision and Pattern Recognition, Maui, HI, USA, 3–6 June 1991.
23. Brox, T.; Bruhn, A.; Papenberg, N.; Weickert, J. High Accuracy Optical Flow Estimation Based on a Theory for Warping. In *Computer Vision—ECCV 2004: 8th European Conference on Computer Vision, Prague, Czech Republic, May 11–14, 2004. Proceedings, Part IV*; Springer Berlin Heidelberg: Berlin, Germany, 2004; pp. 25–36.
24. Jepson, A.; Black, M.J. Mixture models for optical flow computation. In Proceedings of IEEE Computer Society Conference on Computer Vision and Pattern Recognition, New York, NY, USA, 15–17 June 1993.
25. Kim, J.; Wang, X.; Wang, H.; Zhu, C.; Kim, D. Fast moving object detection with non-stationary background. *Multimedia Tools Appl.* **2013**, *67*, 311–335. [CrossRef]
26. Sheikh, Y.; Javed, O.; Kanade, T. Background subtraction for freely moving cameras. In Proceedings of the IEEE 12th International Conference on Computer Vision, Kyoto, Japan, 29 September–2 October 2009.
27. Black, M.J.; Anandan, P. The robust estimation of multiple motions: Parametric and piecewise-smooth flow fields. *Comput. Vis. Image Underst.* **1996**, *63*, 75–104. [CrossRef]
28. Sand, P.; Teller, S. Particle Video: Long-Range Motion Estimation Using Point Trajectories. *Int. J. Comput. Vis.* **2008**, *80*, 72–91. [CrossRef]
29. Guo, G.; Kaye, M.E.; Zhang, Y. Enhancement of Gaussian background modelling algorithm for moving object detection & its implementation on FPGA. In Proceedings of the IEEE 28th Canadian Conference on Electrical and Computer Engineering, Halifax, NS, Canada, 3–6 May 2015.
30. Botella, G.; Martín, J.A.; Santos, M.; Meyer-Baese, U. FPGA-based multimodal embedded sensor system integrating low- and mid-level vision. *Sensors* **2011**, *11*, 8164–8179. [CrossRef] [PubMed]

8

Underwater 3D Surface Measurement Using Fringe Projection Based Scanning Devices

Christian Bräuer-Burchardt [1,*], Matthias Heinze [1], Ingo Schmidt [1], Peter Kühmstedt [1] and Gunther Notni [1,2]

Academic Editors: Fabio Menna, Fabio Remondino and Hans-Gerd Maas

1 Fraunhofer Institute Applied Optics and Precision Engineering, Albert-Einstein-Str. 7, D-07745 Jena, Germany; matthias.heinze@iof.fraunhofer.de (M.H.); ingo.schmidt@iof.fraunhofer.de (I.S.); peter.kuehmstedt@iof.fraunhofer.de (P.K.); gunther.notni@iof.fraunhofer.de (G.N.)

2 Technical University Ilmenau, Ehrenbergstraße 29, D-98693 Ilmenau, Germany

* Correspondence: Christian.braeuer-burchardt@iof.fraunhofer.de

Abstract: In this work we show the principle of optical 3D surface measurements based on the fringe projection technique for underwater applications. The challenges of underwater use of this technique are shown and discussed in comparison with the classical application. We describe an extended camera model which takes refraction effects into account as well as a proposal of an effective, low-effort calibration procedure for underwater optical stereo scanners. This calibration technique combines a classical air calibration based on the pinhole model with ray-based modeling and requires only a few underwater recordings of an object of known length and a planar surface. We demonstrate a new underwater 3D scanning device based on the fringe projection technique. It has a weight of about 10 kg and the maximal water depth for application of the scanner is 40 m. It covers an underwater measurement volume of 250 mm $\times$ 200 mm $\times$ 120 mm. The surface of the measurement objects is captured with a lateral resolution of 150 μm in a third of a second. Calibration evaluation results are presented and examples of first underwater measurements are given.

Keywords: underwater 3D-scanner; fringe projection; underwater stereo camera calibration

---

1. Introduction

Recently, the acquisition of 3D surface geometry of underwater objects has attracted increasing interest in various research fields, including archaeology [1–4], biological applications [5–8], or industrial facility inspection tasks [9]. Typical 3D reconstruction methodologies for underwater applications are underwater photogrammetry [5,7,10–13] or laser scanning techniques [9,14–17]. These techniques are time consuming and have limited precision. The increasing requirements of modern application scenarios concerning accuracy and data acquisition speed, however, require new approaches. One such approach is the use of structured light for the illumination of the underwater scene and to capture 3D measurement data using a stereo camera rig. Structured light illumination should be conducted by projection of a certain pattern (for example fringe pattern, stochastic pattern, other patterns) onto the object to be measured.

Recently, structured light (fringe) projection-based techniques have been increasingly used for underwater 3D measurements [18–20]. Some advantages of this technique in contrast to classical photogrammetric measurements are obvious: more measurement points can be generated and a higher 3D point density can be obtained in the measurement result. Additionally, measurement data can be obtained quickly and automatically. Due to the imaging principle, improved measurement accuracy can be expected. The main disadvantage, however, is the restricted measurement field because of the limited illumination conditions under water. Although the small size of the measurement field is a real limitation for certain potential applications we propose a fringe projection based method for underwater 3D data acquisition. Additionally, we introduce a prototype of a handheld underwater 3D scanning device in order to show the power of the structured light projection technique for underwater 3D measurements.

The main motivation for our developments is high precision inspection of industrial facility parts located under water (see e.g., [21]) such as pipeline systems, tanks, and pylons of off-shore windmills. Here a structured light based inspection methodology can help to carry out the maintenance of such facilities with considerably lower effort.

However, underwater 3D measurements using the structured light projection technique are more complicated than measurements in air. Reasons are the light absorbing property of the water, possible water pollution, and light ray fraction at the interfaces of different media (water, glass, air). Other causes are the more sophisticated requirements of the mechanics concerning robustness against impacts, temperature, humidity, and other environmental influences.

Our first goal was the construction of a new, handheld, practicable 3D scanning device for underwater measurements including algorithms for the attainment of accurate measurement results, and the development of a calibration procedure which can be done with low effort, but makes accurate measurement results possible. The second goal was to show the suitability of the structured light projection technique for sophisticated measurement tasks in underwater environments.

In this paper, an overview over the state of the art in optical underwater 3D measurement techniques is given, the challenges of the application of the structured light projection technique are outlined, the principles of accurate underwater 3D measurements are introduced, the development of a new handheld scanning device is described, and the first experimental results are presented and discussed.

2. State of the Art

Photogrammetry has been applied to underwater measurements with increasing importance and frequency for more than 40 years [22,23]. Documentation of archaeological sites [1,2] or sunken objects like boat parts or ship wrecks [10] are some application fields, as well as inspection and surface measurement of industrial facilities (for example pipeline systems [9]) or the measurement of biological objects such as coral reefs [8] or fishes [5–7].

The challenges for photogrammetric underwater applications are mainly the robustness and the accuracy of the captured 3D data. This is due to potentially bad image recording conditions (polluted water, insufficient light sources, difference in absorption behavior of distinct wavelengths, *etc.*) on the one hand and the possibly limited quality of the photogrammetric sensor calibration on the other hand. Water quality and illumination conditions have considerable influence on the measurement result and have been examined by Sedlazek *et al.* [24], Schechner and Karpel [25], and Bianco *et al.* [20].

Many works are concerned with calibration of photogrammetric underwater stereo scanners. An overview is given by Sedlazeck and Koch [12]. There are works considering the significant refraction effects using plane interface glasses of the underwater sensor housing [11,13,26–29]. Other authors neglect the refraction effects and try to approximate the underwater situation using the common pinhole camera model [5,7,30–32]. They propose modifications concerning the principal distance and the lens distortion [33,34].

A third situation appears when spherical dome ports are used in the underwater housing and the projection centers of the cameras are placed exactly into the sphere center points. Then, theoretically, no refraction occurs because of the perpendicular crossing of the interfaces of all vision rays. In this case, the common pinhole model may be applied for underwater applications in the same way as in the normal case. Furthermore, the calibration procedure can be performed under "normal" laboratory conditions in air. However, if the necessary preconditions concerning the exact spherical form of the dome-ports, the homogeneity of the glass material, and the exact placement of the cameras are denied, then measurement errors occur which should be corrected by a suitable method, as for instance additional distortion correction. Authors who have used this model describe a considerable deviation of the expected parameters [20].

The decision concerning the camera model used should depend on the requirements of the measurement accuracy and the accepted calibration effort. Here application of a suitable evaluation strategy seems meaningful which could be the determination of the measurement error using a certain reference measurement object with known geometry.

The use of structured light for photogrammetric underwater measurements has been recently proposed by Bruno *et al.* [18], Zhang *et al.* [19], and Bianco *et al.* [20], respectively. They showed that this technique can be successfully applied to underwater measurements. However, there are still some restrictions (small measurement fields and short measurement distances) of the structured light technique which should be overcome in the near future.

The present work describes the application of the structured light technique using projection of fringe sequences (Gray-code and sinusoidal sequences) in two orthogonal projection directions and the development of the prototype of an underwater measurement device which could be useful for certain applications and should be developed further for more sophisticated applications.

3. Measurement Principles and Underwater Challenges

3.1. Basic Principles

Contactless optical 3D measurement using active or passive photogrammetric methods employing a stereo camera arrangement will be performed as follows: the measurement object will be recorded by two cameras with known intrinsic parameters (principal point, principal distance, and distortion function) and known relative orientation to each other (extrinsic parameters). The intrinsic and extrinsic camera parameters are obtained by calibration according to the selected camera model (see next paragraph).

Measured 3D points are computed by triangulation (see e.g., [35]) which means the intersection of two rays in the 3D space. If the two rays are not parallel, there exists one unique 3D point which is either a proper intersection point of the two rays or that point having the shortest Euclidean distance to both rays. The vision rays can be obtained by the image co-ordinate and the intrinsic and extrinsic camera parameters. In common air applications typically the well-known pinhole camera model (PM) will be used [35]. However, also other, more general camera models can be applied such as the "raxel" model [36] or the ray-based camera model (RM) as described by Bothe *et al.* [37]. Similarly, whether the ray-based model is used or the pinhole model, calculation of resulting 3D points is obtained by triangulation.

3.2. Correspondence Finding and 3D Point Calculation

The determination of point correspondences may be achieved with the help of markers on the measurement object or by finding some characteristic structure(s) in the images. By using structured illumination one projects such characteristic structures onto the measurement object and increases significantly the number of correctly identifiable corresponding points. It additionally provides the possibility of better sub-pixel exact determination of image co-ordinates of the corresponding points. The projected patterns may be sequences of Gray-codes, sinusoidal fringe patterns, or series of stochastic patterns (see e.g., [38–42]).

The principle of fringe projection based sensors for the 3D surface acquisition is called phasogrammetry [39] or fringe projection profilometry [40] and means the mathematical conjunction of fringe projection and photogrammetry. A projection unit produces sequences of fringe pattern which will be observed by one or more cameras and transformed into so called phase images. Using stereo camera equipment as in our case, the phase images may serve purely for the identification of corresponding points.

From the sequence of sinusoidal fringe images the rough phase will be calculated which is periodically distributed according to the period length over the image. The Gray-code enables the unique identification of the period number and leads together with the rough phase to the fine or unwrapped phase value. Determination of the phase values in two orthogonal projection directions [39] or the use of epipolar geometry information [35] make the phase values unique. Because the epipolar geometry can be applied only approximatively in case of refracted rays in underwater application, two perpendicular sequences of fringes should be used. In the case of underwater application the triangulation is performed between the vision rays in the water together with the calibration parameters of a virtual camera (see Section 3.3).

3.3. Extended Camera Model

In contrast to classical photogrammetric modeling the pinhole camera model is not valid for underwater use because of the fact that there is no single viewpoint (the projection center) where all the image points representing rays pass through. This happens because of the refraction of the vision rays at the interfaces between air and housing glass (viewport) inside and between viewport and water outside the scanner housing (see Figure 1) following Snell's law. An exception would be the usage of spherical dome ports and an exact positioning of the camera's projection center into the center-point of the sphere (see Figure 1). Here (theoretically) no refraction occurs, and the pinhole model can be used as well as the same camera parameters as in air. However, in practice the exact positioning of the cameras is difficult to achieve. Experiments of other researchers [10,20] have showed considerable differences in the calibration parameters when performed both in air and under water. These differences lead to additional errors and distortion effects which must be corrected correspondingly.

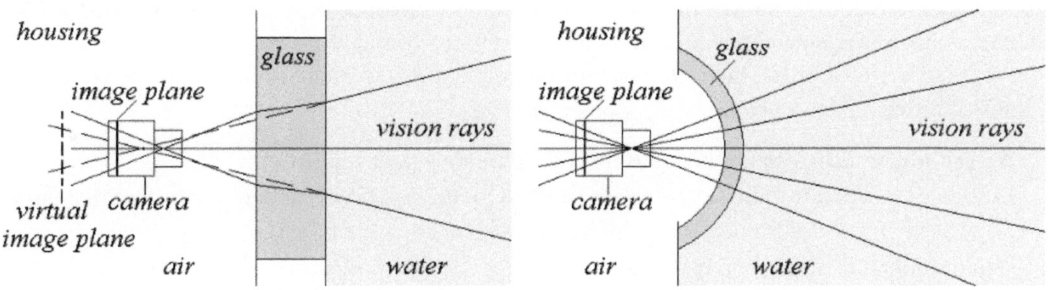

Figure 1. Geometry of rays in case of plane glasses (**left**) and spherical dome ports (**right**).

Let us consider the case of plane interface ports again. The refraction at the two interfaces leads to the effect that the vision rays corresponding to the different image points do not intersect in one single point (the projection center according to PM). Additionally, the principal distance c becomes image point dependent. For illustration see Figure 2. The refraction of the rays is shown for the case of perpendicular orientation of the camera according to the glass interface.

In this case the pinhole camera model can be applied only as an erroneous approximation. In order to avoid these errors the refraction should be taken into account. The description of the camera model can be given accordingly to the ray-based model given by Bothe *et al.* [37] or corresponding to other proposals (see e.g., [11,13,26,28,33,43]).

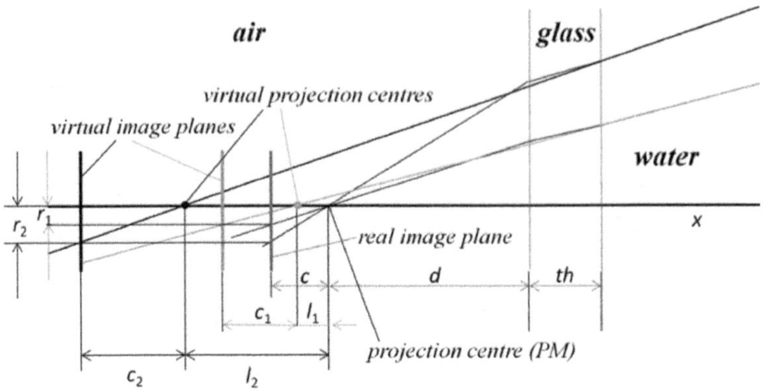

Figure 2. Ray geometry for two image points with radial distances r_1 and r_2 from the principal point using plane glasses (according to [44]).

Our approach is to use the parameters of the air calibration and describe the direction of the vision rays according to refraction and the additional parameters glass thickness *th* and interface distance *d*. Hence, one task of the underwater calibration is to find the values for the additional parameters. The other one is the formulation of the correct calculation of 3D measurement points based on found corresponding image points in the two cameras.

3.4. Approach for Underwater 3D Measurement

Our new approach to obtain accurate underwater 3D measurements using structured light illumination is the following. The exact modeling of the geometric situation of the cameras should be performed by consideration of the refraction effects and use of an accurate air calibration of the stereo scanner. An additional requirement is a relative low effort of the calibration procedure. This means that only few recordings of certain calibration objects should be acquired under water as input of the calibration procedure. The calibration for the underwater stereo scanner will be performed by the following steps:

1. Air calibration using the pinhole model without underwater housing
2. Determination of additional parameters (refraction indices, glass thickness, interface distance)
3. Ray-based model description of the underwater scanner inside housing
4. Determination of additional distortion matrices

When calibration is complete, 3D measurements can be performed under water. The 3D surface data of the measurement objects are obtained by finding corresponding points in the two camera images and performing a triangulation using the extended camera model and the parameters obtained by calibration. In the next section we will describe the calibration process in detail.

3.5. Calibration Procedure

The calibration of the underwater scanner can be performed according to the steps described in the previous section. For the first step any suitable calibration procedure for stereo camera scanners using the PM including distortion description can be used. For the second step we suggest the following procedure: our approach is to use the parameters of the air calibration and describe the direction of the vision rays according to refraction and the additional parameters interface distance *d* and glass thickness *th*. The indices of refraction for air n_a, water n_w, and glass (acrylic) n_g are assumed to be sufficiently exact known. The camera orientations concerning the glass interfaces are supposed to be perpendicular. Possible deviations should be compensated by final application of the additional distortion matrix.

Hence, one task of the underwater calibration is to find the values for the additional parameters. The other one is the formulation of the correct calculation of 3D measurement points based on found corresponding image points in the two cameras.

We assume for simplification an orthogonal normal angle regarding the glass surface. By consideration of the geometry as depicted in Figure 2 we get a shift l of the projection center in direction of the optical axis and a modified principal distance c' depending on the Euclidean distance r of the image points to the principal point:

$$l(r) = \frac{d \cdot r}{c \cdot \tan\left(\arcsin\left(\frac{\sin(\arctan(r/c))}{n_w}\right)\right)} - d - th \cdot \frac{r}{c} \cdot \left(\frac{r}{c} - \tan\left(\arcsin\left(\frac{\sin(\arctan(r/c))}{n_g}\right)\right)\right) \quad (1)$$

$$c\prime(r) = \frac{r}{\tan\left(\arcsin\left(\frac{\sin(\arctan(r/c))}{n_w}\right)\right)} \quad (2)$$

Equations (1) and (2) are applied at calculation of the 3D points by triangulation complementing the known formulas of the pinhole camera case. By this extension we, in fact, obtain a ray-based camera model (RM).

With Equations (1) and (2) we can uniquely describe the rays corresponding to the image points $q = (x, y)$ according to the RM by the two 3D points $O_i(x, y)$ and $Q_i(x, y)$.

$$O_i(x,y) = O_i - l(r) \cdot \mathrm{R} \cdot \mathrm{e} \quad (3)$$

$$Q_i(x,y) = O_i(x,y) - c\prime(r) \cdot \mathrm{R} \cdot \begin{pmatrix} x \\ y \\ -c\prime(r) \end{pmatrix} \quad (4)$$

where O_i is the initially determined projection center of the camera outside housing according to the pinhole model, e is the unit vector, and R is the rotation matrix of the air calibration (see also [45]). For simplification we assume without loss of generality no distortion here.

The next step is the determination of the glass thickness *th*. In our case we measured it tactile before mounting the underwater housing. If this would not be possible, e.g., the method proposed by Chen and Yang [46] can be applied.

Finally, the interface distances d_1 and d_2 for both cameras had to be determined. In order to obtain this, the following algorithm for underwater interface distance determination (UIDD) using four measurements (M_1, M_2, M_3, and M_4) is applied:

- Underwater recording (M_1, M_2) of specimen with known (calibrated) length information L_r (ball bar—see Figure 3) in two different positions
- Underwater recording (M_3, M_4) of a plane surface in two different distances (minimum and maximum distance in the measurement volume) according to the scanner
- Definition of the error function for the test statistics T utilizing length measurement error and flatness error of the plane according to Equations (6) and (7)
- Determination of the searched parameters by minimization of T

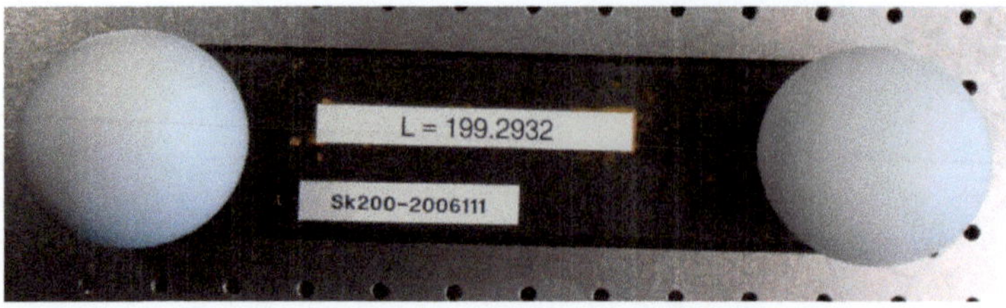

Figure 3. Specimen for calibration: calibrated ball bar.

Minimization of T can be achieved by a systematic search in the parameter space of d_1 and d_2 with meaningful search interval limits and step-widths. Having only two parameters, systematic search may be considerable more effective than trying to formulate an optimization function, because of the trigonometric functions in the Equations (1) and (2). The test quantity T is defined as:

$$T(d_1, d_2) = (E_{L,M1})^2 + (E_{L,M2})^2 + (E_{F,M3})^2 + (E_{F,M4})^2 \tag{5}$$

with relative length error E_L and relative flatness error E_F defined as:

$$E\left|\frac{L_a - L_r}{L_r}\right| \tag{6}$$

and:

$$E_F = \frac{|d_{\max}| + |d_{\min}|}{d_r} \tag{7}$$

Here, L_r is the calibrated reference distance between the two spheres of the ball bar and L_a is the measured distance. Quantity d_r represents the maximal possible distance in the observed region (e.g., diameter of a circle area or diagonal of a rectangular region) at the plane measurement. It is a constant which is fixed according to the measurement field size and which should have a length of at least half the diagonal of the measurement volume. The term $|d_{max}| + |d_{min}|$ denotes the flatness measurement error, which is defined as the range of the signed distances of the measured points from the best fit plane as defined in [47]. Application of this algorithm leads to the searched parameters d_1 and d_2. A more detailed description of the calibration procedure can be found in [44]. For all measurements it is allowed to remove a maximum of 0.3% outliers according to [46].

Application of this calibration procedure yielded good and meaningful results (see Section 5). However, the E_F error seemed to be too high, and the shape deviation of the plane was quite systematic (bowl-shape). This led to the introduction of an additional or alternative, respectively, distortion function and a corresponding matrix operator D for distortion correction. This operator will be obtained as follows. The two plane measurements (M_3 and M_4) are used for the determination of D. For both a fitting of a plane to the 3D points is performed leading to the plane parameters $E_1 = \{A_1, B_1, C_1, D_1\}$ and $E_2 = \{A_2, B_2, C_2, D_2\}$, respectively. Now, every calculated 3D point P is replaced by the nearest point P' in the fitted plane. Using P', new residuals for the corresponding image points are determined. Finally, the new distortion function is obtained by approximation of the residuals (from both planes) by a polynomial function of third degree leading to ten parameters (see [47] for more details).

In order to compensate new arising scaling errors after application of the new distortion compensation operator, last step of UIDD algorithm is performed again. If necessary, this procedure can be performed iteratively. The improvement in the resulting calibration is described in Section 5. Additionally, first results of the underwater measurements are presented.

4. New Underwater 3D Scanner

The paragon for the inner parts of the scanner was the handheld "kolibri Cordless" mobile scanning device [48]. For the underwater application, however, certain functional units such as the PC technique, the included display, and considerably more robust mechanics had to be developed completely anew. Additionally, an underwater resistant housing was necessary. The desired technical system parameters were similar to those of the "kolibri Cordless" and are listed in Table 1.

Table 1. Desired system parameters.

Property	Desired Parameter
Measurement volume (MV)	$250 \times 200 \times 150$ mm^3
Working distance	500 mm
Camera resolution	1600×1200 pixel
Lateral resolution in the MV	150 µm
Noise of the 3D points	10 µm ... 50 µm
Frame rate	60 Hz
Recording time per scan	350 ms
Maximal water depth	40 m
Sensor weight (without housing)	2 kg
Sensor weight with housing	10 kg

A new principle is the direct connection and insertion of the control and analysis PC and the display into the underwater housing. This allows the omission of a connection between the scanner and an external PC, which would be otherwise carried in a backpack or external case. Additionally, underwater handling of the scanner becomes easier. The consequently higher weight of the device is no disadvantage in underwater use because it is only important that neither sinking nor upwelling of the scanner occur. The following criteria were essential for the scanner construction:

- Desired technical system parameters (see Table 1)
- Easy handling at use and also at the process of mounting into the housing
- Compactness of the scanner including housing
- Low weight
- Suitable heat exhausting at underwater use
- Easy navigation of the scanner under water

The main components of the scanner, which should be connected compactly for mounting into the housing, are the projection unit, two cameras, the PC, the display, cooling elements, and mechanical elements. The projection unit including the lens causes the structured illumination of the scene which is observed by the two cameras. The PC controls the projection and observation, stores the recorded image data, and performs data processing. A rough presentation of the resulting 3D point cloud on the display allows the user to evaluate the quality of the measurement. Cooling elements such as heatsinks, heatpipes, and cooling ribs are responsible for heat dissipation from the housing. Mechanical elements are necessary for the connection of the principal parts.

The projection unit was built up using a commercially available beamer (HX301G, LG, Seoul, Korea) with a frame rate of 60 Hz. The power consumption of the beamer is less than 100 W and the luminous flux is 300 Lm. The maximal pixel resolution is 1680×1050, but for fringe pattern projection the native resolution of 1024×768 pixels was used.

A sequence of fringes (Gray-code and sinusoidal fringes) in two orthogonal directions is projected onto the scene by the projection unit (see Figure 4). Typically, a fringe period length of 16, 32, or 64 projector pixels is used. This leads to 64, 32, or 16 vertical and 48, 24, or 12 horizontal projected fringes and seven, six, or five Gray-code images per projection direction. Using a 90° phase shift four sinusoidal fringe images per projection direction are necessary. Hence, one complete projection sequence consists of 24, 22, or 20 single fringe images taking 400, 367, or 333 ms time, respectively.

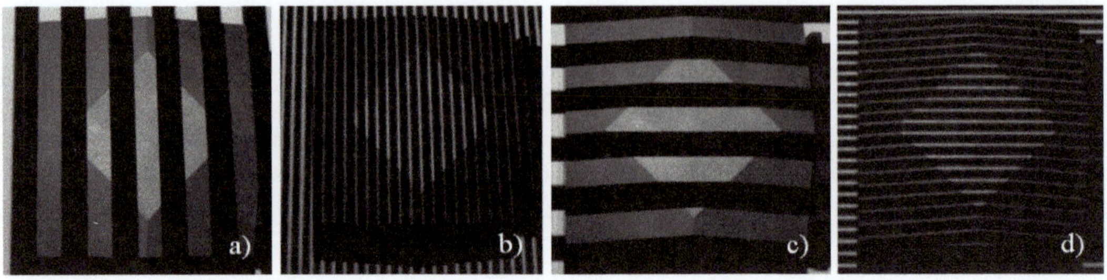

Figure 4. Selected images from fringe sequence: parts of Gray-code sequence (**a**,**c**) and sinusoidal fringes (**b**,**d**) taken from air measurements of a frustum of a pyramid.

The fringe image sequence is synchronously recorded by both cameras. From these image sequences the so called phase images are calculated, which are used for the determination of the 3D data (see e.g., [39]). The 3D calculation is performed on the PC, which also has the task to control the image recording. The measurement results are also indicated on the display (see Figure 5). Additional components are two laser pointers for checking the current measurement distance. Figure 5 shows the main components of the scanner in two construction drawings.

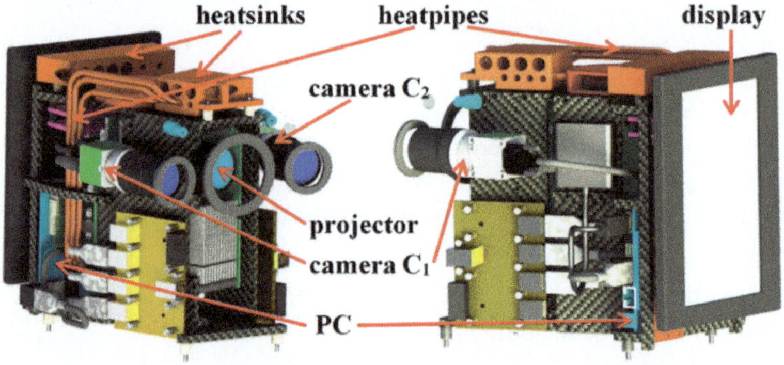

Figure 5. Principal components in two views of a construction drawing.

The underwater housing was produced using the synthetic material PA 2200 (see [49]) and developed by the 4h Jena Engineering GmbH (see [50]). It can be used in both fresh and salt water. It can distinctly withstand the water pressure at a diving depth of 40 m. The interfaces for the projector, the cameras, and the laser beams were made from sapphire glass, whereas the window for the display is made from polycarbonate. The planar windows for the cameras and the projector do not lie in a common plane but are tilted according to the directions of the optical axis. This should simplify the calibration procedure (see Section 3.5).

One major problem of the scanner was the adequate heat dissipation from the housing. A base plate with cooling ribs and appropriate heat sinks were constructed. For power supply and signal lines a separable under water plug-in connector cable was selected and employed. Separable inductive switching boards including control keys and interfaces to the scanner were developed in order to provide a correct handling also with diver gloves (see Figure 6). A construction drawing of the scanner is shown in Figure 7 as well as a photograph of the scanner in the housing without back panel with the display.

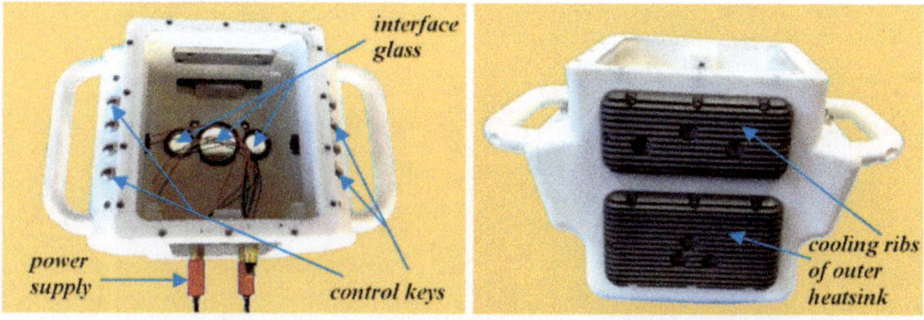

Figure 6. Two housing views showing certain sensor elements.

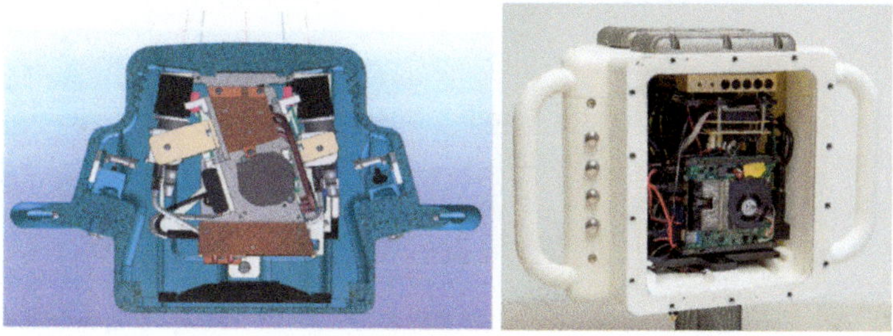

Figure 7. Housing views: construction draw, view from above (**left**), sensor inside housing without back-panel (**right**).

5. Measurements

5.1. Calibration Evaluation

After the four measurements which were used for the calibration of the underwater parameters we made several underwater measurements of certain specimen in order to evaluate the quality of the calibration. According to the suggestions of the VDI/VDE guidelines [46] the measurement objects ball bar and plane were placed in different positions in the measurement volume. Additionally, a frustum of a pyramid (see Figure 8) was measured. All measurements were performed in clear water using a rain barrel (see Figure 8). Figure 9 shows an example for the projected fringe patterns under water. Compared to Figure 4 differences in the image quality can be hardly detected.

For evaluation the quantities relative length deviation and relative flatness deviation defined by Equations (5) and (6) were used. For comparison, all measurements and calculations were performed

- With the sensor outside the housing in the laboratory (Air)
- Underwater with the pinhole camera model with modified parameters (PM)
- Underwater with the extended model, but without additional distortion matrices (RM)
- Underwater with the extended model including the additional distortion matrices (RMD)

Figure 8. Frustum of pyramid in a rain barrel (**left**) and scanner in underwater use (**right**).

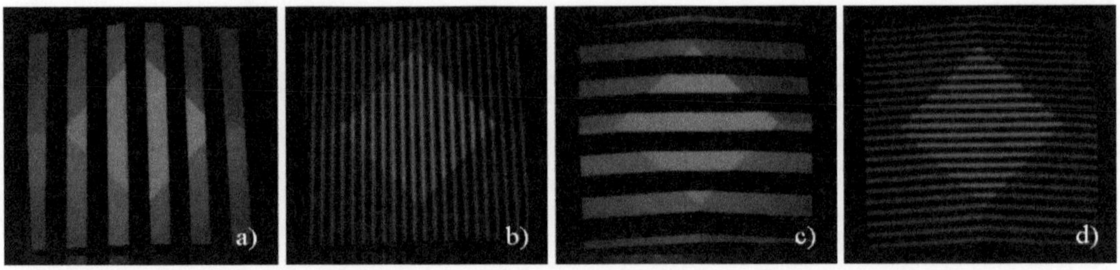

Figure 9. Selected images from fringe sequence: parts of Gray-code sequence (**a**,**c**) and sinusoidal fringes (**b**,**d**) taken from underwater measurements of a frustum of a pyramid.

In all cases at least seven measurements were performed. Noise determination was performed additionally. It was defined as standard deviation of the measured 3D points from a small locally fitted plane (or sphere, respectively). The obtained maximum error results are documented by Table 2. Note that for E_L and E_F the percentage values are given in contrast to Equations (6) and (7), and *noise* is an averaged value over all noise measurements.

Table 2. Calibration evaluation results (given in %) using specimen ball bar (E_{Lb}), frustum of pyramid (E_{Lp}), and plane (E_F), and averaged 3D noise values.

LocationQuantity	E_{Lb} (%)	E_{Lp} (%)	E_F (%)	*noise* (mm)
Air, PM	0.1	0.4	0.15	0.02
Water, PM	1.0	1.45	1.7	0.05
Water, RM	0.4	0.6	0.65	0.04
Water, RMD	0.35	0.35	0.4	0.03

The results show that pinhole approximation in water is not sufficient to obtain acceptable measurement accuracy. Although the modeling without additional distortion correction should completely describe the ray geometry, the results are not yet as precise as desired. RMD results are acceptable, but should also be improved at future measurements after certain enhancements which can be obtained, for instance, by using a free ray-based camera model. However, for improved ray-based model representation an additional calibration step must be performed. An approach of this idea will be given in the outlook section.

5.2. Examples of Underwater Measurements

Underwater measurements were performed next by application of the scanner in a water basin. It was handled by a diver (Figure 8). The first measurement objects were a pipe (Figures 10 and 11), a fossil sea shell (Figure 12), and stones (Figure 13).

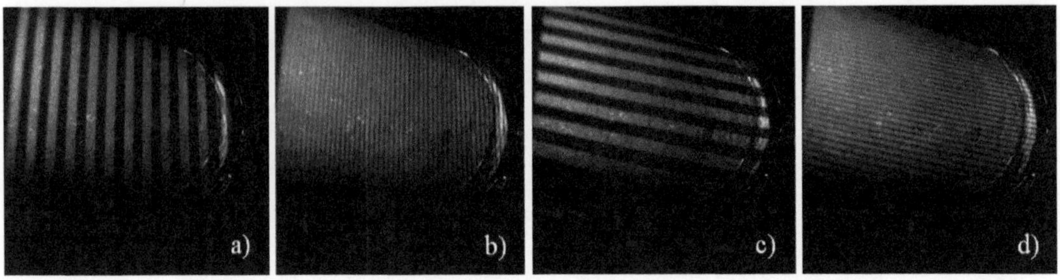

Figure 10. Selected images from fringe sequence: parts of Gray-code sequence (**a**,**c**) and sinusoidal fringes (**b**,**d**) taken from underwater measurements of a pipe.

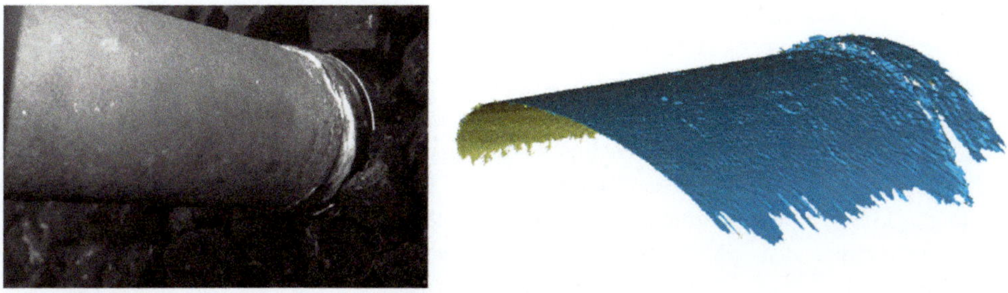

Figure 11. Underwater measurement of a pipe (**left**) and 3D result representation (**right**).

Figure 12. Fossil seashell: photograph (**left**), color-coded 3D- (**middle**), and 3D-surface presentation (**right**).

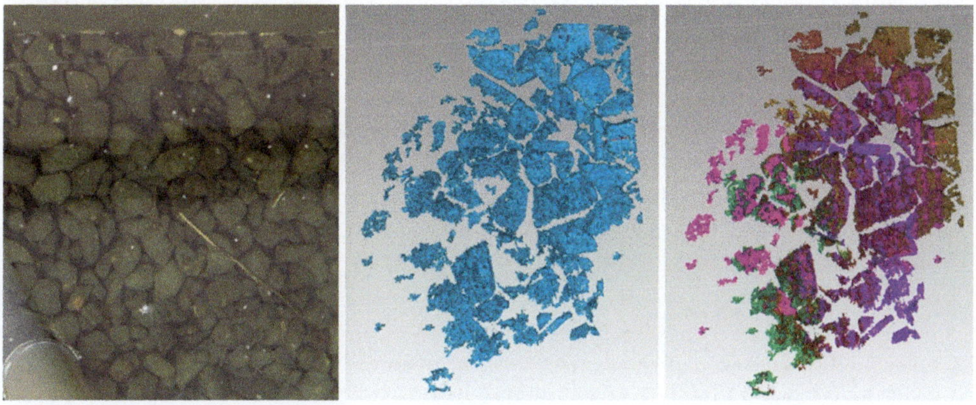

Figure 13. Example underwater measurement stones: photograph (**left**), surface 3D representation (**middle**), and color coding of six different scans (**right**).

The examples can give a first impression of the capabilities of the underwater scanner. The scan results are obtained few seconds after the recording. It can be assumed that the measurement precision is sufficient for a number of application scenarios. However, for bigger objects the 3D point clouds must be merged. Here, additional software is necessary.

6. Summary, Discussion and Outlook

The principle of optical 3D surface measurement based on the fringe projection technique for underwater applications has been described in this work and the challenges of the underwater use of this technique are shown and discussed in contrast to the classical application.

A new fringe projection based 3D underwater scanner which covers a measurement volume of 250 mm × 200 mm × 120 mm for diving depths up to 40 m was introduced. It can record one 3D scan in approximately 350 ms with a lateral resolution of 150 μm in the object space. Measurement objects larger than the measurement volume should be scanned in several views and merged subsequently.

The presented scanning device for 3D surface measurements is one of the first underwater scanners based on fringe projection technique. It is quite complicate to perform a comparison of technical details and measurement capability to other systems based on this technique [18–20]. These systems are laboratory setups and the projector is not included in the camera housing. Certainly, what can be assessed is that our scanner has a smaller measurement field. Measurement accuracy, although also difficult to compare, seems to be similar. Compared to photogrammetric underwater systems (e.g., [13,26]) the measurement precision has the same magnitude. The main differences with photogrammetric systems are the smaller measurement volume and the higher measurement point density.

During the image recording time of 0.33 or 0.4 s operator movements may disturb the image recording. Effective movement correction algorithms could manage this, but, unfortunately such methods are not available for this scanner. However, we have developed an algorithm which detects vibrations and movements during the image recording and suggests the rejection of the affected current measurement and its repetition.

If moving dust clouds or dirt particles disturb the image acquisition, it must be decided whether to wait for better recording conditions or to take the images and to accept decreased image quality. In order to make a more exact statement possible, appropriate experiments should be performed in the future.

The first quantitative measurements showed acceptable measurement accuracy. However, best results could be achieved only by application of additional (heuristically found) distortion corrections which cannot be explained by the model. They perhaps compensate partly the effects of the deviations of the current modelling from the real situation. The search of the origins of these effects may be a part of our future work.

First measurements and experiments were performed in clear freshwater in the laboratory and in an outdoor water basin with slightly more pollution. This, however, is not a practice-oriented situation. It is expected that in a sea environment with certain turbidity the robustness of the object surface acquisition will decrease and the noise of the measurement values will increase. Measurement accuracy, however, will be affected only by a slightly higher random error. The corresponding experiments should confirm and quantify these expectations.

The main part of our future work should be occupied by experiments concerning the dependence of the measurement accuracy on the water quality. Here the influence of difference levels of turbidity to the robustness and accuracy of the measurements must be analyzed. Additionally, comparisons between measurements in fresh and salt water will be made.

In order to avoid the necessity of additional distortion compensation by special operator, the complete ray-based representation of the observation system should be used in the future. Here, an additional calibration step with underwater recordings of a plane must be applied similarly to the method described by Bothe *et al.* [37]. With this final transition to the ray-based camera model a further reduction of the remaining length error and flatness deviation should be achieved.

As it was already mentioned before, the size of the measurement volume is not very large and seems to be too small for certain applications such as pipeline system inspection or survey of archaeological sites. The possible stitching of resulting 3D datasets is sophisticated, time consuming, and prone to errors. Hence, scanning devices with considerable larger measurement volume are desired. This requires stronger illumination power of the projection unit with coincident more efficient heat dissipation. The design of such a more powerful underwater 3D scanner will be also a part of our future work. Perhaps, the underwater housing for the projector should be separated from the housing for the cameras because of the heat dissipation.

Acknowledgments: This work was supported by the state Thuringia and the European Union (EFRE) under grant label TNA VIII-1/2012.

Conflicts of Interest: The authors declare no conflict of interest.

References

1. Roman, C.; Inglis, G.; Rutter, J. Application of structured light imaging for high resolution mapping of underwater archaeological sites. In Proceedings of the 2010 IEEE OCEANS, Sydney, Australia, 24–27 May 2010; pp. 1–9.
2. Drap, P. Underwater Photogrammetry for Archaeology. In *Special Applications of Photogrammetry*; Carneiro Da Silva, D., Ed.; InTech: Rijeka, Croatia, 2012; pp. 111–136.
3. Eric, M.; Kovacic, R.; Berginc, G.; Pugelj, M.; Stopinsek, Z.; Solina, F. The Impact of the Latest 3D Technologies on the Documentation of Underwater Heritage Sites. In Proceedings of the IEEE Digital Heritage International Congress 2013, Marseille, France, 28 October–1 November 2013; Volume 2, pp. 281–288.
4. Canciani, M.; Gambogi, P.; Romano, F.G.; Cannata, G.; Drap, P. Low cost digital photogrammetry for underwater archaeological site survey and artifact insertion. The case study of the Dolia wreck in secche della Meloria-Livorno-Italia. *Int. Arch. Photogramm. Remote Sens. Spat. Inf. Sci.* **2003**, *34 Pt 5*, 95–100.
5. Harvey, E.; Cappo, M.; Shortis, M.; Robson, S.; Buchanan, J.; Speare, P. The accuracy and precision of underwater measurements of length and maximum body depth of southern bluefin tuna (Thunnus maccoyii) with a stereo–video camera system. *Fish. Res.* **2003**, *63*, 315–326. [CrossRef]
6. Dunbrack, R.L. *In situ* measurement of fish body length using perspective-based remote stereo-video. *Fish. Res.* **2006**, *82*, 327–331. [CrossRef]
7. Costa, C.; Loy, A.; Cataudella, S.; Davis, D.; Scardi, M. Extracting fish size using dual underwater cameras. *Aquac. Eng.* **2006**, *35*, 218–227. [CrossRef]
8. Bythell, J.C.; Pan, P.; Lee, J. Three-dimensional morphometric measurements of reef corals using underwater photogrammetry techniques. *Coral Reefs* **2001**, *20*, 193–199.
9. Tetlow, S.; Allwood, R.L. The use of a laser stripe illuminator for enhanced underwater viewing. *Proc. SPIE* **1994**, *2258*, 547–555.
10. Korduan, P.; Förster, T.; Obst, R. Unterwasser-Photogrammetrie zur 3D-Rekonstruktion des Schiffswracks "Darßer Kogge". *Photogramm. Fernerkund. Geoinf.* **2003**, *5*, 373–381.
11. Kwon, Y.H.; Casebolt, J. Effects of light refraction on the accuracy of camera calibration and reconstruction in underwater motion analysis. *Sports Biomech.* **2006**, *5*, 315–340. [CrossRef] [PubMed]
12. Sedlazeck, A.; Koch, R. Perspective and non-perspective camera models in underwater imaging—Overview and error analysis. In *Theoretical Foundations of Computer Vision*; Springer: Berlin, Germany, 2011; Volume 7474, pp. 212–242.
13. Telem, G.; Filin, S. Photogrammetric modeling of underwater environments. *ISPRS J. Photogramm. Remote Sens.* **2010**, *65*, 433–444. [CrossRef]
14. Moore, K.D. Intercalibration method for underwater three-dimensional mapping laser line scan systems. *Appl. Opt.* **2001**, *40*, 5991–6004. [CrossRef] [PubMed]
15. Narasimhan, S.G.; Nayar, S.K. Structured Light Methods for Underwater Imaging: Light Stripe Scanning and Photometric Stereo. In Proceedings of the 2005 MTS/IEEE OCEANS, Washington, DC, USA, 17–23 September 2005; Volume 3, pp. 2610–2617.
16. Tan, C.S.; Seet, G.; Sluzek, A.; He, D.M. A novel application of range-gated underwater laser imaging system (ULIS) in near-target turbid medium. *Opt. Lasers Eng.* **2005**, *43*, 995–1009. [CrossRef]
17. Massot-Campos, M.; Oliver-Codina, G. Underwater laser-based structured light system for one-shot 3D reconstruction. In Proceedings of the 5th Martech International Workshop on Marine Technology, Girona, Spain, 2–5 November 2014.
18. Bruno, F.; Bianco, G.; Muzzupappa, M.; Barone, S.; Razionale, A.V. Experimentation of structured light and stereo vision for underwater 3D reconstruction. *ISPRS J. Photogramm. Remote Sens.* **2011**, *66*, 508–518. [CrossRef]
19. Zhang, Q.; Wang, Q.; Hou, Z.; Liu, Y.; Su, X. Three-dimensional shape measurement for an underwater object based on two-dimensional grating pattern projection. *Opt. Laser Technol.* **2011**, *43*, 801–805. [CrossRef]
20. Bianco, G.; Gallo, A.; Bruno, F.; Muzzupappa, M. A comparative analysis between active and passive techniques for underwater 3D reconstruction of close-range objects. *Sensors* **2013**, *13*, 11007–11031. [CrossRef] [PubMed]

21. McLeod, D.; Jacobson, J.; Hardy, M.; Embry, C. Autonomous inspection using an underwater 3D LiDAR. In Proceedings of the Ocean in Common, San Diego, CA, USA, 23–27 September 2013.
22. Höhle, J. Zur Theorie und Praxis der Unterwasser-Photogrammetrie. Ph.D. Thesis, Bayerische Akademie der Wissenschaften, München, Germany, 1971.
23. Moore, E. J. Underwater photogrammetry. *Photogramm. Rec.* **1976**, *8*, 748–163. [CrossRef]
24. Sedlazeck, A.; Koser, K.; Koch, R. 3D reconstruction based on underwater video from rov kiel 6000 considering underwater imaging conditions. In Proceedings of the 2009 OCEANS—Europe, Bremen, Germany, 11–14 May 2009; pp. 1–10.
25. Schechner, Y.Y.; Karpel, N. Clear underwater vision. In Proceedings of the 2004 IEEE Computer Society Conference on Computer Vision and Pattern Recognition, Washington, DC, USA, 27 June–2 July 2004; Volume 1, pp. 536–543.
26. Li, R.; Tao, C.; Curran, T.; Smith, R. Digital underwater photogrammetric system for large scale underwater spatial information acquisition. *Mar. Geod.* **1996**, *20*, 163–173. [CrossRef]
27. Maas, H.G. New developments in multimedia photogrammetry. In *Optical 3-D Measurement Techniques III*; Grün, A., Kahmen, H., Eds.; Wichmann Verlag: Karlsruhe, Germany, 1995.
28. Sedlazeck, A.; Koch, R. Calibration of housing parameters for underwater stereo-camera rigs. In Proceedings of the 22nd British Machine Vision Conference, Dundee, UK, 29 August–2 September 2011.
29. Kawahara, R.; Nobuhara, S.; Matsuyama, T. A Pixel-wise Varifocal Camera Model for Efficient Forward Projection and Linear Extrinsic Calibration of Underwater Cameras with Flat Housings. In Proceedings of the 2013 IEEE International Conference on Computer Vision Workshops, Sydney, Australia, 2–8 December 2013; pp. 819–824.
30. Shortis, M.R.; Harvey, E.S. Design and calibration of an underwater stereo-video system for the monitoring of marine fauna populations. *International Archives Photogramm. Remote Sens.* **1998**, *32*, 792–799.
31. Fryer, J.G.; Fraser, C.S. On the calibration of underwater cameras. *Photogramm. Rec.* **1986**, *12*, 73–85. [CrossRef]
32. Bryant, M.; Wettergreen, D.; Abdallah, S.; Zelinsky, A. Robust camera calibration for an autonomous underwater vehicle. In Proceedings of the Australian Conference on Robotics and Automation (ACRA 2000), Melbourne, Australia, 30 August–1 September 2000.
33. Lavest, J.M.; Rives, G.; Lapreste, J.T. Underwater camera calibration. In Proceedings of the 6th European Conference on Computer Vision (ECCV 2000), Dublin, Ireland, 26 June–1 July 2000; pp. 654–668.
34. Lavest, J.M.; Rives, G.; Lapreste, J.T. Dry camera calibration for underwater applications. *Mach. Vis. Appl.* **2003**, *13*, 245–253. [CrossRef]
35. Luhmann, T.; Robson, S.; Kyle, S.; Harley, I. *Close Range Photogrammetry*; Wiley Whittles Publishing: Caithness, UK, 2006.
36. Grossberg, M.D.; Nayar, S.K. The raxel imaging model and ray-based calibration. *Int. J. Comput. Vis.* **2005**, *61*, 119–137. [CrossRef]
37. Bothe, T.; Li, W.; Schulte, M.; von Kopylow, C.; Bergmann, R.B.; Jüptner, W. Vision ray calibration for the quantitative geometric description of general imaging and projection optics in metrology. *Appl. Opt.* **2010**, *49*, 5851–5860. [CrossRef] [PubMed]
38. Sansoni, G.; Carocci, M.; Rodella, R. Three-Dimensional Vision Based on a Combination of Gray-Code and Phase-Shift Light Projection: Analysis and Compensation of the Systematic Errors. *Appl. Opt.* **1999**, *38*, 6565–6573. [CrossRef] [PubMed]
39. Schreiber, W.; Notni, G. Theory and arrangements of self-calibrating whole-body three-dimensional measurement systems using fringe projection techniques. *Opt. Eng.* **2000**, *39*, 159–169. [CrossRef]
40. Zhang, S. Recent progresses on real-time 3D shape measurement using digital fringe projection techniques. *Opt. Lasers Eng.* **2010**, *48*, 149–158. [CrossRef]
41. Schaffer, M.; Große, M.; Harendt, B.; Kowarschik, R. Statistical patterns: An approach for high-speed and high-accuracy shape measurements. *Opt. Eng.* **2014**, *53*. [CrossRef]
42. Salvi, J.; Fernandez, S.; Pribanic, T.; Llado, X. A state of the art in structured light patterns for surface profilometry. *Pattern Recognit.* **2010**, *43*, 2666–2680. [CrossRef]
43. Yau, T.; Gong, M.; Yang, Y.H. Underwater Camera Calibration Using Wavelength Triangulation. In Proceedings of the 2013 IEEE Conference on Computer Vision and Pattern Recognition (CVPR), Portland, OR, USA, 23–28 June 2013; pp. 2499–2506.

44. Bräuer-Burchardt, C.; Kühmstedt, P.; Notni, G. Combination of air- and water-calibration for a fringe projection based underwater 3D-Scanner. In Proceedings of the 16th International Conference (CAIP 2015), Valletta, Malta, 2–4 September 2015; pp. 49–60.
45. Chen, X.; Yang, Y.H. Two view camera housing parameters calibration for multi-layer flat refractive interface. In Proceedings of the 2014 IEEE Conference on Computer Vision and Pattern Recognition (CVPR), Columbus, OH, USA, 23–28 June 2014; pp. 524–531.
46. The Association of German Engineers (VDI). Optical 3D-Measuring Systems. In *VDI/VDE Guidelines, Parts 1–3*; VDI/VDE 2634; VDI: Duesseldorf, Germany, 2008.
47. Bräuer-Burchardt, C.; Kühmstedt, P.; Notni, G. Ultra-Precise Hybrid Lens Distortion Correction. In Proceedings of the International Conference on Image and Vision Computing (ICIVC 2012), Venice, Italy, 14–16 November 2012; Volume 71, pp. 179–184.
48. Munkelt, C.; Bräuer-Burchardt, C.; Kühmstedt, P.; Schmidt, I.; Notni, G. Cordless hand-held optical 3D sensor. In Proceedings of the SPIE Optical Metrology 2007, Munich, Germany, 18–22 June 2007; Volume 6618.
49. Eos e-Manufacturing Solutions. Available online: http://www.eos.info/material-p (accessed on 3 December 2015).
50. 4h Jena. 2015. Available online: http://www.4h-jena.de/ (accessed on 3 December 2015).

9

AUV Underwater Positioning Algorithm Based on Interactive Assistance of SINS and LBL

Tao Zhang [1,2,*], Liping Chen [1,2] and Yao Li [1,2]

Academic Editor: Jaime Lloret Mauri

1 Key Laboratory of Micro-Inertial Instrument and Advanced Navigation Technology, Ministry of Education, Southeast University, Nanjing 210096, China; 220132597@seu.edu.cn (L.C.); liyao@seu.edu.cn (Y.L.)
2 School of Instrument Science and Engineering, Southeast University, Nanjing 210096, China
* Correspondence: 101011356@seu.edu.cn

Abstract: This paper studies an underwater positioning algorithm based on the interactive assistance of a strapdown inertial navigation system (SINS) and LBL, and this algorithm mainly includes an optimal correlation algorithm with aided tracking of an SINS/Doppler velocity log (DVL)/magnetic compass pilot (MCP), a three-dimensional TDOA positioning algorithm of Taylor series expansion and a multi-sensor information fusion algorithm. The final simulation results show that compared to traditional underwater positioning algorithms, this scheme can not only directly correct accumulative errors caused by a dead reckoning algorithm, but also solves the problem of ambiguous correlation peaks caused by multipath transmission of underwater acoustic signals. The proposed method can calibrate the accumulative error of the AUV position more directly and effectively, which prolongs the underwater operating duration of the AUV.

Keywords: LBL; TDOA; underwater positioning; information fusion

---

1. Introduction

An AUV (autonomous underwater vehicle) is applied to execute all kinds of underwater tasks, including ocean exploration, underwater mine clearance and collecting bathymetry data of ocean and rivers [1–4]. In order to guarantee that underwater tasks will be completed smoothly and accurate underwater measurement data will be acquired, the AUV is required to have long-term autonomous high-precision positioning and navigation abilities and invisibility [5]. In an underwater environment, electromagnetic wave signals have the characteristic of serious attenuation. In deep sea or under an ice surface, adopting GPS and other radio positioning means cannot achieve ideal positioning effects. In order to meet the navigation requirements, DVL (Doppler velocity log) and SINS (strapdown inertial navigation system) are often used to integrate navigation [6], and the position will be estimated by dead reckoning. However, when this means is used for positioning, positioning errors will accumulate as time goes on [7]. When the AUV is performing tasks in shallow sea, it can adopt the navigation mode of "submerge, water surface calibration, submerge" to launch positioning and navigation; in other words, the AUV relies on SINS/DVL to launch positioning and navigation when navigating under water. After the AUV has been submerged under water for a certain time, in order to calibrate the accumulative errors, the AUV must emerge from the water, and the SINS/GPS integrated navigation system must be used to do the calibration [8]. Adopting this scheme can reach the goal of calibrating accumulative errors, but the AUV must be required to travel to and fro between the underwater operation position and the water surface, which will not only influence the working efficiency and increase the energy consumption, but also expose the position of the AUV. Especially when the AUV is operating in deep sea or under an ice surface, this scheme will be more impractical. Hence, it is very important to study a method in which reliable assistance positioning can be conducted for a long time underwater. This paper suggests an interactive assistance positioning method that integrates an LBL (long base line) underwater acoustic positioning system and an SINS/DVL/MCP (magnetic compass pilot) integrated navigation system, and this is a new idea for solving the above problems.

An LBL underwater acoustic positioning system, usually consisting of a seabed transponder matrix and an interrogation responder with a base length of hundreds to thousands of meters [9,10], adopts distance information between the underwater objective and the seabed matrix element to solve the target position. It can provide accurate positioning of an underwater vehicle within a local area without accumulative errors. Hence, an LBL underwater acoustic positioning system is very applicable to an underwater AUV to launch assistance positioning. Among some of the research of predecessors, Liu, Y. puts forward an underwater AUV positioning and navigation algorithm, which adopts an LBL underwater acoustic positioning system, an ADCP (acoustic Doppler current profiler) and depthometer-assisting INS [11]. Miller, P.A. *et al.* puts forward a tight integrated system based on LBL/DVL/INS [3]. Cheng,W.H. proposes a modification method, which is based on the periodically-measured actual navigation distance and is associated with the TOA positioning algorithm [12]. Jakuba, M.V. *et al.* report results for LBL acoustic navigation during autonomous under-ice surveys near the seafloor and adaptation of the LBL concept for several typical operational situations, including navigation in proximity to the ship during vehicle recoveries [13]. Eustice, R.M. *et al.* report recent experimental results in the development and deployment of a synchronous-clock acoustic navigation system suitable for the simultaneous navigation of multiple underwater vehicles [14]. Chen, Y.M. *et al.* propose a near-real-time approach to underwater inertial navigation with LBL, which uses a ping-response protocol, resulting in asynchronous measurements [15]. Although these systems have reached a certain positioning effect, there are some deficiencies. Firstly, these systems do not explain how to solve the positioning difficulty brought by the multi-path transmission of the underwater acoustic signal. In addition, acoustic velocity is distributed unevenly with the change of underwater depth, and sound ray transmission is curved, which will result in big positioning errors; additionally, the above systems have not proposed any solution.

When solving the target position, the LBL underwater acoustic positioning system can adopt the TOA (time of arrival) positioning algorithm and the TDOA (time difference of arrival) positioning algorithm. The equation set formulated by the TOA positioning algorithm can be directly transformed into a simple linear system of equations with a simple solution. However, strict time synchronization between the hydrophones and the sound source is required to measure a relatively accurate TOA value. It is very hard to do so in reality. The TDOA positioning algorithm acquires the TDOA value by conducting a generalized cross-correlation calculation of the signal received from one hydrophone and another, and then makes the positioning calculation. This method does not have to assure synchronization of the sound source and hydrophones, and the communication between them is quite simple. Hence, it is often adopted in wireless positioning. As acoustic signals will finally be in a coherence stack at one hydrophone through different paths, there will be multiple correlation peaks with approximate amplitudes in the generalized cross-correlation results, thus forming a phenomenon of ambiguous correlation peaks. Then, An, L. *et al.* came up with an ambiguity-solving algorithm based on underwater acoustic propagation characteristics [16]. This method, by studying the distribution rule of cross-correlation peaks forming the multipath transmission of underwater signal channels, tracks stable correlation peaks, which can effectively correct the miscalculation of the TDOA value caused by the ambiguity of the correlation peaks. However, under the condition that distances between fake peaks and the main peak do not differ that much, it is still very hard to accurately track, and the tracking error will be enlarged and finally diverge. In addition, underwater acoustic propagation channels will change as the underwater environment changes, so will the distribution rule of correlation peaks: if the former tracking strategy is still adopted, there will also be errors. Hence, the adaption of this tracking algorithm is not that strong.

In this paper, we propose an underwater positioning method based on the interactive assistance of LBL/SINS/DVL for an AUV. This positioning system consists of SINS, DVL and a sound source installed on the AUV and an LBL underwater matrix located at the seabed. The hydrophones of LBL receive signals from the sound source and conduct cross-correlation calculation and acquire the TDOA value. We adopt the hyperbolic model to solve the position of the sound source (namely, the position of the AUV), correct accumulative errors of SINS/DVL, use resolving results of SINS/DVL to assist in solving ambiguous correlation peaks when LBL is launching underwater acoustic positioning, estimate the TDOA value and improve the solution accuracy of LBL positioning. This scheme can not only directly correct accumulative errors caused by the dead reckoning algorithm, but also solves the problem of ambiguous correlation peaks caused by multipath transmission of underwater acoustic signals. Therefore, it is quite applicable to underwater positioning and navigation of an AUV.

The structure of this paper is as follows: Firstly, we introduce the principle and structure of the underwater assistance positioning system and then introduce key technologies of the system, such as underwater acoustic propagation channel modeling, the calculation of time delay differences, the Taylor series expansion algorithm, the TDOA position solution method and the interactive assistance algorithm. Finally, we verify the effectiveness of the algorithm through a simulation experiment.

2. Principle and Structure of the System

2.1. Placement and Positioning of the Hydrophone

The underwater LBL system needs to use a seabed hydrophone to confirm the position of the vehicle. The calculated position coordinates are the ones corresponding to the seabed hydrophone matrix. Hence, the hydrophone fixed on the seabed should be positioned firstly, and then, its absolute geographic position should be calculated.

As shown in Figure 1, we install the hydrophone reception matrix at the bottom of mother ship (at least three hydrophones, usually more than three), which will receive signals from the hydrophone (with the sound source) underwater and then calculate the three-dimensional position coordinates of each underwater hydrophone corresponding to the hydrophone matrix at the bottom of the mother ship according to short base line positioning principles. The GPS, IMU and compass are installed on the mother ship to provide the accurate geographic position (longitude, latitude and depth) of the mother ship, as well as the attitude angle. We combine factors, such as this information of the mother ship and the installation errors, and calculate the absolute geographic position of each hydrophone under geodetic coordinates. The underwater AUV can adopt these hydrophones (their accurate positions are already acquired) to launch the local area positioning of itself.

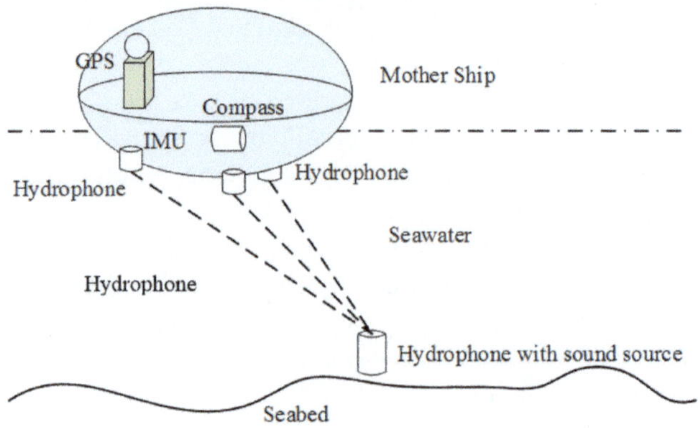

Figure 1. Positioning of the seabed hydrophone.

2.2. LBL Underwater Positioning Model Based on TDOA

TDOA positioning is a method that adopts delay inequality to perform the positioning. By measuring the time difference of a signal reaching different hydrophones, the distance difference between the signal source and different hydrophones can be acquired. As shown in Figure 2, suppose: hydrophones $T_i (i = 0, 1, 2)$ located at three different positions; a sound source on the AUV sends a signal, and the transmission time of the signal reaching three hydrophones is $t_i (i = 0, 1, 2)$; the sound velocity is a steady-state value (suppose it is c); then:

$$R_2 - R_0 = c(t_2 - t_0) = c\Delta t_{20} \tag{1}$$

$$R_1 - R_0 = c(t_1 - t_0) = c\Delta t_{10} \tag{2}$$

$R_i (i = 0, 1, 2)$ represents the distance between the sound source and the hydrophone T_i. The above two equations respectively represent a hyperbolic curve n, which takes T_0 and T_2 as focal points, and hyperbolic curve m, which takes T_0 and T_1 as focal points; their point of intersection is the position of the sound source. As a certain error exists in the measured distance difference, there may be a condition with no solution. Hence, in view of this condition of multiple hydrophones placed on the seabed, redundant information is usually used to acquire the position closest to the actual position.

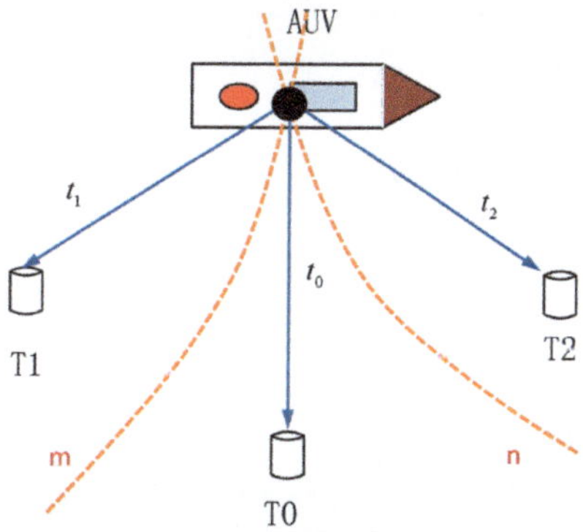

Figure 2. Schematic diagram of the TDOA positioning model.

2.3. Working Process of the System

Figure 3 is a schematic diagram of AUV positioning based on LBL, with multiple fixed hydrophones on the seabed (there are four hydrophones in the diagram); a sound source fixed at the bottom of the AUV will send sound signals; first, accurate positioning of the hydrophones through sensors, like the GPS, IMU and compass, is performed, and the absolute geographic coordinates of each hydrophone are acquired, which is in preparation for solving the position of the sound source; then, a generalized correlation calculation of the sound source signals received by each hydrophone is done. As signals will be refracted and reflected during the transmission, multiple correlation peaks will be generated, resulting in the ambiguity of correlation peaks. Directed at this problem, this paper adopts SINS position assistance to estimate the time difference of sound source signals reaching each hydrophone, solves the distance difference according to the time difference and the equivalent transmission velocity of the signal and, finally, calculates the position of the sound source according to the hyperbolic positioning model. Hence, the interactive assistance positioning technology of SINS and LBL is an innovation point of this paper.

Figure 4 is the operating block diagram of the system. The positioning system mainly consists of LBL, an SINS/DVL/MCP integrated system and a data processing unit. The solution will be made according to the sequence number in the box; firstly, the hydrophones in the LBL system receive signals (Box 1) from the sound source on the AUV and conduct a generalized correlation calculation of received signal ($x_i(t)$,$x_j(t)$) of hydrophone i and hydrophone j, and the calculation result is a group of ambiguous correlation peaks (Box 2). Then, the current AUV position information P_{SINS} and the absolute position of the hydrophones (Box 3) according to the SINS/DVL/MCP integrated system are acquired, and the delay inequality t'_{ij} (Box 4) of the sound source signal reaching hydrophone i and hydrophone j is calculated. We adopt the correlation peak screening module again to screen the former acquired ambiguous correlation peaks and acquire the actual delay inequality t_{ij} (Box 5), then we acquire the distance difference by combining the sound velocity correction algorithm, formulate a positioning solution equation, take P_{SINS} as the initial iterative position, adopt the Taylor series expansion algorithm to solve AUV position P_{LBL} (Box 6) in LBL positioning and, finally, input the difference value of P_{LBL} and P_{SINS} as the external observation information into a Kalman filter; velocity information provided by the DVL and heading information provided by the MCP are also taken as an observed quantity. Filtering results will correct the errors of SINS, and navigation results, such as the accurate position P_{AUV}, velocity and attitude of the AUV, will finally be acquired.

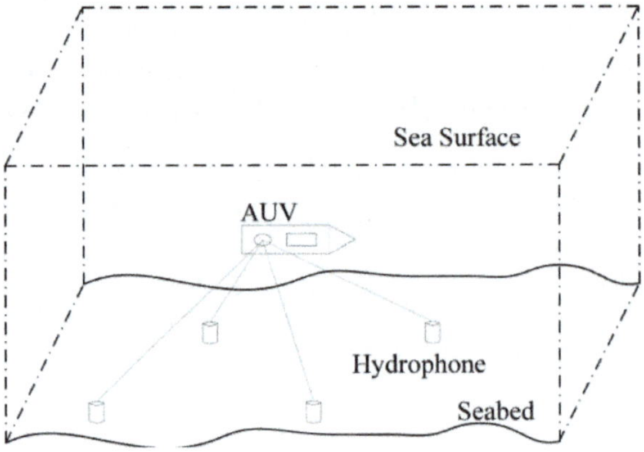

Figure 3. Schematic diagram of AUV underwater LBL positioning.

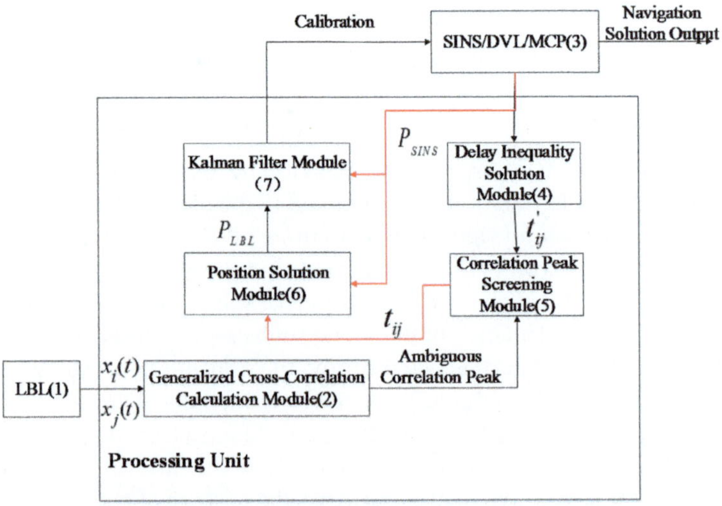

Figure 4. Operation principle block diagram of the system. SINS, strapdown inertial navigation system; DVL, Doppler velocity log; MCP, magnetic compass pilot.

3. Principle of the Interactive Assistance Positioning Algorithm of SINS/DVL/MCP/LBL

This section introduces the realization principles of the algorithms in Figure 4, including generalized cross-correlation calculation of hydrophone receiving signals; SINS assists in seeking the ideal time differences and AUV position calculation based on TDOA.

3.1. Generalized Cross-Correlation Calculation of Hydrophone Receiving Signals

$x(t)$ represents the sound source signal; suppose that the signal received by No. i hydrophone is:

$$x_i(t) = \alpha_i x(t - \tau_i) + n_i(t) \tag{3}$$

The signal received by No. j hydrophone is:

$$x_j(t) = \alpha_j x(t - \tau_j) + n_j(t) \tag{4}$$

α_i and α_j are attenuation coefficients of sound signals propagating underwater; $n_i(t)$ and $n_j(t)$ are non-correlative noise signals; τ_i and τ_j are propagation time.

The ross-correlation function of $x_i(t)$ and $x_j(t)$ is:

$$R_{x_i x_j}(\tau) = E[x_1(t)x_2^*(t-\tau)] = \frac{1}{T-\tau}\int_{\tau}^{T} x_i(t)x_j(t-\tau)dt \tag{5}$$

$\tau = \tau_j - \tau_i$ represents TDOA; T is observation time. According to the characteristics of the correlation function, if the peak value of $R_{x_i x_j}(\tau)$ is found, then the corresponding τ is the right time difference.

3.2. Multi-Path Effect of Sound Signals Underwater

Figure 5 is a simplified multi-path underwater sound propagation model. Place a sound source and two hydrophones R1 and R2; simply consider nonstop path (P*i*d, *i*=1,2), sea surface reflection path (P*i*s, *i*=1,2) and seabed reflection path (P*i*b, *i*=1,2); set the sound source signal as $x(t)$; then, the reception model of the hydrophone is as shown in Equation (6):

$$\begin{cases} x_1(t) = \alpha_{1D}x(t-\tau_{1D}) + \alpha_{1S}x(t-\tau_{1S}) + \alpha_{1B}x(t-\tau_{1B}) + n_1(t) \\ x_2(t) = \alpha_{2D}x(t-\tau_{2D}) + \alpha_{2S}x(t-\tau_{2S}) + \alpha_{2B}x(t-\tau_{2B}) + n_2(t) \end{cases} \tag{6}$$

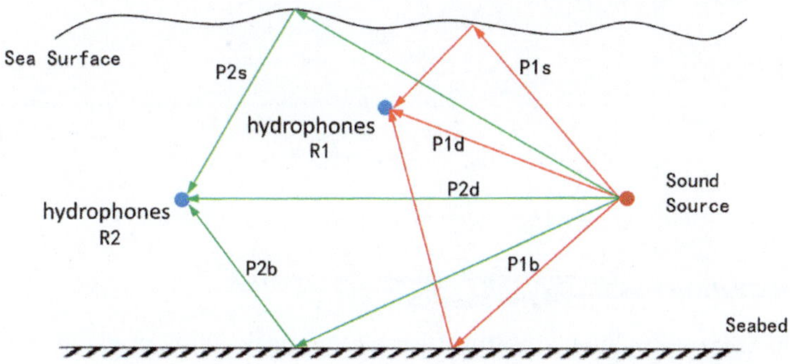

Figure 5. Simplified model of the underwater multi-path effect.

α_{iD}, α_{iS} and α_{iB} are respectively attenuation coefficients of P*i*d path, P*i*s path and P*i*b path (i = 1,2). τ_{iD}, τ_{iS} and τ_{iB} are respectively the propagation time of P*i*d path, P*i*s path and P*i*b path. Suppose that sound source signal $x(t)$ is irrelevant to noise $n_1(t)$ and noise $n_2(t)$ and that $n_1(t)$ is irrelevant to noise $n_2(t)$, then the cross-correlation function of $x_1(t)$ and $x_2(t)$ is as shown in Equation (7). $R_{xx}(\tau)$ is a self-correlation function of $x(t)$. It can be seen from Equation (7) that: peak values of the cross-correlation functions of $x_1(t)$ and $x_2(t)$ occur respectively on nine time delay of arrival points, such as $\tau_{1D} - \tau_{2D}$, $\tau_{1D} - \tau_{2S}$ and $\tau_{1D} - \tau_{2B}$ (nine peak values will occur when the nine points are unequal; if an equality situation among nine points exists, then there will be an overlapping phenomenon, and the number of peak values will reduce); the peak value will be decided by the corresponding attenuation coefficient. The specific effect is as shown in Figure 6. Under a practical situation, we only need to calculate the time difference of arrival of a nonstop path (main peak), so other peak values will interfere with confirming the main peak, which makes it impossible to accurately estimate the TDOA of signals.

$$\begin{aligned} R_{x_1x_2}(\tau) &= E[x_1(t)x_2^*(t-\tau)] \\ &= \alpha_{1D}\alpha_{2D}R_{xx}(\tau_{1D}-\tau_{2D}-\tau) \\ &+\alpha_{1D}\alpha_{2S}R_{xx}(\tau_{1D}-\tau_{2S}-\tau) \\ &+\alpha_{1D}\alpha_{2B}R_{xx}(\tau_{1D}-\tau_{2B}-\tau) \\ &+\alpha_{1S}\alpha_{2D}R_{xx}(\tau_{1S}-\tau_{2D}-\tau) \\ &+\alpha_{1S}\alpha_{2S}R_{xx}(\tau_{1S}-\tau_{2S}-\tau) \\ &+\alpha_{1S}\alpha_{2B}R_{xx}(\tau_{1S}-\tau_{2B}-\tau) \\ &+\alpha_{1B}\alpha_{2D}R_{xx}(\tau_{1B}-\tau_{2D}-\tau) \\ &+\alpha_{1B}\alpha_{2S}R_{xx}(\tau_{1B}-\tau_{2S}-\tau) \\ &+\alpha_{1B}\alpha_{2B}R_{xx}(\tau_{1B}-\tau_{2B}-\tau) \end{aligned} \tag{7}$$

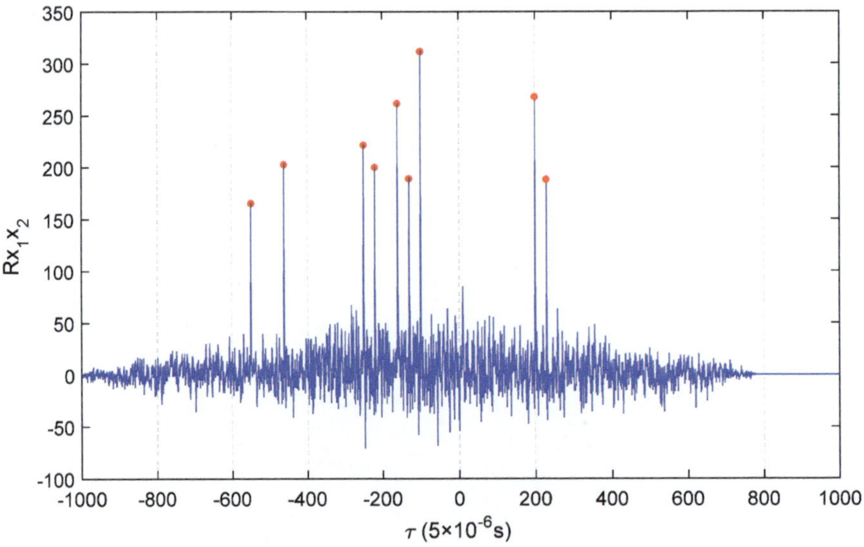

Figure 6. Ambiguity phenomenon of correlation peaks.

3.3. SINS/DVL Assistance in Seeking the Ideal Delay Inequality

Because of the multi-path effect, multiple correlation peaks will appear in the cross-correlation function, and these peaks differ a little in value; it is very difficult to judge which one corresponds to the most ideal time difference. In multi-path time delay estimation, the methods that can be adopted are usually the self-adaptation method, the generalized cross-correlation method, the auto-correlation method, the cepstrum method, *etc.* Although the self-adaptation method is of high estimation accuracy and a strong resolution ratio, its search scope is quite broad, which makes it hard to guarantee convergence and results in a large calculated quantity and poor instantaneity; besides, it has a certain requirement for the signal-to-noise ratio. The generalized cross-correlation method is of simple calculation, but the main correlation peak is not obvious; and there are certain deficiencies in its performance. Resolution ratios of the auto-correlation method and the cepstrum method are used for calculating the time delay, as they need a relatively larger signal bandwidth [17–20]. This paper opens up a new path by putting forward a method that selects the main correlation peak based on

SINS/DVL positioning assistance after combining the advantages and disadvantages of the generalized cross-correlation method. This algorithm is simple and easy to realize; besides, it can estimate the time differences of multiple paths. Even with a low signal-to-noise ratio, it can effectively do the estimation; in the meantime, it can solve the problem of selecting the main correlation peak with high accuracy when correlation peaks are ambiguous. Now, the principle of the algorithm will be introduced in detail:

In the LBL underwater acoustics positioning system, set the position of No. i hydrophone as $P_i(x_i, y_i, z_i)$ and the AUV position output by the SINS integrated system at time k as $P_{SINS}(x_{SINS}(k), y_{SINS}(k), z_{SINS}(k))$; adopt $P_{SINS}(x_{SINS}(k), y_{SINS}(k), z_{SINS}(k))$ to calculate the distance between the hydrophone and AUV at time k as:

$$D_i(k) = \sqrt{(x_i - x_{SINS}(k))^2 + (y_i - y_{SINS}(k))^2 + (z_i - z_{SINS}(k))^2} \tag{8}$$

The distance difference between two random hydrophones i and j and the AUV at time k is:

$$D_{ij}(k) = D_i(k) - D_j(k)(i \neq j) \tag{9}$$

Then, the calculation of the time difference of the two hydrophones receiving signals is:

$$\Delta t'_{ij}(k) = \frac{D_{ij}(k)}{c_{ij}(k)} \tag{10}$$

$c_{ij}(k)$ is the equivalent sound velocity of signals corresponding to time difference $\Delta t'_{ij}(k)$ at time k.

As at time $k-1$, the surrounding environment of the AUV is not changed that much at time k, the change of the sound ray structure is little. Hence, the equivalent sound velocity of the last time cycle can be used as the current equivalent acoustic velocity; in other words, the velocity can be acquired by using the distance difference between hydrophones i and j and the sound source at time $k-1$ to divide the time difference. The specific calculation is as follows:

At time $k - 1$, the corrected position of AUV by LBL is $P_{LBL/SINS}(x_{LBL/SINS}(k-1), y_{LBL/SINS}(k-1), z_{LBL/SINS}(k-1))$, and the distance of hydrophone i and the sound source is:

$$R_i(k-1) = \sqrt{(x_i - x_{LBL/SINS}(k-1))^2 + (y_i - y_{LBL/SINS}(k-1))^2 + (z_i - z_{LBL/SINS}(k-1))^2} \tag{11}$$

The distance difference between hydrophones i and j and the sound source is:

$$\Delta R_{ij}(k-1) = R_i(k-1) - R_j(k-1) \tag{12}$$

If the time difference of hydrophones i and j (which have been screened out) receiving the signal at time t_{k-1} is $\Delta t_{ij(k-1)}$, then the current equivalent acoustic velocity is:

$$c_{ij}(k) = \frac{\Delta R_{ij}(k-1)}{\Delta t_{ij}(k-1)} \tag{13}$$

Substitute the computed results of Equations (9) and (13) into Equation (10), and time difference $\Delta t'_{ij}(k)$ can be calculated. Seek the peak value that is the most proximate to $\Delta t'_{ij}(k)$ among a group of ambiguous correlation peaks in Equation (7), and take the time difference corresponding to this peak value as the more accurate time difference $\Delta t_{ij}(k)$.

3.4. TDOA Three-Dimensional Positioning Algorithm Based on SINS/DVL Assistance

Three-Dimensional Positioning Algorithm Base on Taylor Series Expansion

For the moment, there are many algorithms that use TDOA measured to perform positioning, such as Chan's algorithm, the Taylor algorithm, the Friedlander algorithm, *etc.* Chan's algorithm has strict requirements for the measurement accuracy of the time difference and is more applicable to a line-of-sight transmission channel environment. Under a non-line-of-sight transmission condition, as for the measurement errors of the time difference of the signal arrival, the positioning errors of the algorithm will be large and will be easily influenced by the effects of reflection, scattering and refraction. What the Friedlander algorithm acquires is only the second-best solution. The Taylor algorithm has no special requirements for the statistic property of measuring errors or *a priori* information, and it can provide a higher positioning degree on a certain Gaussian noise level. However, this algorithm is an iterative one without a final expression solution, and algorithmic convergence needs to be guaranteed by an initial position that is not far from the actual position [21,22]. Hence, this paper suggests the TDOA positioning algorithm based on SINS/DVL assistance, and the algorithm takes the positioning results of SINS/DVL as the iterative initial value of the Taylor algorithm, which not only guarantees algorithmic convergence, but also reduces the iterations. It is very applicable to underwater positioning.

This algorithm will be introduced in detail as follows:

Suppose that there are n hydrophones in the matrix, then formulate (n – 1) equations according to the hyperbolic positioning model:

$$R_{i1} = R_i - R_1 = c_{i1}\Delta t_{i1}(i = 2, 3, 4, \ldots, n) \tag{14}$$

R_{i1} is the function of x, y, z, x_i, y_i, z_i; (x, y, z) represents the position of the AUV; (x_i, y_i, z_i) represents the position of No. i hydrophone; then, it can be expressed as $f_i(x, y, z, x_i, y_i, z_i) = R_i - R_1$; set it as the objective function. Suppose that the measured value of objective function $f_i(*)$ (namely $f_i(x, y, z, x_i, y_i, z_i)$) is $m_i = c_{i1}\Delta t_{i1}$, and the actual value is u_i, $u_i = m_i - e_i$, e_i is the measuring error. Adopt initial value $(\hat{x}, \hat{y}, \hat{z})$, and they meet $x = \hat{x} + \Delta x$, $y = \hat{y} + \Delta y$, $z = \hat{z} + \Delta z$; then, expand objective function $f_i(*)$ in $(\hat{x}, \hat{y}, \hat{z})$ according to the Taylor series as the following Equation (15):

$$\begin{aligned} f_i(x, y, z, x_i, y_i, z_i) &= f_i(\hat{x}, \hat{y}, \hat{z}, x_i, y_i, z_i) + (\Delta x\frac{\partial}{\partial x} + \Delta y\frac{\partial}{\partial y} + \Delta z\frac{\partial}{\partial z})f_i(\hat{x}, \hat{y}, \hat{z}, x_i, y_i, z_i) \\ &+ \frac{1}{2!}(\Delta x\frac{\partial}{\partial x} + \Delta y\frac{\partial}{\partial y} + \Delta z\frac{\partial}{\partial z})^2 f_i(\hat{x}, \hat{y}, \hat{z}, x_i, y_i, z_i) + \cdots \\ &+ \frac{1}{n!}(\Delta x\frac{\partial}{\partial x} + \Delta y\frac{\partial}{\partial y} + \Delta z\frac{\partial}{\partial z})^n f_i(\hat{x}, \hat{y}, \hat{z}, x_i, y_i, z_i) \\ + \frac{1}{(n+1)!}&(\Delta x\frac{\partial}{\partial x} + \Delta y\frac{\partial}{\partial y} + \Delta z\frac{\partial}{\partial z})^{n+1} f_i(\hat{x} + \xi\Delta x, \hat{y} + \xi\Delta y, \hat{z} + \xi\Delta z, x_i, y_i, z_i), (0 < \xi < 1) \end{aligned} \tag{15}$$

Ignore high-order terms above the quadratic term in the expanded section, then the above equation can be expressed as:

$$f_i(x, y, z, x_i, y_i, z_i) = u_i \approx f_i(\hat{x}, \hat{y}, \hat{z}, x_i, y_i, z_i) + (\Delta x\frac{\partial}{\partial x} + \Delta y\frac{\partial}{\partial y} + \Delta z\frac{\partial}{\partial z})f_i(\hat{x}, \hat{y}, \hat{z}, x_i, y_i, z_i) \tag{16}$$

It can be acquired according to the hyperbolic positioning model that:

$$f_i(x, y, z, x_i, y_i, z_i) = \sqrt{(x - x_i)^2 + (y - y_i)^2 + (z - z_i)^2} - \sqrt{(x - x_1)^2 + (y - y_1)^2 + (z - z_1)^2} \tag{17}$$

Expand Equation (17) according to Equation (16) and acquire:

$$f_i(x,y,z,x_i,y_i,z_i) \approx \hat{R}_i - \hat{R}_1 + \Delta x(\frac{\hat{x}-x_i}{\hat{R}_i} - \frac{\hat{x}-x_1}{\hat{R}_1}) + \Delta y(\frac{\hat{y}-y_i}{\hat{R}_i} - \frac{\hat{y}-y_1}{\hat{R}_1}) + \Delta z(\frac{\hat{z}-z_i}{\hat{R}_i} - \frac{\hat{z}-z_1}{\hat{R}_1}) \tag{18}$$

where:

$$\hat{R}_1 = \sqrt{(\hat{x}-x_1)^2 + (\hat{y}-y_1)^2 + (\hat{z}-z_1)^2} \tag{19}$$

$$\hat{R}_i = \sqrt{(\hat{x}-x_i)^2 + (\hat{y}-y_i)^2 + (\hat{z}-z_i)^2} \tag{20}$$

Set $\hat{f}_i = f_i(\hat{x},\hat{y},\hat{z},x_i,y_i,z_i)$, then:

$$a_{i1} = \frac{\partial f_i(x,y,z,x_i,y_i,z_i)}{\partial x}\Big|_{\substack{x=\hat{x}\\ y=\hat{y}\\ z=\hat{z}}} = \frac{\hat{x}-x_i}{\hat{R}_i} - \frac{\hat{x}-x_1}{\hat{R}_1} \tag{21}$$

$$a_{i2} = \frac{\partial f_i(x,y,z,x_i,y_i,z_i)}{\partial y}\Big|_{\substack{x=\hat{x}\\ y=\hat{y}\\ z=\hat{z}}} = \frac{\hat{y}-y_i}{\hat{R}_i} - \frac{\hat{y}-y_1}{\hat{R}_1} \tag{22}$$

$$a_{i3} = \frac{\partial f_i(x,y,z,x_i,y_i,z_i)}{\partial z}\Big|_{\substack{x=\hat{x}\\ y=\hat{y}\\ z=\hat{z}}} = \frac{\hat{z}-z_i}{\hat{R}_i} - \frac{\hat{z}-z_1}{\hat{R}_1} \tag{23}$$

and then:

$$\hat{f}_i + a_{i1}\Delta x + a_{i2}\Delta y + a_{i3}\Delta z \approx m_i - e_i \tag{24}$$

For the n matrix elements, there will be:

$$\varepsilon \approx h - G\delta \tag{25}$$

where:

$$\varepsilon = \begin{bmatrix} e_2 \\ e_3 \\ \vdots \\ e_n \end{bmatrix} \tag{26}$$

$$h = \begin{bmatrix} m_2 - \hat{f}_2 \\ m_3 - \hat{f}_3 \\ \vdots \\ m_n - \hat{f}_n \end{bmatrix} = \begin{bmatrix} c_{21}\Delta t_{21} - (\hat{R}_2 - \hat{R}_1) \\ c_{31}\Delta t_{31} - (\hat{R}_3 - \hat{R}_1) \\ \vdots \\ c_{n1}\Delta t_{n1} - (\hat{R}_n - \hat{R}_1) \end{bmatrix} \tag{27}$$

$$G = \begin{bmatrix} a_{21} & a_{22} & a_{33} \\ a_{31} & a_{32} & a_{33} \\ \vdots & \vdots & \vdots \\ a_{n1} & a_{n2} & a_{n3} \end{bmatrix} == \begin{bmatrix} \frac{\hat{x}-x_2}{\hat{R}_2} - \frac{\hat{x}-x_1}{\hat{R}_1} & \frac{\hat{y}-y_2}{\hat{R}_2} - \frac{\hat{y}-y_1}{\hat{R}_1} & \frac{\hat{z}-z_2}{\hat{R}_2} - \frac{\hat{z}-z_1}{\hat{R}_1} \\ \frac{\hat{x}-x_3}{\hat{R}_3} - \frac{\hat{x}-x_1}{\hat{R}_1} & \frac{\hat{y}-y_3}{\hat{R}_3} - \frac{\hat{y}-y_1}{\hat{R}_1} & \frac{\hat{z}-z_3}{\hat{R}_3} - \frac{\hat{z}-z_1}{\hat{R}_1} \\ \vdots & \vdots & \vdots \\ \frac{\hat{x}-x_n}{\hat{R}_n} - \frac{\hat{x}-x_1}{\hat{R}_1} & \frac{\hat{y}-y_n}{\hat{R}_n} - \frac{\hat{y}-y_1}{\hat{R}_1} & \frac{\hat{z}-z_3}{\hat{R}_n} - \frac{\hat{z}-z_1}{\hat{R}_1} \end{bmatrix} \tag{28}$$

$$\delta = \begin{bmatrix} \Delta x \\ \Delta y \\ \Delta z \end{bmatrix} \tag{29}$$

Take δ as an unknown variable; suppose that Q is the covariance matrix of ε; then adopt the method of weighing least squares estimation; it can be acquired that:

$$\begin{bmatrix} \Delta x \\ \Delta y \\ \Delta z \end{bmatrix} = \delta = \left[G^T Q^{-1} G\right]^{-1} G^T Q^{-1} h \tag{30}$$

The calculation process of the Taylor algorithm can be concluded as follows:

(1) Select an initial value $(\hat{x}, \hat{y}, \hat{z})$;
(2) Substitute $(\hat{x}, \hat{y}, \hat{z})$ into Equations (20) and (28) to calculate G and $\hat{R}_i$;
(3) Substitute c_{i1}, Δt_{21} and $\hat{R}_i$ into Equation (27) to calculate h;
(4) Substitute G, h and Q into Equation (30) and update δ; if $\sqrt{(\Delta x)^2 + (\Delta y)^2 + (\Delta z)^2} < \eta$, η is a very small threshold value, then the iteration ends; $(\hat{x}, \hat{y}, \hat{z})$ is the final positioning result. Otherwise, update $(\hat{x}, \hat{y}, \hat{z})$ according to Equation (31), and repeat $(\hat{x}, \hat{y}, \hat{z})$ until the above conditions are met.

$$\begin{cases} \hat{x} \leftarrow \hat{x} + \Delta x \\ \hat{y} \leftarrow \hat{y} + \Delta y \\ \hat{z} \leftarrow \hat{z} + \Delta z \end{cases} \tag{31}$$

The algorithm needs an estimated position value as an initial value to make iterations. The accuracy of the initial position value has great influence on the convergence of the algorithm. As shown in Figure 7, the convergence error threshold for the iterative algorithm is 10^{-7} m. When the initial position error is 80 m, the iterative result cannot converge. When the initial position error is 35 m, the iterative result converges, but the number of iterations is 468 steps. While the initial position error is set as 10 m, the number of iterations requires only 364 steps. Thus, the higher the accuracy of the initial position value, the faster the convergence rate. This paper selects the output position P_{SINS} of SINS/DVL as the initial position value of the iteration, which can not only satisfy the convergence of the algorithm, but also greatly reduce the iteration steps.

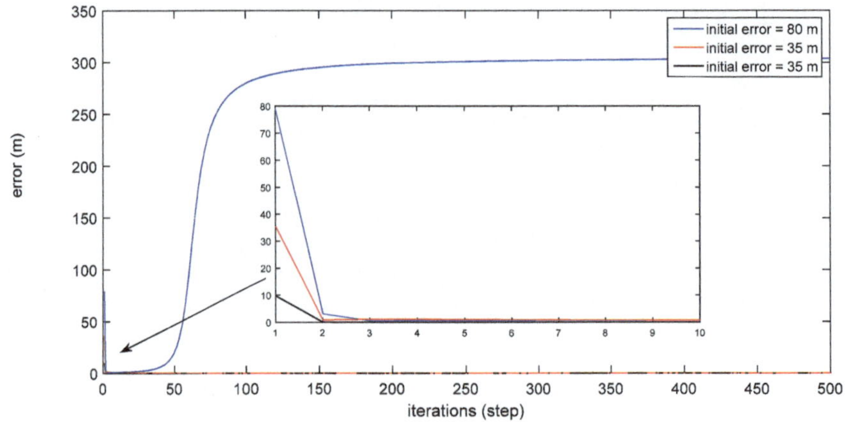

Figure 7. The iteration results with different initial position values.

3.5. *SINS/DVL/MCP/LBL Integrated System Modeling*

When the AUV does not enter the effective signal scope of LBL, adopt the SINS/DVL/MCP integrated navigation system as shown in Figure 8 to perform the navigation and positioning. When the AUV enters the effective signal scope of LBL, adopt the SINS/DVL/MCP/LBL integrated navigation system as shown in Figure 9 to perform the navigation and positioning.

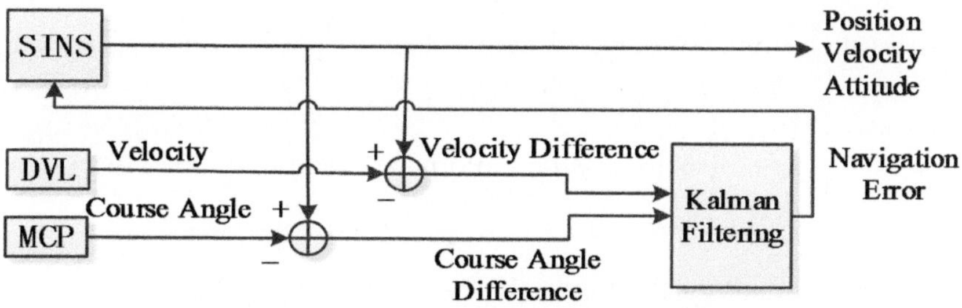

Figure 8. SINS/DVL/MCP integrated navigation system.

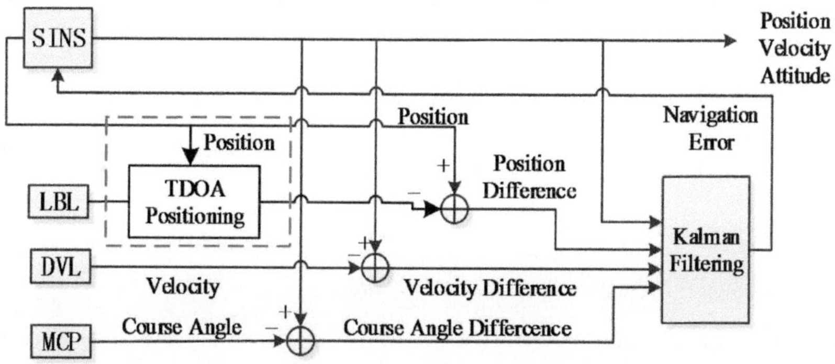

Figure 9. SINS/DVL/MCP/LBL integrated navigation system.

The state equation of the integrated system is:

$$\dot{\mathbf{X}} = \mathbf{FX} + \mathbf{GW} \tag{32}$$

where **X** is the state variable, **F** is the state-transition matrix, **G** is the transition matrix of the process noise and **W** is systematic noise.

Select the velocity error, attitude error, accelerometer zero offset and gyroscopic drift as state vector **X**:

$$\mathbf{X} = \begin{bmatrix} \delta V_E & \delta V_N & \delta V_U & \phi_E & \phi_N & \phi_U & \delta L & \delta\lambda & \delta h & \nabla_{bx} & \nabla_{by} & \nabla_{bz} & \varepsilon_{bx} & \varepsilon_{by} & \varepsilon_{bz} \end{bmatrix}^T \tag{33}$$

$\delta V_E, \delta V_N, \delta V_U$ are respectively the velocity errors of the directions of east, north and the local vertical(up). ϕ_E, ϕ_N, ϕ_U are respectively the misalignment angles of the directions of east, north and the local vertical(up). $\delta L, \delta\lambda, \delta h$ are respectively the errors of latitude, longitude and altitude. $\nabla_{bx}, \nabla_{by}, \nabla_{bz}$ are respectively biased errors of the three axial directions of the accelerator. $\varepsilon_{bx}, \varepsilon_{by}, \varepsilon_{bz}$ are respectively the drifts of the three axial directions of the gyroscopes. **F** can be confirmed by the SINS error equation.

The measuring equation of the integrated system is:

$$\mathbf{V_{SINS}} = \begin{bmatrix} V_{SINSE} \\ V_{SINSN} \\ V_{SINSU} \end{bmatrix} \tag{34}$$

Z is the observation vector; $\mathbf{P_{SINS}} = \begin{bmatrix} L_{SINS} \\ \lambda_{SINS} \\ h_{SINS} \end{bmatrix}$ is the position information output of the SINS system; $\mathbf{P}_{LBL} = \begin{bmatrix} L_{LBL} \\ \lambda_{LBL} \\ h_{LBL} \end{bmatrix}$ is the position information output by the LBL system; $\mathbf{V_{SINS}} = \begin{bmatrix} V_{SINSE} \\ \lambda_{SINSN} \\ h_{SINSU} \end{bmatrix}$ is the velocity information output by the SINS system; $\mathbf{V}_{DVL} = \begin{bmatrix} V_{DVLE} \\ V_{DVLN} \\ V_{DVLU} \end{bmatrix}$ is the velocity information output by the DVL system; φ_U is the heading information output by the SINS system; φ_{MCP} is the heading information output by the MCP system; **V** is the observation noise vector; and **H** is the measurement matrix, which satisfies:

$$\mathbf{H} = \begin{bmatrix} 1 & 0 & 0 & 0 & 0 & 0 & 0 & 1 & 0 & 0 & 0 & 0 & 0 & 0 & 0 \\ 0 & 1 & 0 & 0 & 0 & 0 & 0 & 0 & 1 & 0 & 0 & 0 & 0 & 0 & 0 \\ 0 & 0 & 1 & 0 & 0 & 1 & 0 & 0 & 0 & 1 & 0 & 0 & 0 & 0 & 0 \end{bmatrix} \tag{35}$$

4. Simulation and Experiment

4.1. Underwater Sound Signal Propagation Channel Modeling

The underwater acoustic channel is a time varying and space varying random channel of high environmental noise, a narrow channel bandwidth, large transmission loss and a serious multi-path effect. Because of the slow underwater movement of the AUV, the acoustic channel can be seen as a slow time-varying system. The system can be approximated as a LTI (linear time-invariant)system. Suppose that there are n paths for the acoustic signal to transmit from the sound source to the hydrophones, and then, the unit impulse response of the multi-path channel from the sound source to the hydrophones will be:

$$h(t) = \sum_{i=1}^{N} a_i \delta(t - \tau_i) \tag{36}$$

a_i is the attenuation coefficient of the No. i transmission path; τ_i is the relative time delay of the transmission along the No. i transmission path. The underwater acoustic signal transmission can be simplified as the model that is shown in Figure 10:

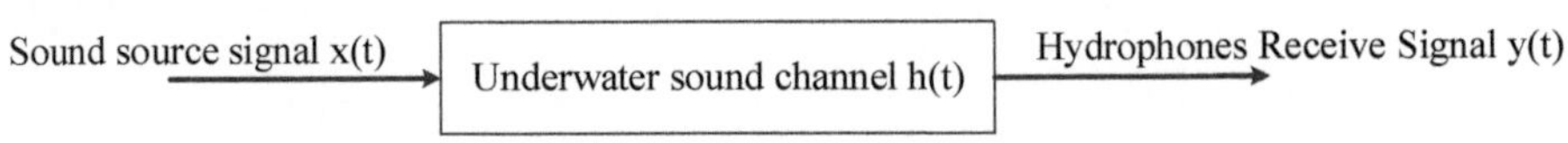

Figure 10. A simplified model of underwater sound transmission.

Then, the hydrophones receive signal *y(t)* as a convolution of sound source signal *x(t)* and the unit impulse response *h(t)*, namely:

$$y(t) = x(t) * h(t) \tag{37}$$

Porter, M.B. *et al.* have developed BELLHOP software, which simulates the marine environment according to this model. This software can acquire amount N, the angle of incidence, the range and the time delay of intrinsic sound rays and provide the unit impulsive response of the system by inputting marine environment parameters [23]. This paper, according to the sound velocity distribution curve under a lake, which is as shown in Figure 11, adopts the BELLHOP software to establish an underwater sound signal transmission channel model, which will describe the sound field of the underwater environment and calculate the sound ray transmission path by setting the positions of the sound source and hydrophones, as shown in Figure 12; then, it will solve the unit impulse response function $h(t)$ of the system and, finally, acquire the received signals of the hydrophones by the convolution operation and simulation.

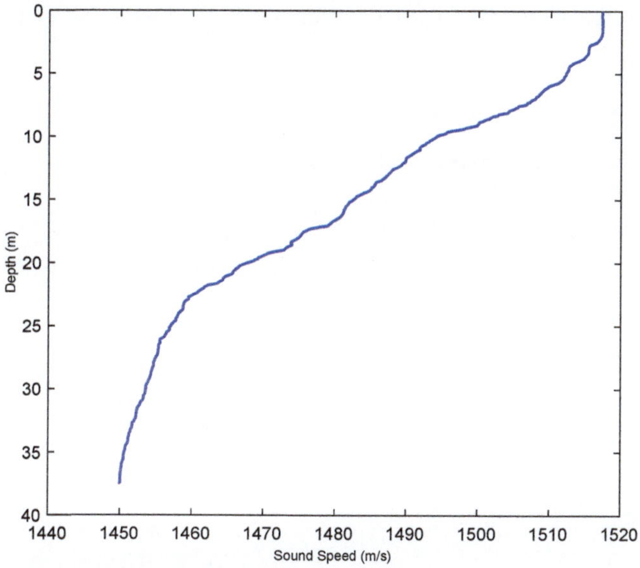

Figure 11. Underwater sound velocity distribution curve of a lake.

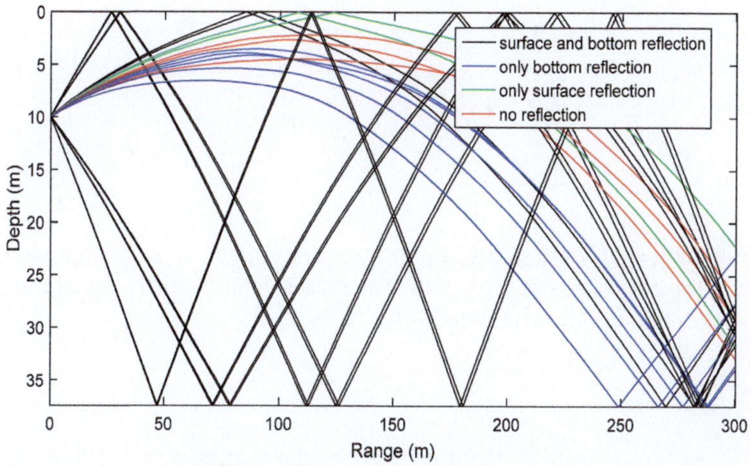

Figure 12. Sound ray transmission path.

The simulated sound source signal adopts the amplitude-modulated signal expressed in Equation (38). Its bandwidth is 50 kHz, and its center frequency is 25 kHz. Suppose that the noise model of the facility is a white noise model; then, the waveform of the sound source signal will be as shown in Figure 13.

$$sigIN = (\sum_{i}^{N} a\sin(2\pi i\omega t))\cos(2\pi\omega_c t) \tag{38}$$

Calculate the unit impulse response as shown in Figure 14 according to the propagating sound rays, and the received signal can be calculated as shown in Figure 15 according to Equation (37). Adopt the sound field model established by BELLHOP and the underwater sound velocity distribution to simulate the situation of reflection and refraction, which may occur in the acoustic signal transmission process, as well as the signals received by the hydrophones, all of which will be used for the simulation and positioning calculation.

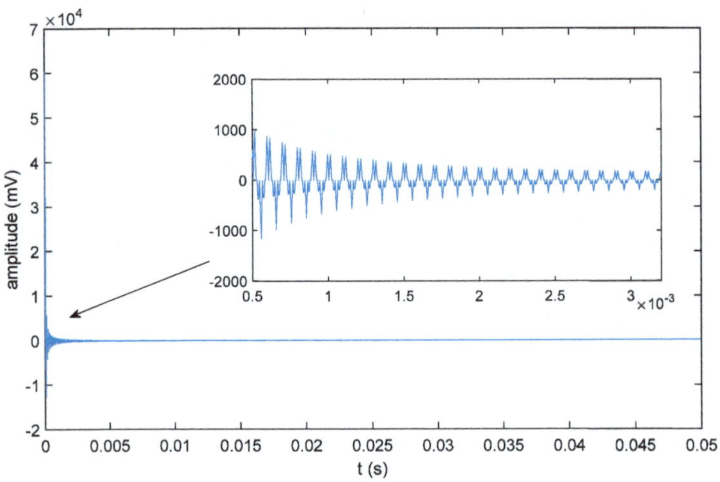

Figure 13. Oscillogram of the sound source signal.

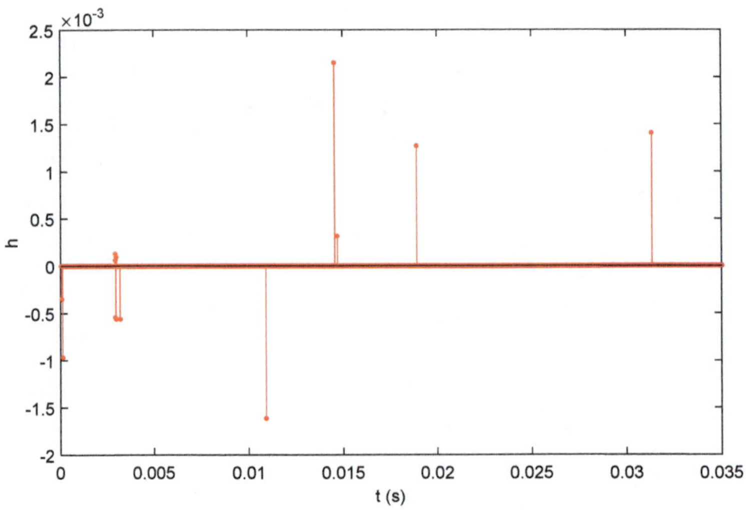

Figure 14. Unit impulse response h(t).

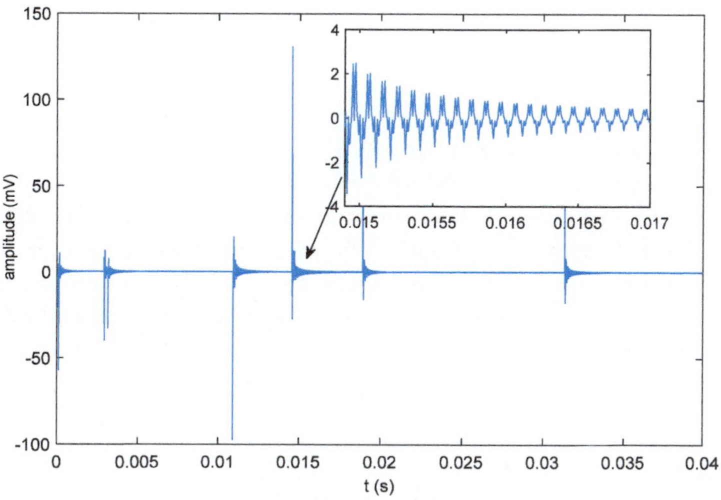

Figure 15. Oscillogram of received signals.

4.2. Simulation of SINS Assistance in the LBL Positioning Algorithm

As shown in Figure 16, set the positions of the hydrophones and the AUV; place five hydrophones underwater; their positions will be expressed with longitude and latitude, and they are respectively(118.01°, 32°), (118.01°, 32.01°), (118.01°, 32.02°), (118.02°, 32.01°), (118°, 32.01°); all of their depths are 30 m. Suppose that the current actual position of the AUV is (118°, 32°) with a depth of 10 m. The underwater acoustic velocity distribution data are acquired by an experiment on this lake. Simulate the acoustic signals received by the hydrophones according to the method introduced in Section 4.1, and calculate the time difference of different hydrophones receiving the signal.

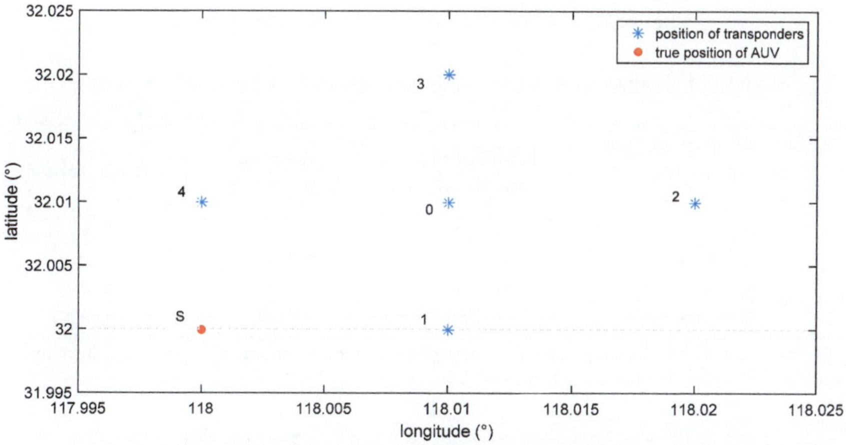

Figure 16. Layout of the hydrophones and the AUV.

4.2.1. Simulation of SINS Assisting LBL in Tracking the Optimal Time Difference

As shown in Figure 17, as the sound source signal will go through multiple paths, there will be also multiple correlation peaks in the generalized cross-correlation result. The correlation peaks whose amplitude is in the top twenty can be selected as alternative correlation peaks, while others are neglected due to excessive signal attenuation. The distance between the time difference with the highest amplitude (Point A in Figure 17) and the true time difference (the red circle in Figure 17) is relatively larger, so Point A cannot be selected as the true time difference. Using the proposed method, the time difference (symbol × in Figure 16), which is calculated through Equation (10) by the aiding of the SINS/DVL integrated navigation, better approaches the truth value. Therefore, the alternative correlation peak (Point B in Figure 17), which is closest to the symbol "×", is selected as the ideal time difference, because it has the smallest distance to Point "A". Traditional algorithms directly take the correlation peak that is corresponding to the maximum peak value as the main correlation peak to calculate the time difference; there will be larger errors under the multi-path effect. The improved algorithm adopted in this paper selects the main correlation peak through the positions obtained by the SINS/DVL integrated navigation system. A comparison of the calculation results and errors of the two algorithms is as shown in Tables 1 and 2 which show that: the improved algorithm reduces the interference of multiple correlation peaks with time difference estimation under the multi-path effect and makes the calculation value accuracy of the final time difference superior to that of the traditional algorithm.

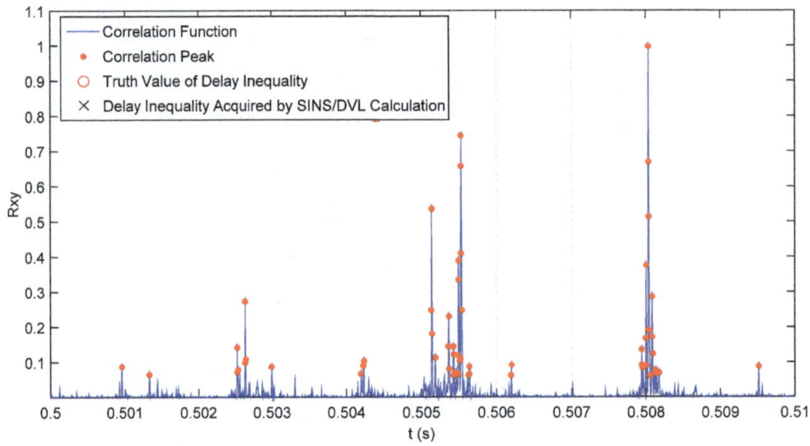

Figure 17. Screening out of the main correlation peak.

Table 1. Comparison of the calculation results of the time differences.

Time Difference of Two Hydrophones Receiving the Signal	Traditional Algorithm/s	Improved Algorithm/s	Truth Value/s
Ts1−Ts0	−0.3532	−0.3436	−0.3470
Ts2−Ts0	0.5080	0.5043	0.5044
Ts3−Ts0	0.6623	0.6499	0.6492
Ts4−Ts0	−0.2388	−0.2297	−0.2297

Note: The expression (Ts1−Ts0) represents the time difference from the sound source from the sound source propagating to Hydrophone 1 and Hydrophone 0, and so on.

Table 2. Error comparison of the calculation values of the time difference.

Time Difference of Two Hydrophones Receiving the Signal	Error of the Traditional Algorithm/s	Error of the Improved Algorithm/s
Ts1−Ts0	0.0062	0.0034
Ts2−Ts0	0.0036	0.0001
Ts3−Ts0	0.0132	0.0007
Ts4−Ts0	0.0091	0.0001

4.2.2. Simulation of the Acoustic Velocity Correction Algorithm

The traditional algorithm directly adopts acoustic velocity in a traditional sense to calculate the distance difference. The improved algorithm adopted in this paper calculates equivalent acoustic velocity, which will be used to calculate the distance difference. A comparison of the calculation results and errors of the two algorithms is as shown in Tables 3 and 4 which show that: as the improved algorithm adopts the equivalent acoustic velocity to calculate the distance difference, it corrects the sound velocity, under situations of the multi-path effect and sound ray curve. It greatly enhances the accuracy of the calculation values of the distance difference. Therefore, adopting the time difference algorithm and distance difference algorithm in this paper can greatly reduce the errors of the time difference and distance difference, which play a significant role in improving the AUV underwater positioning accuracy.

Table 3. Comparison of the calculation results of the distance difference.

Distance Difference of Two Hydrophones Receiving the Signal	Traditional Algorithm/m	Improved Algorithm/m	Ideal Distance Difference/m
Ds1−Ds0	−523.8619	−509.5792	−510.0840
Ds2−Ds0	739.6344	734.1894	734.3538
Ds3−Ds0	973.0620	954.7788	954.1334
Ds4−Ds0	−360.3024	−346.6323	−346.1796

Note: The expression (Ds1−Ds0) represents the distance difference from the sound source propagating to Hydrophone 1 and Hydrophone 0, and so on.

Table 4. Error comparison of the calculation values of the distance difference.

Distance Difference of Two Hydrophones Receiving the Signal	Error of the Traditional Algorithm/m	Error of the Improved Algorithm/m
Ds1−Ds0	13.7779	0.5048
Ds2−Ds0	5.2806	0.1644
Ds3−Ds0	18.9286	0.6454
Ds4−Ds0	14.1228	0.4527

4.2.3. Simulation of the TDOA Positioning Algorithm

In order to verify the positioning effect of the algorithm, we conduct a simulation under the MATLAB environment. Adopt the BELLHOP model to simulate the acoustic signal receiving from the hydrophones, and calculate the time difference and distance difference; finally, adopt the position obtained from SINS/DVL integrated navigation system as the iterative initial value of the Taylor series expansion method to solve the positioning results; then obtain the longitude, latitude and depth of AUV, and make a comparison with the positioning results of the traditional algorithm. The result is as shown in Figure 17.

It can be seen from Figure 18 that: the positioning result of the traditional algorithm has far deviated from the actual position, while that of the improved algorithm is very proximate to the actual value. This is because the error of the distance difference of the traditional algorithm is larger, which results in the non-convergence of the positioning result or big error; then, an acceptable positioning result cannot be acquired. However, the error of the distance difference of the improved algorithm is less than 1 m, and a more accurate positioning result can be acquired finally.

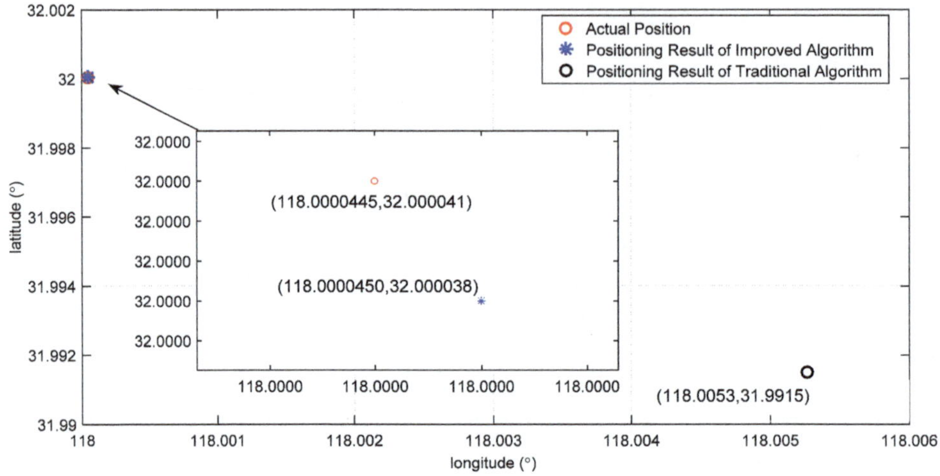

Figure 18. Comparison of the positioning results.

4.3. Dynamic Simulation of AUV Integrated Navigation System Based on SINS/LBL/DVL/MCP

In order to further verify the effectiveness of this algorithm when the AUV is in dynamic operation, a dynamic simulation has been performed on this algorithm. Place five hydrophones underwater; their positions are the same as the above simulation. Suppose that the AUV starts from (118°, 32°) and moves along north by east 45°, which is as shown in Figure 19. The random shift and constant value shift of the gyroscope is 50 μg; the constant value bias is 50 μg. The initial misalignment angles are respectively: pitching angle 1.5°, roll angle 1.5° and course angle 1.5°. The velocity error of DVL is 0.2 m/s. The heading error of MCP is 0.3°. The velocity of the AUV is 1 m/s. Adopt the algorithm in this paper to perform the positioning calculation, and integrate the positioning results with the SINS/DVL/MCP integrated system to calibrate accumulative errors; simulation time is three hours. In order to enhance the fault-tolerant ability of the system, if the positioning results are not converged, then the positioning results of LBL will not be used for the calibration. Figure 19 indicates that: when the AUV enters the scope of the hydrophones, the trajectory of the improved algorithm is fundamentally overlapping with the ideal trajectory, and that of the traditional algorithm deviates from the ideal trajectory. When the AUV leaves the scope of the hydrophones, the trajectory of both the traditional and improved algorithms deviates from the ideal trajectory. Figure 19 is a comparison graph of the positioning errors of the traditional and improved algorithms. The positioning errors will be expressed by the distance between the positioning results and the actual position. Figure 20 indicates that: within 0~1000 s, the positioning errors of both the traditional and improved algorithms will gradually enlarge, because within this period, the AUV is somewhat far away from the scope of the hydrophones; it takes a long time for the hydrophones to receive the sound source signal, and there will be great delay for the positioning results. Hence, LBL will not be used to calibrate within this period; instead, the SINS/DVL/MCP integrated navigation system will be used to perform the positioning. When the AUV approaches the scope of hydrophones, namely after 1000 s, we adopt LBL to perform the calibration. As the errors of adopting the traditional algorithm to calculate are larger, the LBL positioning results will not converge, which has nearly no contribution to calibrating for the SINS/DVL/MCP integrated navigation system. Within 1000 s to 3000 s, the positioning errors enlarge gradually, with the maximum reaching 30 m or so. However, the improved algorithm in this paper greatly reduces the errors of the calculation values of the distance difference. The LBL positioning results can calibrate the accumulative errors of the integrated navigation system, which will control the positioning errors within 2 m. When the AUV leaves the scope of the hydrophones, the LBL positioning system will lose its effect; however, in general, the precision is superior to the system that has not been calibrated by LBL. The above analysis shows that: under actual dynamic operation, when the AUV approaches the scope of the hydrophones, the improved algorithm can be used to perform the positioning calculation, and the calculation results can be used to calibrate the SINS/DVL/MCP integrated navigation system, which will greatly improve the positioning accuracy of the AUV when it navigates underwater.

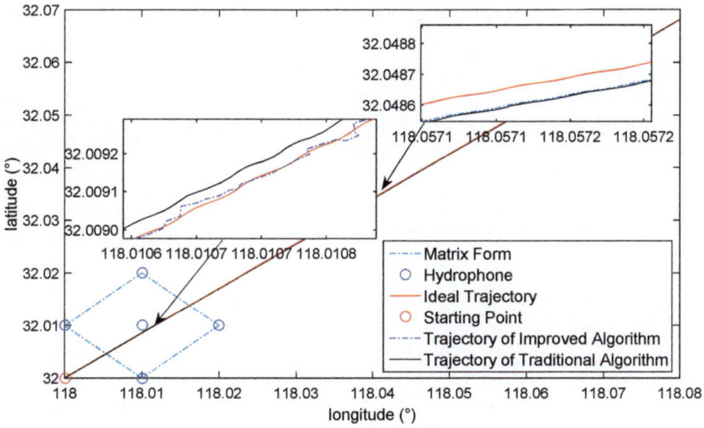

Figure 19. Dynamic simulation of the AUV integrated navigation system.

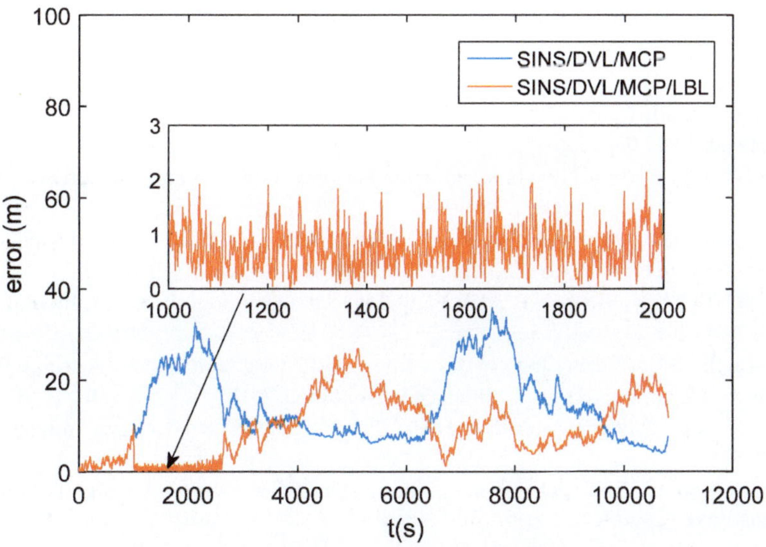

Figure 20. Error comparison of the positioning results.

5. Conclusions

This paper, directed at deficiencies in existing underwater positioning technology, proposes an underwater positioning system based on the mutual assistance of SINS/DVL/MCP and LBL. The whole system consists of an SINS/DVL/MCP integrated navigation system and an LBL underwater acoustic positioning system. The latter adopts the TDOA positioning algorithm based on Taylor series expansion, as well as the positioning results provided by the SINS/DVL/MCP integrated navigation system, which will assist in calculating the delay inequality and distance difference. In the meantime, the positioning results will also be used as the iterative initial value of the positioning algorithm. The final positioning results will be used to calibrate the accumulative errors of the position of the SINS/DVL/MCP integrated navigation system.

The final simulation results show that: compared to the traditional algorithm, this scheme can not only directly correct accumulative errors caused by the dead reckoning algorithm, but also solves the problem of ambiguous correlation peaks caused by the multipath transmission of underwater acoustic signals. This algorithm can greatly improve the AUV underwater positioning accuracy. Only when the AUV installed with the SINS/DVL integrated navigation system approaches the scope of the hydrophones can accumulative errors be effectively calibrated. Hence, it has strong practicability.

Acknowledgments: This study is supported in part by the National Natural Science Foundation of China (Grant No. 51375088), the Foundation of Key Laboratory of Micro-Inertial Instrument and Advanced Navigation Technology of the Ministry of Education of China (201403), the Fundamental Research Funds for the Central Universities (2242015R30031) and the Key Laboratory fund of the Ministry of Public Security based on large data structure (2015DSJSYS002).

Author Contributions: Tao Zhang: Design of system structure and interactive assistance and writing of the paper. Liping Chen: Implementation of algorithm and simulation analysis and writing of the paper. Yao Li: Scientific advising and writing.

Conflicts of Interest: The authors declare no conflict of interest.

References

1. Paull, L.; Saeedi, S.; Seto, M.; Li, H. AUV Navigation and Localization: A Review. *IEEE J. Ocean. Eng.* **2014**, *39*, 131–149. [CrossRef]
2. Tan, Y.T.; Gao, R.; Chitre, M. Cooperative Path Planning for Range-Only Localization Using a Single Moving Beacon. *IEEE J. Ocean. Eng.* **2014**, *39*, 371–385. [CrossRef]
3. Miller, P.A.; Farrell, J.A.; Zhao, Y.; Djapic, V. Autonomous Underwater Vehicle Navigation. *IEEE J. Ocean. Eng.* **2010**, *35*, 663–678. [CrossRef]
4. Kinsey, J.C.; Eustice, R.M.; Whitcomb, L.L. A Survey of Underwater Vehicle Navigation: Recent Advances and New Challenges. In Proceedings of the IFAC Conference of Manoeuvering and Control of Marine Craft, Lisbon, Portugal, 20–22 September 2006.
5. Li, P.; Xu, X.; Zhang, T. Application of Information Fusion to Integrated Navigation System of Underwater Vehicle. *J. Chin. Inert. Technol.* **2009**, *17*, 344–349.
6. Morgado, M.; Batista, P.; Oliveira, P.; Silvestre, C. Position USBL/DVL Sensor-Based Navigation Filter in the Presence of Unknown Ocean Currents. *Automatica* **2011**, *47*, 2604–2614. [CrossRef]
7. Zhang, T.; Xu, X.; Li, Y.; Gong, S.P. AUV Fault-Tolerant Technology Based on Inertial Navigation and Underwater Acoustics Assisted Navigation System. *J. Chin. Inert. Technol.* **2013**, *21*, 512–516.
8. Pan, X. Research on Key Technology of AUV Integrated Navigation System Based on SINS/DVL/GPS. Master's Thesis, Ocean University of China, Qingdao, China, 2011.
9. Tian, T. *Underwater Positioning and Navigation Technology*; National Defense Industry Press: Beijing, China, 2007.
10. Ning, J.; Wu, Y.; Sun, D. The Development of LBL Acoustic Positioning System and Its Application. *Hydrogr. Surv. Charting* **2014**, *1*, 72–75.
11. Liu, Y.; Li, X.R. Aided Strapdown Inertial Navigation for Autonomous Underwater Vehicles. In Proceedings of the SPIE Conference on Signal and Data Processing of Small Targets, Orlando, FL, USA, 5–8 April 2010; Volume 7698.
12. Cheng, W.H. A Study of Increasing the Precision of Navigation Position for Submerged Body. *Ocean Eng.* **2004**, *31*, 693–707. [CrossRef]
13. Jakuba, M.V.; Roman, C.N.; Singh, H.; Murphy, C.; Kunz, C.; Willis, C.; Sato, T.; Sohn, R.A. Long-Baseline Acoustic Navigation for Under-Ice Autonomous Underwater Vehicle Operations. *J. Field Robot.* **2008**, *25*, 861–879. [CrossRef]
14. Eustice, R.M.; Whitcomb, L.L.; Singh, H.; Grund, M. Experimental Results in Synchronous-Clock One-Way-Travel-Time Acoustic Navigation for Autonomous Underwater Vehicles. In Proceedings of the 2007 IEEE International Conference on Robotics and Automation, Roma, Italy, 10–14 April 2007; pp. 4257–4264.
15. Chen, Y.M.; Zheng, D.; Miller, P.A.; Farrell, J.A. Underwater inertial navigation with long base line transceivers: A Near-Real-Time Approach. In Proceedings of the 2013 IEEE 52nd Annual Conference on Decision and Control (CDC), Firenze, Italy, 10–13 December 2013; pp. 5042–5047.

16. An, L.; Chen, L.; Fang, S. Investigation on Correlation Peaks Ambiguity and Ambiguity Elimination Algorithm in Underwater Acoustic Passive Localization. *J. Electron. Inf. Technol.* **2013**, *35*, 2948–2953. [CrossRef]
17. Hertz, D. Time delay estimation by combining efficient algorithms and generalized cross-correlation methods. *IEEE Trans. Acoust. Speech Signal Process.* **1986**, *34*, 1–7. [CrossRef]
18. Knapp, C.; Carter, G.C. The generalized correlation method for estimation of time delay. *IEEE Trans. Acoust. Speech Signal Process.* **1976**, *24*, 320–327. [CrossRef]
19. Krolik, J.; Joy, M.; Pasupathy, S.; Eizenman, M. A comparative study of the LMS adaptive filter *versus* generalized correlation method for time delay estimation. In Proceedings of the IEEE International Conference on Acoustics, Speech, and Signal Processing (ICASSP '84), San Diego, CA, USA, 19–21 March 1984.
20. Zhong, S.; Xia, W.; He, Z. Multipath Time Delay Estimation Based on Gibbs Sampling under Incoherent Reception Environment. *IEICE Trans. Fundam.* **2015**, *E98A*, 1300–1304. [CrossRef]
21. Chen, D.; Tang, H.; Wu, J. Research of TDOA Cooperative Location Algorithm Based on Chan and Taylor. *Comput. Sci.* **2011**, *38*, 406–407.
22. Jiao, X.; Li, J.; Men, L. Accuracy Analysis of Two TDOA Algorithms in Passive Underwater Acoustic Positioning. *Audio Eng.* **2013**, *1*, 73–75.
23. Porter, M.B. *The BELLHOP Manual and User's Guide: Preliminary Draft*; Heat, Light, and Sound Research, Inc.: La Jolla, CA, USA.

10

Secure and Efficient Reactive Video Surveillance for Patient Monitoring

An Braeken [1], Pawani Porambage [2,*,†], Andrei Gurtov [3,4,†] and Mika Ylianttila [2,†]

Academic Editor: Rongxing Lu

1 Department of Industrial Sciences and Technology (INDI), Vrije Universiteit Brussel, Brussels 1000, Belgium; an.braeken@vub.ac.be
2 Centre for Wireless Communication (CWC), University of Oulu, Oulu 90014, Finland; mika.ylianttila@ee.oulu.fi
3 Helsinki Institute of Information Technology (HIIT), Aalto University, Espoo 00076, Finland; gurtov@hiit.fi
4 Department of "Secure Information Technologies" (SIT), ITMO University, Kronverkskiy prospekt 49, St. Petersburg 197101, Russia
* Correspondence: pporamba@ee.oulu.fi
† These authors contributed equally to this work.

Abstract: Video surveillance is widely deployed for many kinds of monitoring applications in healthcare and assisted living systems. Security and privacy are two promising factors that align the quality and validity of video surveillance systems with the caliber of patient monitoring applications. In this paper, we propose a symmetric key-based security framework for the reactive video surveillance of patients based on the inputs coming from data measured by a wireless body area network attached to the human body. Only authenticated patients are able to activate the video cameras, whereas the patient and authorized people can consult the video data. User and location privacy are at each moment guaranteed for the patient. A tradeoff between security and quality of service is defined in order to ensure that the surveillance system gets activated even in emergency situations. In addition, the solution includes resistance against tampering with the device on the patient's side.

Keywords: patient monitoring; visual sensor networks; security; privacy; authentication

---

1. Introduction

The rapid advancements of communication and sensing technologies have been deployed in many healthcare applications and ambient assisted living (AAL) systems. Due to the increasing elderly population and the desire of the general public for independent living, most of these systems are expected to be operated autonomously with minimum or no human intervention. Under such a circumstance, the monitoring applications of the caliber of video surveillance that are deployed in hospitals or retirement homes for patients and elderly people should always ensure their safety and privacy while maintaining operational efficiency and accuracy. Therefore, our proposed solution is applicable to those patients who are not immediately in a critical health situation, however being interested in following up on their health in a secure way, taking into account their privacy. From the moment a potential health issue occurs, the patient gets the opportunity to react. If the patient is not reacting (e.g., due to temporary inability), he or she is aware of the absence of that particular segment of security. However, the patient can still check and validate the procedures followed afterwards.

In this paper, we consider a conceptual design for the network architecture of a reactive video surveillance system in a medical environment, as illustrated in Figure 1. Accordingly, each patient *U* wears a wireless body area network (WBAN), which contains the medical sensors to acquire health records for different aspects and a body gateway (BG) node to aggregate and analyze the sensed data. The BG node also performs as a communicating entity from the WBAN to the outside network. Once a health-concerned or security threshold is notified by a particular medical sensor in the WBAN, the BG or the patient is responsible for invoking the video camera (VC) in closest proximity. The communication between the BG and VC should be activated by means of a private and authenticated request. The patient is in control of requests with a minor priority, while for urgent requests, no interaction of the patient is required. The privacy includes that no outsider can derive the identity of the person sending the request, whereas the authentication allows the submission of requests limited to people known to the system. Having a positive authentication validation, the VC captures the videos of the patient and sends them in the encrypted format to the cloud server (CS). The CS ensures the freshness of the data, stores the encrypted data and sends a notification about the newly-received encrypted video to the authorized persons (D) (e.g., doctors, nurses and caretakers). After that, the doctors with the corresponding credentials can decrypt the video and consult the patients accordingly.

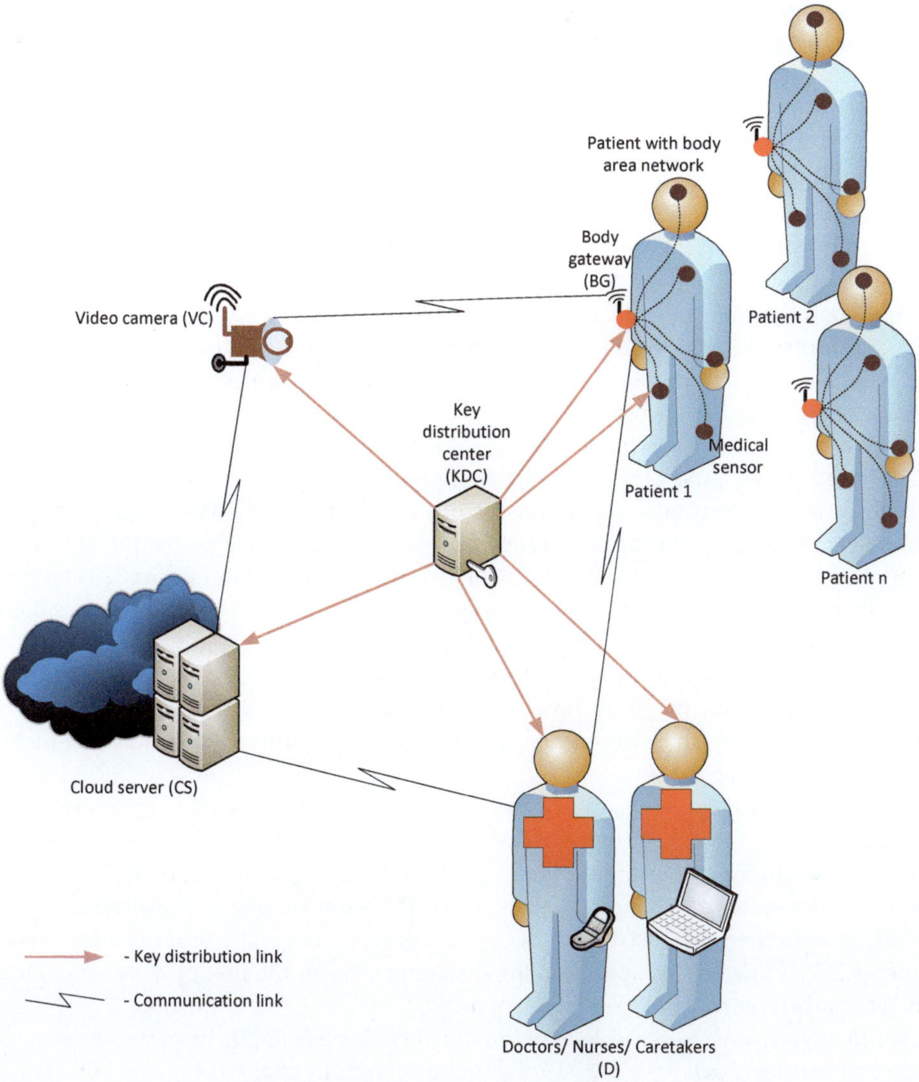

Figure 1. Network system architecture.

For this particular network setting, we propose a secure video surveillance scheme that satisfies the following criteria:

- The scheme ensures the efficient involvement of all of the entities in the system with minimum power consumption and high performance.
- The patients' user privacy and location privacy are guaranteed along with the data confidentiality.
- The VCs can be triggered manually by the patient or autonomously by the BG when the patient is unable to react. Since the second scenario does not guarantee the patient's authentication, the VC and BG mutually authenticate and create user awareness and feedback.
- The solution is completely tamper proof at the patient level due to the construction of the key material. A leakage of security materials at the level of the VC is also detectable.
- The solution is flexible in terms of the patients' password updates and the changes in doctor-patient relations.

The remainder of the paper is organized as follows: Section 2 provides a brief overview about the related previous work. Section 3 describes the threat model and the notations used. Section 4 explains the secure video surveillance scheme. Sections 5 and 6 respectively discuss the security and performance of the proposed scheme. Finally, Section 7 summarizes the work and draws the conclusions.

2. Background and Related Work

In [1–3], the authors provide an extensive analysis of the wireless multimedia sensor networks and visual sensor networks along with their security and privacy threats and the protection mechanisms. According to [1], visual sensor network (VSN) security requirements can be classified into four key areas in terms of data, node, network and user-centric security. In data-centric security, it is expected to protect all forms of data (*i.e.*, raw and processed images or videos) made available by the visual sensors and ensure non-repudiation and confidentiality. Under data-centric security, privacy is identified as a sub-property of confidentiality, which indicates the protection of sensitive data against misuse by legitimate users.

In [4], Winkler and Rinner propose a user-specific and location-aware privacy awareness system integrated as an Android smartphone application. A distributed approach with the hybrid cryptosystem that ensures the confidentiality in a video surveillance system is presented in [5]. However, most of the security solutions proposed in the state of the art are for general purpose applications of video surveillance. Recently, numerous techniques have been proposed for the coordination and the control of multiple cameras, taking into account the potential loss of privacy [6–8]. Nevertheless, the key objectives of these mechanisms slightly deviate from the key scope of this paper. A complete security architecture for patient or elderly monitoring use cases is not particularly addressed in the previous literature. Although [9–12] specifically describe the VSNs in the context of assisted living and elderly care, they do not provide sufficient solutions for the security vulnerabilities, as discussed in [1].

A multitude of research on WBAN security and privacy already exists in the literature [13,14]. However, since the internal security of the WBAN is not the explicit focus of this paper, we here use a simplified security solution of a common shared group key among the sensors of the WBAN. Security solutions of eHealth systems, where the data of a WBAN is transmitted to the electronic records of the hospitals, has also been extensively studied. The most recent results propose adaptive security measurements [15]. However, by intuition, these solutions are in the first place too complicated to derive and secondly require complex implementations.

Our solution gives an answer to several identified challenges in [1]. First, we are able to create user awareness and feedback. We did not include user control of the data, since we believe that because of safety reasons, a doctor should be able to consult the data without waiting for patient approval. Second, we also define a solution that inherently satisfies tamper resistance from the patient side. The security mechanisms used in this paper are inspired from a protocol for biometric-based authentication with multi-server login using a smart card [16].

3. Threats and Notations

According to the network architecture, explained in Figure 1, several security threats can be encountered in the system as follows:

3.1. Security Threats

In [1], the following security threats in a video surveillance network can be distinguished.

- Illegitimate data access: An attacker tries to eavesdrop on the information exchanged in the network by, for instance overhearing the communication channel in order to use it for its own purposes.
- Illegitimate control access: An attacker tries to alter, insert, delete or change data sent in the network. In the worst case, an attacker is able to compromise one or several video cameras or body gateways in the network.
- Service degradation and denial of service: An attacker tries to reduce the availability of the network by, for instance jamming the wireless communication or by injection of a large amount of invalid requests. These attacks are often facilitated if the attacker has compromised one or several video cameras or body gateways in the network, as described in the previous attack.
- Malicious inside users: Here, special care needs to be taken, such that legitimate users do not leak any confidential information to outsiders. Minimal measurements to limit or trace this type of attack are the installation of a logging infrastructure.
- HW/SWattacks: An attacker tries to modify the configuration of the video camera or the body gateway in order to compromise the device.

It must be mentioned that complete protection against these types of attacks is inherently difficult. A minimal requirement is that detection mechanisms should be incorporated.

3.2. Notations

The most important notations and abbreviations used in the security scheme are described in Table 1.

Table 1. Notations used in the proposed security scheme.

Notation	Description
x	Master key of key distribution center (KDC)
y	Secret key shared among D's and KDC
ID_i, T_i	Identity and nonce of patient i
ID_D	Identity of the doctor D
K_G	Group key in WBAN
K_D	Secret key between D and KDC
K	Key to encrypt video by VC
V	Identifier for all VCs
c	Control status of message
$RPW_i, A_i, B_i, C_i, D_i, E_i$	Security material related to ID_i
$x\|y$	Concatenation of x and y
$H()$	One-way hash function, e.g., Secure Hash Algorithm 3 (SHA3)
$E_K()$	Symmetric authenticated key encryption using key K, e.g., AES-CCM

4. Overview of the Secure Video Surveillance Scheme

The proposed video surveillance scheme consists of six different phases. First, it starts with the initialization phase followed by a registration phase of the patients. For the sake of simplicity and the clarity of notations used, we restrict the explanations to one patient. The system architecture can accommodate multiple patients, where each patient will undergo the same procedure and receive similar forms of security materials with a unique identity. Then, the videos are captured, stored and processed respectively in the video recording, storage and decryption phases. Finally, the security materials are renewed in the update phase. Each phase is described in detail in the following sections.

4.1. Phase 1: Installation

The KDC contains one master key x and shares the following secret information with the different entities in the system, by means of physical contact or a secured channel.

- D: A common secret key y is shared with all of the doctors, together with distinct private shared secret keys $K_D = H(x\|y\|ID_D)$. The identity of doctor D corresponds to ID_D.
- VC: The values $H(x\|y)$, $H(V\|H(x))$ and $H(V\|0)$ are shared with all VCs. The common identifier V is identical to every VC.
- Sensors in WBAN: A common group key K_G is securely delivered to all of the medical sensors in the WBAN.
- BG: The group key K_G and the identifier V are securely delivered to the BG.

4.2. Phase 2: Registration

This phase involves the registration of the patient and his or her WBAN into the system. First, the security materials should be derived for the BG based on the patient's password. Then, the patient should establish the secure links with one or more doctors responsible for following up.

4.2.1. Security Materials for BG

This phase requires physical contact between the patient and the key distribution center (KDC). The patient chooses a random value b and a password PW_i and registers with the KDC using the information ID_i, $RPW_i = H(b \oplus PW_i)$. The KDC computes the parameters A_i, B_i, C_i, D_i and E_i as follows:

$$
\begin{aligned}
A_i &= H(y\|ID_i\|T_i) \\
B_i &= H(x\|y) \oplus A_i \\
C_i &= H(RPW_i\|ID_i) \oplus H(A_i) \\
D_i &= H(x) \oplus H(ID_i) \\
E_i &= RPW_i \oplus ID_i
\end{aligned}
$$

Then, the BG stores the B_i, C_i, D_i, E_i parameters and the b value. Note that an attacker has no additional benefit when it is in possession of B_i, C_i, D_i, E_i.

4.2.2. Linking the Patient with the Doctor

The doctor(s) responsible for the care of patient ID_i receives the parameter ID_i together with a random value T_i at the registration phase from the KDC. This information can be remotely sent as $E_{K_D}(ID_i\|T_I)$ using the shared key K_D with the doctor. After that, the KDC eliminates all of the data derived for patient ID_i.

4.3. Phase 3: Video Request

The medical sensors in the WBAN sense data, encrypt them with K_G and deliver them to the BG. Secure communication within the WBAN is not discussed in this paper, since it is out of the scope of the main scheme.

The BG analyzes the incoming data and invokes the alarms in two states according to predefined rules or threshold values. These states correspond to two different scenarios. In the first scenario (*i.e.*, medium alarm), the system requires the input of the identity and password of the patient. The second scenario (*i.e.*, urgent alarm) immediately starts sending a request. The medium alarm does not strictly require an input from the patient after two reminders. For the urgent alarms, the system can autonomously initiate the communication. Note that a patient might also have the possibility to request a video and/or an audio recording, even when there is no alarm. These situations all belong to Scenario 1. Both scenarios ensure the patient's privacy, whereas the authentication of the request is only guaranteed in the first scenario. Finally, the last possibility is that there are three consecutive false inputs of the user ID and password, without the existence of an alarm. This situation might occur in the case of an attack and requires a change of password of the BG before any further action. The steps to perform here are discussed in Section 4.6.1. We now explain in detail the video request phase for both Scenarios 1 and 2 into detail.

4.3.1. Scenario 1

Once the patient enters identity ID_i and the password into the BG, it first computes RPW_i using the inputs and the value b and then obtains $RPW_i \oplus ID_i$. If this corresponds to the stored E_i, the patient is authenticated on the BG, and the process can be continued. Otherwise, a new trial should be undertaken by the patient using a random nonce N_i generated by the BG as follows:

$$\begin{aligned} H(x) &= D_i \oplus H(ID_i) \\ H(A_i) &= C_i \oplus H(RPW_i \| ID_i) \\ C_1 &= H(V \| H(x)) \oplus H(ID_i \| N_i) \\ C_2 &= H(A_i) \oplus N_i \\ V_1 &= H(N_i \oplus B_i) \\ CID_i &= B_i \oplus H(H(V \| H(x)) \| H(ID_i \| N_i)) \end{aligned}$$

The request to the VC consists of the message:

$$c \| CID_i \| V_1 \| C_1 \| C_2 \| H(c \| CID_i \| V_1 \| C_1 \| C_2).$$

Note that control status c is a three-bit value, which is expected to be distinguished by the following possibilities.

- $c = 0$: The user gave a correct input to the BG for a medium alarm.
- $c = 1$: The user requests video and audio recording without a particular alarm situation.
- $c = 2$: The user requests video recording without a particular alarm situation.
- $c = 3$: The user requests audio recording without a particular alarm situation.
- $c = 4$: The user did not react to the two alarms sent by the BG.
- $c = 5$: The user gave the wrong input to the alarm of the BG.
- $c = 6$: The alarm is urgent.
- $c = 7$: The user gave the wrong input values more than three times.

Consequently, Scenario 1 corresponds to Control Statuses 1 to 3.

4.3.2. Scenario 2

For Control Statuses 5 to 7, Scenario 2 is applied. As the patient does not enter the identity and the password, the authentication of the patient is not executed. Moreover, its security mainly depends on the stored value V, which means that the request can also be sent by compromised devices. Next, the same computations as before are performed, but now with $H(x) = 0$, $ID_i = 0$ and $H(A_i) = 0$. Consequently:

$$
\begin{aligned}
H(x) &= 0 \\
H(A_i) &= 0 \\
C_1 &= H(V\|0) \oplus H(0\|N_i) \\
C_2 &= N_i \\
V_1 &= H(N_i \oplus B_i) \\
CID_i &= B_i \oplus H(H(V\|0), H(0\|N_i))
\end{aligned}
$$

The transmitted message also contains c, CID_i, V_1, C_1, C_2 and its corresponding hash.

4.4. *Phase 4: Video Recording*

When the message is received, the video cameras first check the integrity of the message by verifying the hash on the first part of the message. If this verification is successful, depending on the value of the control status, one of the two scenarios should be followed to check the authenticity of the message. We first describe the steps to be executed in the two scenarios as mentioned in Phase 3. Next, we explain which message should be sent from the VC to the CS.

4.4.1. Scenario 1

The stored values $H(V\|H(x))$ and $H(x\|y)$ are required to execute the following computations:

$$
\begin{aligned}
H(ID_i\|N_i) &= H(V\|H(x)) \oplus C_1 \\
B_i &= CID_i \oplus H(H(V\|H(x))\|H(ID_i\|N_i)) \\
A_i &= B_i \oplus H(x\|y) \\
N_i &= H(A_i) \oplus C_2 \\
V_1^* &= H(N_i \oplus B_i)
\end{aligned}
$$

If V_1^* equals the transmitted V_1, the message is authenticated. A confirmation message, allowing mutual authentication, is sent to the BG. This message contains $(C_3\|V_2)$ and is calculated as:

$$
\begin{aligned}
C_3 &= N_j \oplus H(ID_i\|N_i) \\
V_2 &= N_i \oplus H(H(V\|H(x))\|B_i\|N_j)
\end{aligned}
$$

Having these values, the BG can derive $N_j = C_3 \oplus H(ID_i\|N_i)$. If $N_i \oplus H(H(V\|H(x))\|B_i\|N_j)$ corresponds to the transmitted V_2, the BG knows that the request is successfully treated. If too many requests are unanswered, the BG sends an alarm to the patient and the system administrator.

4.4.2. Scenario 2

For Scenario 2, similar operations are performed in a different order, taking into account some simplifications, such as $A_i = H(A_i) = 0$ with the stored value $H(V\|0)$.

$$
\begin{aligned}
H(ID_i\|N_i) &= H(V\|0) \oplus C_1 \\
B_i &= CID_i \oplus H(H(V\|0)\|H(ID_i\|N_i)) \\
A_i &= 0 \\
N_i &= C_2 \\
V_1^* &= H(N_i \oplus B_i)
\end{aligned}
$$

If V_1^* equals the transmitted V_1, the validity of the request can be obtained. As long as the value V is shared only among valid and honest users, the user can be considered as a registered user (although he is not authenticated at the BG level). In order to detect the misbehavior of the system, a list containing B_i, C_2 and c is stored at the VC, and the detection of a problem is sent to the KDC from the moment abnormal incidents occur.

Furthermore, a confirmation message, allowing mutual authentication, is sent to the BG. This message contains $(C_3 \| V_2)$ and is calculated as:

$$\begin{aligned} C_3 &= N_j \oplus H(B_i \| N_i) \\ V_2 &= N_i \oplus H(H(V \| 0) \| B_i \| N_j) \end{aligned}$$

The BG can now derive $N_j = C_3 \oplus H(B_i \| N_i)$. If $N_i \oplus H(H(V \| 0) \| B_i \| N_j)$ corresponds to the transmitted V_2, the BG knows that the request is successfully treated.

4.4.3. Recording and Submission

The recording can be started when the request is accepted. The video m is encrypted using the key $K = H(C_2 \| H(A_i))$. Note that in the second scenario, only B_i is derived. In order to obtain the corresponding A_i, the BG computes $A_i = B_i \oplus H(x \| y)$, using its stored secret $H(x \| y)$.

The following message $C_2 \| H(H(C_2 \| H(A_i))) \| E_K(m)$ is then sent to the cloud.

4.5. Phase 5: Storage and Decryption of Video

The server first checks for the uniqueness of C_2. All messages with the same C_2 should arrive in a limited time frame, as otherwise, it can come from a replay attack. Moreover, thanks to the usage of a fixed parameter, C_2, the videos related to the same event can be grouped.

The patients can check the video as they have temporarily stored the value C_2, and after entering the password and identity, also $H(A_i)$ can be derived. In this way, the patient receives feedback on the recorded video. Note that even if the request was sent as an emergency alarm, where only B_i is used, the video can only be encrypted after a proper authentication of the patient with the BG.

On the other hand, a notification of video reception is sent by the CS to all D's of the system, together with the challenge $C_2 \| H(H(C_2 \| H(A_i)))$. Any D who has the information ID_i, T_i can derive A_i using its secret information y. Consequently, this authorized D can then compute $H(H(C_2 \| H(A_i)))$. If the challenge matches, the D is able to decode the message with key $H(C_2 \| H(A_i))$.

4.6. Phase 6: Updates of Secure Materials

4.6.1. User Password Update

This can be easily executed by the user, without the involvement of the KDC or other entities in the system. New values of C_i and E_i are stored at the BG.

4.6.2. Patient Update with Doctor

If the patient changes doctors, the patient should be removed from the doctor's list. Moreover, new values of A_i, B_i, C_i, D_i, E_i should be computed at the KDC since the registration random value T_i is changed. The updated values of B_i, C_i, D_i, E_i for the BG can be remotely sent, as an attacker does not gain any knowledge with solely these values. The other doctors only need an update on the value of T_i, which can also be remotely sent in the encrypted format using the secret shared key K_D.

4.6.3. Change in Patient and Doctor Relation

In fact, it is sufficient to update the list of patients, corresponding to the doctor. However, as the doctor also holds a long time secret y, it might be time to update all of the parameters. On the other hand, note that a doctor with knowledge of solely y is still unable to perform any disturbing activities in the system.

4.6.4. Corrupted BG and Patient

In this case, the parameters $H(x)$ and V are revealed. The only impact is that now, the video requests, corresponding to alarm Phase 2, can be sent without the guarantee that the user is a registered one. The privacy of the user in the request is still present, only the authentication cannot be verified any longer. If there is abuse of this situation, which is determined thanks to the registration of these events by the video camera or perhaps by the central server in the case of increased inactivity on certain requests, the system operator can still decide to update the parameter V.

4.6.5. Corrupted Video Camera

If the system parameters $H(x\|y)$, $H(V\|0)$ and $H(V\|H(x))$ are leaked, any request can be monitored, and the data submitted can be manipulated. In addition, also requests with an urgent alarm can be sent. However, as the video camera has no direct link to the identity of the user, the privacy of the patient is still guaranteed.

Moreover, in most of the cases, the request is taken by several video servers in the neighborhood. Consequently, the origin of the corruption can be revealed when comparing the different inputs. Once detection is noticed, it is advocated to update the system parameters for all entities of the system.

5. Security Analysis

This section discusses in detail the security and privacy properties of the proposed video surveillance scheme and how it overcomes the threats mentioned in Section 3.

5.1. Illegitimate Data Access

The only entities in the system that are able to decrypt the message are the patient himself/herself and the corresponding doctors (or care takers). This follows from the fact that the data are encrypted by the video cameras by using a key only derivable by the involved parties.

5.2. Illegitimate Control Access

In both scenarios of medium and urgent alarm phases, only the registered users can send a valid video request. Even when the identifier V of the videos is revealed, the requests of a medium alarm phase can be solely derived by the registered users. Synchronized timers are not exploited due to the high level of resource consumption. As a result, there can be a possibility of replaying requests. Notification of replay with the same VC will be noticed on account of the storage of the temporary list of B_i, C_2, c in the VC. However, since all of the VC's share the same nonce V, the message might also be replayed to another VC.

Although, even if the messages that can create replay attacks are not detected by the VC, there is still the powerful server that will immediately detect a replay and continue with further actions. Due to the mutual authentication of the video request, the BG will receive an acceptance message from the video camera. If this is not followed by any request, the alarm can be notified by the BG, as well. Furthermore, without knowledge of the security material (e.g., $H(x\|y)$ to derive A_i), no harm can come to the encrypted video.

5.3. *Service Degradations and Denial of Service*

Denial of service would be a potential danger for the video camera. However, the camera starts recording only when a valid request is sent. Due to the usage of a random nonce, the requests cannot be replayed without a notification by one of the entities, as mentioned before. Moreover, no heavy cryptographic operations are exploited in the system. On the other hand, suppose that many invalid messages are sent to the CS; the CS will notice that no actions are followed by either patients or doctors after forwarding the message and notifying of the problem.

5.4. *Malicious Inside Users*

Here, we can distinguish the BG in combination with a malicious user or a malicious doctor. None of these entities have enough knowledge to generate new accounts. Moreover a malicious doctor does not have the knowledge to generate a valid video request of another user. When a video is accidentally leaked, the log files need to be investigated in order to find the source.

5.5. *HW/SW Attacks*

The system finds its origin in providing security for smart cards without tamper-proof requirements. The same ideas can be applied here to the side of the BG. Even with knowledge of B_i, C_i, D_i, E_i, an attacker has no further advantage, since the input of the identity and password of the user is required. As mentioned before, breaking into a video camera has a minor direct security impact on the users. Moreover, it can be quite easily detected. However, a compromised video camera requires a complete update of the system.

5.6. *Privacy*

Note that the video request contains the parameter CID_i. This is a dynamic reference, related to the hidden identity B_i of the user. Consequently, no outsider can ever link the request to a certain patient, nor link those with other previous or pending requests. This also guarantees the location privacy of the patient. Moreover, as the video camera has no knowledge about the link between the real identity and the parameter B_i, even a malicious video camera cannot break the privacy of the patient.

6. Performance Analysis

The performance of the proposed security scheme is presented on behalf of the BG and the VC, as those are considered the most resource-constrained devices. Since all of the operations in the system are limited to XORs, hashes and symmetric encryption, the computational complexity of the proposed security solution is reasonably low.

The analysis is limited to the first scenario, which has slightly higher complexity. Denote the number of XORs by N_X, the number of hashes by N_H and the number of authenticated encryptions by N_E. Table 2 summarizes the number of computations at the BG and the VC, for each valid request, leading to the recording and submission of the encrypted video.

Table 2. Performance of the body gateway (BG) and video camera (VC).

Phase	BG	VC
Video request	$6N_X + 7N_H$	0
Video recording	$2N_X + 3N_H$	$7N_X + 8N_H + 1N_E$

Moreover, for the required communication phases, the length of the messages in each phase is moderate enough. For instance, assuming the length of the parameters in the system to be 128 bits,

Table 3 summarizes the length of the transmitted and received messages, corresponding to the video request and video recording phase at the BG and VC. Let $\|m\|$ denote the length of the video.

Table 3. Communication length at the BG and VC.

	BG		VC	
	Phase 1	**Phase 2**	**Phase 1**	**Phase 2**
Transmission	80 bytes	–	–	64 bytes + $\|m\|$
Reception	–	32 bytes	80 bytes	–

In order to provide further insight into the performance of the proposed surveillance scheme, we have quantified the energy consumptions on behalf of BG and VC entities, which are considered the most critical devices for power optimization. We implemented the corresponding cryptographic and control operations on a Libelium Waspmote platform [17] using Waspmote cryptographic libraries. Waspmote has an Atmega1281 microcontroller running at 8 MHz with 8 KB SRAM, 4 KB EEPROM and 128 KB flash memory. We measured the execution time (t) for individual operations on the BG and VC and thereby calculate the computation energy cost using formula $U \times I \times t$ based on the execution time (t), the nominal voltage (U) and the current draw in active mode (I) on Waspmote sensors. The communication between the BG and VC was performed by the Bluetooth Low Energy (BLE) protocol. The captured video is uploaded by the VC over a 3G network via the Secure File Transfer Protocol (FTPS). Table 4 shows the energy consumption values for different operations at the BG and VC.

Table 4. Energy consumption at the BG and VC.

	Operation	Time (ms)	Energy (mJ)
BG	Process video request	25	1.575
	Transmit video process request to VC	25	4.095
	Receive confirmation from VC	21	1.793
	Authenticate VC	45	2.835
	Total at BG	116	10.298
VC	Receive request from BG	25	2.135
	Authenticate BG	47	2.961
	Transmit ACK to BG	21	3.440
	Total at VC	93	8.536
VC	Transmit 2-s video to cloud	19,438	38,876

We could observe that a single elliptic curve cryptographic (ECC) multiplication operation consumed approximately 13.365 mJ on the Waspmote platform (for secp160r1 elliptic curve (EC) domain parameters). However, the total energy consumptions (*i.e.*, including computation and communication) at each entity (*i.e.*, the BG and VC) are less than this value. This gives an implicit assurance that the proposed scheme is efficient, and it will outperform resource-less devices compared to the ECC-related schemes.

7. Conclusions and Future Work

A patient monitoring system by means of reactive video surveillance strictly requires dedicated security mechanisms in order to guarantee the privacy and confidentiality of the patient's data. In this paper, we proposed a complete and highly efficient solution solely using symmetric key cryptographic primitives to establish this goal. The patient is aware of the video monitoring and receives the required feedback.

The proposed security architecture can be easily extended to obtain extra performance features or even stronger security properties. Below, we present some examples.

- The data measured in the WBAN can be securely sent to the central server by using the VC's as gateways or edge routers. These edge routers could share the same security mechanisms as the VC's, however, then with a distinct common identifier E.
- In many commercial systems, video surveillance is combined with voice communication. Consequently, if a smartphone takes the role of the BG, the doctor can call the patient if required.
- We did not add watermarking techniques to the side of the VC's, since we assumed that one request is taken by different VC's, and thus, the combined information reveals the location of recording. In order to guarantee the integrity and authentication of the data from a particular VC, a watermark could be added to the image encryptions.
- As mentioned before, the usage of time stamps could speed up the detection of replay attacks. However, special care needs to be given to secure the synchronization between the clocks of the different entities.
- An additional communication layer can be added while applying the multi-camera fusion techniques. This may increase the performance of the system drastically.

As future work, we intend to implement the same conceptual design in a real system to retrieve the opinion of patients and healthcare professionals.

Acknowledgments: This work is supported by Finnish funding agency TEKES and the European Celtic-Plus project CONVINcE which was partially funded by Finland, France, Sweden and Turkey. The authors would also like to acknowledge networking support by the COST Action IC1303 Architectures, Algorithms and Platforms for Enhanced Living Environments (AAPELE).

Author Contributions: An Braeken and Pawani Porambage conceived and designed the experiments, analyzed the data, contributed reagents/materials/analysis tools, and wrote the paper. Pawani Porambage performed the experiments. Andrei Gurtov and Mika Ylianttila are the supervisors of Pawani Porambage and contributed to the organization of the paper. All authors have contributed to the discussion of the proposed framework, and have read and approved the final version of the manuscript.

Conflicts of Interest: The authors declare no conflict of interest. The founding sponsors had no role in the design of the study; in the collection, analyses, or interpretation of data; in the writing of the manuscript, and in the decision to publish the results.

References

1. Winkler, T.; Rinner, B. Security and Privacy Protection in Visual Sensor Networks: A Survey. *ACM Comput. Surv.* **2014**, *47*, 1–42.
2. Guerrero-Zapata, M.; Zilan, R.; BarcelÃş-Ordinas, J.; Bicakci, K.; Tavli, B. The Future of Security in Wireless Multimedia Sensor Networks. *Telecommun. Syst.* **2010**, *45*, 77–91.
3. Akyildiz, I.F.; Melodia, T.; Chowdhury, K.R. A Survey on Wireless Multimedia Sensor Networks. *Comput. Netw.* **2007**, *51*, 921–960.
4. Winkler, T.; Rinner, B. User-centric Privacy Awareness in Video Surveillance. *Multimedia Syst.* **2012**, *18*, 99–121.
5. Schaffer, M.; Schartner, P. Video Surveillance: A Distributed Approach to Protect Privacy. In Proceedings of the Communications and Multimedia Security–9th IFIP TC-6 TC-11International Conference, CMS 2005, Salzburg, Austria, 19–21 September 2005.
6. Saini, M.; Atrey, P.K.; Mehrotra, S.; Kankanhalli, M. W3-Privacy: Understanding What, When, and Where Inference Channels in Multi-camera Surveillance Video. *Multimedia Tools Appl.* **2014**, *68*, 135–158.
7. Natarajan, P.; Atrey, P.K.; Kankanhalli, M. Multi-Camera Coordination and Control in Surveillance Systems: A Survey. *ACM Trans. Multimedia Comput. Commun. Appl.* **2015**, *11*, 1–30.
8. Meliones, A.; Kokkovos, S. Privacy-preserving Intelligent Networked Video Surveillance for Patient Monitoring and Alarm Detection. In Proceedings of the 8th ACM International Conference on PErvasive Technologies Related to Assistive Environments, PETRA 2015, Corfu, Greece, 1–3 July 2015; pp. 1–8.
9. Aghajan, H.; Augusto, J.C.; Wu, C.; McCullagh, P.; Walkden, J.-A. Distributed Vision-Based Accident Management for Assisted Living. In Proceedings of the International Conference on Smart Homes and Health Telematics, ICOST 2007, Nara, Japan, 21–23 June 2007; pp. 196–205.

10. Fleck, S.; Straßer, W. Smart Camera Based Monitoring System and its Application to Assisted Living. *IEEE Proc.* **2008**, *96*, 1698–1714.
11. Bamis, A.; Lymberopoulos, D.; Teixeira, T.; Savvides, A. The behavior scope framework for enabling ambient assisted living. *Personal Ubiquitous Comput.* **2010**, *14*, 473–487.
12. Pinto, A. Wireless Embedded Smart Cameras: Performance Analysis and their Application to Fall Detection for Eldercare. Ph.D. Thesis, University of Nebraska–Lincoln, Lincoln, NE, USA, 2011.
13. Latré, B.; Braem, B.; Moerman, I.; Blondia, C.; Demeester, P. A Survey on Wireless Body Area Networks. *Wirel. Netw.* **2011**, *17*, 1–18.
14. Li, M.; Lou, W.; Ren, K. Data Security and Privacy in Wireless Body Area Networks. *IEEE Wirel. Commun.* **2010**, *17*, 51–58.
15. Leister, W.; Hamdi, M.; Abie, H.; Poslad, S. An Evaluation Scenario for Adaptive Security in eHealth. In Proceedings of Fourth International Conference on Performance, Safety and Robustness in Complex Systems and Applications, Nice, France, 23–27 February 2014.
16. Baruah, K.C.; Banerjee, S.; Dutta, M.P.; Bhunia, C.T. An Improved Biometric-Based Multi-Server Authentication Scheme Using Smart Card. *Int. J. Secur. Its Appl.* **2015**, doi:10.14257/ijsia.2015.9.1.38.
17. Libelium Inc. Waspmote Sensor Boards. Available online: http://www.libelium.com/products/waspmote/ (accessed on 10 December 2015).

11

Segmentation of a Vibro-Shock Cantilever-Type Piezoelectric Energy Harvester Operating in Higher Transverse Vibration Modes

Darius Zizys [1,*], Rimvydas Gaidys [1], Rolanas Dauksevicius [2], Vytautas Ostasevicius [2] and Vytautas Daniulaitis [3]

Academic Editor: Davide Brunelli

1 Faculty of Mechanical Engineering and Design, Kaunas University of Technology, Studentu 56, Kaunas LT-51368, Lithuania; rimvydas.gaidys@ktu.lt

2 Institute of Mechatronics, Kaunas University of Technology, Studentu 56-123, Kaunas LT-51368, Lithuania; rolanas.dauksevicius@ktu.lt (R.D.); vytautas.ostasevicius@ktu.lt (V.O.)

3 Faculty of Informatics, Kaunas University of Technology, Studentu 50, Kaunas LT-51368, Lithuania; vytautas.daniulaitis@ktu.lt

* Correspondence: darius.zizys@ktu.edu

Abstract: The piezoelectric transduction mechanism is a common vibration-to-electric energy harvesting approach. Piezoelectric energy harvesters are typically mounted on a vibrating host structure, whereby alternating voltage output is generated by a dynamic strain field. A design target in this case is to match the natural frequency of the harvester to the ambient excitation frequency for the device to operate in resonance mode, thus significantly increasing vibration amplitudes and, as a result, energy output. Other fundamental vibration modes have strain nodes, where the dynamic strain field changes sign in the direction of the cantilever length. The paper reports on a dimensionless numerical transient analysis of a cantilever of a constant cross-section and an optimally-shaped cantilever with the objective to accurately predict the position of a strain node. Total effective strain produced by both cantilevers segmented at the strain node is calculated via transient analysis and compared to the strain output produced by the cantilevers segmented at strain nodes obtained from modal analysis, demonstrating a 7% increase in energy output. Theoretical results were experimentally verified by using open-circuit voltage values measured for the cantilevers segmented at optimal and suboptimal segmentation lines.

Keywords: piezoelectric; optimal segmentation; vibration energy harvesting; resonant frequency; strain node; numerical modelling

---

1. Introduction

Various different approaches exist for vibration energy harvesting via piezoelectric, electromagnetic and electrostatic transduction mechanisms, which have been widely discussed and compared by [1–3]. The piezoelectric transduction mechanism was extensively studied in recent years [3] and proven to be a prime choice for MEMS energy harvesting from harmonic ambient vibrations [4]. A piezoelectric energy harvester usually constitutes a cantilevered transducer with single or multiple layers of piezoelectric material bonded on its surface. The transducer is mounted on a vibrating structure for voltage generation via a direct piezoelectric effect. Vibration-based energy harvesters are usually designed to exhibit natural frequencies that match ambient vibration frequencies. Various authors have focused on modeling mechanical [5] and electronical [6] aspects of piezoelectric energy harvesting, as well as on different optimization techniques to increase the efficiency of the electromechanical conversion [7–10]. The qualitative factors of a vibration energy harvesting system have been recognized by [2,11]. A narrow bandwidth is a major issue for transducers with a high quality factor, and this issue has been addressed by the authors of [12], who developed a design of plate structures for vibration energy harvesting from two or more modes of vibration to respond to variable frequency sources of base excitation. Different approaches to multi-modal harvesters were further investigated by [13,14].

Mechanical-to-electrical energy conversion can be investigated based on piezoelectric constitutive laws [4] and fundamental relations of mechanics of materials [15]. The electric charge collected at the electrodes is the integral of the normal component of electric displacement over the electrode area, and the electric displacement field generated in the piezoelectric layer during vibration is a function of the strain distribution over transducer length. If the strain distribution and the corresponding electric displacement component changes sign under full-sheet electrodes, a cancellation effect manifests, leading to a substantial reduction in piezoelectric charge output [16–18]. In higher vibration modes (second and above), a certain strain node is present, where the strain field changes sign, meaning that if a continuous electrode is applied on the surface, a significant cancelation in collected charge occurs, resulting in losses in harvested energy. Several works were published investigating the normal strain nodes in higher modes [19] and proper segmentation techniques [16,17] for the maximization of harvested energy. Besides the conventional vibration energy harvesting from harmonic vibrations in resonant or off-resonant mode, there were a series of harvesters dedicated to impact energy harvesting [20,21]. Modeling of contact dynamics aspects was thoroughly investigated in [22].

During an impact, not only the first natural vibration mode is excited, but also higher natural modes, so the cantilever shape after the impact may be represented as a superposition of the first and higher natural modes. The number of modes will participate in cantilever vibration as separate components of periodic vibration [23]. Therefore, in practice, higher modes of the harvester can be excited due to the random, varying frequency or impulse-type excitations generated by ambient vibration sources [17]. Bearing in mind the latter, [24] focused on the dynamic efficiency of the cantilever vibrating in its third natural mode. The authors proposed a few approaches of the excitation of the third natural mode, namely vibro-impact or forced excitation. In [16,17,24], the need for proper segmentation at the strain nodes of the cantilevers vibrating in higher natural modes was highlighted. The authors of [10,25] proposed a multi-beam piezoelectric energy harvester that exploits impact to transform low-frequency ambient mechanical vibrations toward higher resonant frequencies of the piezoelectric transducers. The system consists of a steel driving beam that is exciting two piezoelectric beams via impact. The authors of [26] proposed a mechanism for achieving frequency up-conversion for low frequency harvesters exploiting impact between end-stop and a cantilever beam: a seven-fold increase in the oscillation frequency of the transducer was induced if compared to the base excitation frequency.

In this paper, the ideas presented by [16,17,19] are further developed. The positions of strain nodes of a cantilevered Euler–Bernoulli beam without a tip mass for second natural frequency is calculated from modal analysis and compared to the strain node found from transient analysis. During the transitional processes, higher transient vibration modes are excited in vibro-impact harvesters. Therefore, in contrast to the conventional piezoelectric vibration energy harvester (PVEH), the piezoelectric layer has to be segmented to maximize the energy output and avoid the generated charge cancelation due to effects related to strain nodes. Two setups of segmentation were investigated in this work: one obtained from modal analysis (in further sections, referred to as suboptimal) and the other from transient analysis, further referred to as optimal segmentation.

2. FEM Modelling of PVEH Segmentation in Higher Vibration Modes

Two types of cantilevers were chosen for investigation: a conventional cantilever of a constant cross-sectional area and a cantilever of optimized shape. The latter refers to a cantilever that has a minimal volume and the same second transverse vibration Eigen frequency ω_2 as the cantilever of a constant cross-section. The procedure for cantilever shape optimization is given in further sections. Both setups had a pair of PVDF (polyvinylidene fluoride) layers attached on their top surface. Cantilevers were modeled as uniform composite beams for linearly elastic deformations and geometrically small oscillations based on the Euler–Bernoulli beam assumption.

The effects of shear deformations and rotary inertia are neglected, and this is a reasonable assumption, since typical piezoelectric cantilevers are designed and manufactured as thin beams, as described in [6]. The principal scheme of the cantilever of a constant cross-section is provided in Figure 1. The mechanical properties and dimensions of both cantilevers are listed in Table 1. The first PVDF layer is mounted from the fixed end to the strain node, the second from the strain node up to the cantilever end. The properties of the PVDF are given in Table 2. The Eigen frequencies of both cantilevers were obtained from the modal analysis.

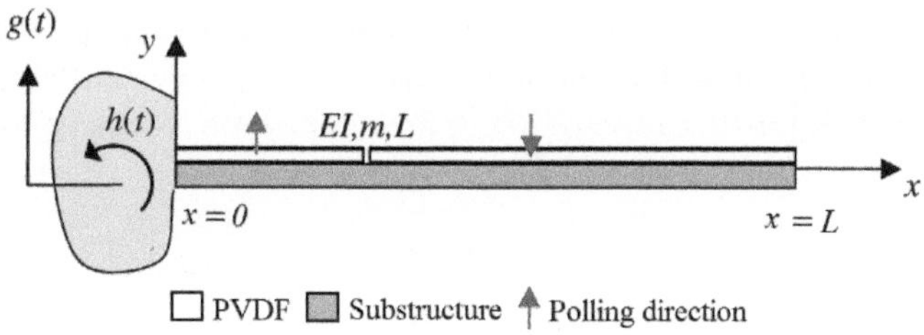

Figure 1. Piezoelectric cantilever under translational and small rotational base motions (adapted from [17]). PVDF, polyvinylidene fluoride.

Table 1. Mechanical and geometrical properties of the considered cantilever setups.

Parameter	Constant Cross-Section Area Cantilever		Optimal Shape
ω_1, Hz	86		66
ω_2, Hz	541		534
Density, kg/m^3		7850	
Elastic modulus, N/m^2		2×10^{10}	
Poisson's ratio		0.33	
Length, m		0.1	
Width a, m		0.01	
Thickness b, m	1×10^{-3}		Varying from 4×10^{-4} to 1×10^{-3}

Table 2. PVDF properties.

Parameter	Name	PVDF	Units
d_{31}	Piezoelectric strain constant	23	(pC/N)
g_{31}	Piezoelectric stress constant	216	(10^{-3} Vm/N)
k_t	Electromechanical coupling factor	12%	
C	Capacitance	1.4–2.8	nF
Y	Young's modulus	4	10^9 N/m^2
ε	Permittivity	110	10^{-12} F/m
ρ	Mass Density	1780	kg/m^3
t	Thickness	64	µm

2.1. Finite Element Model

The differential equations were solved numerically using the finite element method. The cantilevers under investigation are considered to be thin, which is important, because in thin beams, the shear deformations in the transverse direction are neglected. The PVEH model is composed of two piezoelectric layers of opposite polarities, covered on both sides by electrode layers, mounted on top of a substructure and connected in a series configuration. The function of the electrode is to provide uniform potential. Electromechanical modeling of cantilever-type piezoelectric transducers has been studied by several authors in the past [4,5]. The cantilever was excited by harmonic base movement with frequencies matching the second natural frequency of the cantilevers (see Table 1) with an acceleration of 1.3 g. Equation (8) was used for modal analysis, while Equation (13) was used to solve the transient analysis problem. Excitation parameters are varied by the frequency ω_n of the time-dependent force *f(t)*. A 2D in-plane strain application was applied for modeling. The element type that was used is a quadratic Lagrange element with second-order polynomial approximation. FE mesh with 350 elements per length of the cantilever was implemented. The Structural Mechanics module of COMSOL was adopted for the calculation of the Eigen frequencies and the strain nodes of the cantilevers. Simulation results were obtained by deriving voltage output from piezo plane strain analysis. The study was performed both for the cantilever of a constant cross-section and the optimally-shaped cantilever (Table 1), which was obtained by solving the shape optimization problem with the objective to obtain the cantilever of minimal volume for a fixed second natural frequency of transverse vibrations.

2.2. Constitutive Equations for Substructure and Piezoelectric Layers

The structure is composed of a load-bearing material and two layers of piezoelectric material. Under the linear elasticity assumption, the constitutive equation for the load-bearing material is given as Equation (1):

$$T = C^H S \tag{1}$$

where T is mechanical stress, C^H is the elasticity matrix of the host layer and S is strain. The electromechanical coupling effect of the piezoelectric material can be described by the following constitutive Equations (2) and (3).

$$T = C^P S - eE\ S \tag{2}$$

$$D = e^T S + \varepsilon^S E\ S \tag{3}$$

where C^P is the elasticity matrix of the piezoelectric layer, D denotes the electric displacement, e and E are the piezoelectricity matrix and applied electric field, respectively, and ε^S is the permittivity matrix. Constitutive equations are further described by [27]. It can be assumed that the electric potential varies linearly across the thickness of the piezoelectric layer. The electrical boundary condition is open circuit, thus $D = 0$. If an axial stress T_1 is applied, it will deform, and hence, the charge will displace toward the electrodes. Under open circuit conditions ($D = 0$), the voltage V is given by:

$$V = g_{31} h T_1 \tag{4}$$

As the substrate vibrates in its second natural frequency, the PVDF layers undergo dynamic strain, thus generating alternating voltage output via direct piezoelectric effect.

2.3. Procedure for Cantilever Shape Optimization

The optimal shape of the cantilever was obtained by subjecting it to the optimization procedure as follows: minimize the volume of the cantilever-type harvester.

$$Min\ Volume\ (x) \tag{5}$$

defined by state equation:

$$K(x)\, v = \omega_i^2 M(x)\, v \tag{6}$$

Here, ω_i is the natural frequency and v is the "mode shape" of the system. With a constraint of the shape:

$$x_{min} < x < x_{max} \tag{7}$$

and Eigen frequency:

$$w = w^* \tag{8}$$

Here, x is the vector of the design parameters, w the Eigen frequency of the cantilever and w^* the desired frequency. The application of this optimization procedure resulted in the optimal shape of the cantilever for the second transverse vibration (Figure 2).

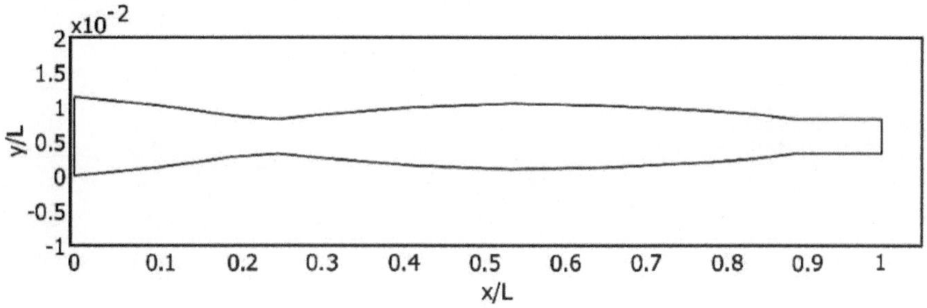

Figure 2. Normalized shape of the optimally-shaped cantilever.

2.4. Determination of Strain Node Position for the Second Vibration Mode via Modal Analysis

For modal analysis, the cantilever is considered undamped and undergoing free vibrations; thus, the governing equation as following.

$$M\ddot{u} + Ku = 0 \tag{9}$$

Internal elastic forces Ku act as an offset to the internal forces $M\ddot{u}$. In this case, K is the stiffness matrix, and M is the mass matrix. Reducing Equation (9), assuming the solution of Equation (10) can be derived and Eigen frequencies of the cantilever obtained.

$$Kv = \omega_i^2 Mv \tag{10}$$

One end of the cantilever is fixed and the second end is free, as shown in Equation (11) for the fixed end and Equation (12) for the free end, where $Y(x)$ is displacement in the y direction at distance x from the fixed end.

$$x = 0,\ Y(x) = 0,\ \frac{dY(x)}{dx} = 0 \tag{11}$$

$$\mathrm{x} = L,\ \frac{d^2Y(x)}{dx^2} = 0,\ \frac{d^3Y(x)}{dx^3} = 0 \tag{12}$$

Figure 3a shows the second transverse vibration mode shape of both cantilevers obtained by modal analysis. Meanwhile, Figure 3b illustrates the strain mode shape for the second natural mode of the cantilever of the constant cross section and optimally shaped cantilever; rough estimation of strain node locations is 0.216 L and 0.238 L, respectively. The location of strain node of a cantilever of constant cross section complies well with [17] that predicted a strain node of second natural mode to be at 0.2165 L. The slope of strain curve of optimally shaped cantilever in region of strain node is higher. This means that interval of cantilever length with different sign strain is shorter. Figure 4 shows the second normal mode shapes of both cantilevers with normal strain distribution along their faces for cantilever of constant cross section and optimally shaped cantilever in Figure 4a,b, respectively.

It can be observed that the upper and lower faces of the cantilever are subjected to strain of different signs; while the upper face is negatively strained (compressed) at x/L = 0, the lower face is under tension.

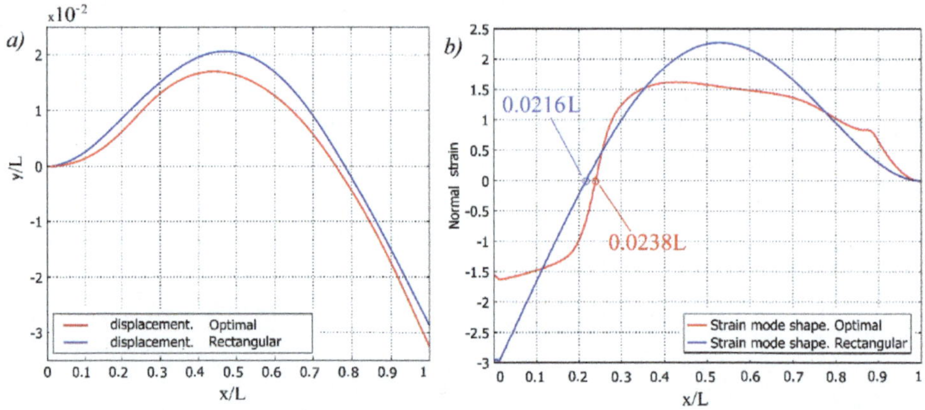

Figure 3. Second transverse vibration mode of a cantilever: (**a**) displacement; (**b**) normal strain.

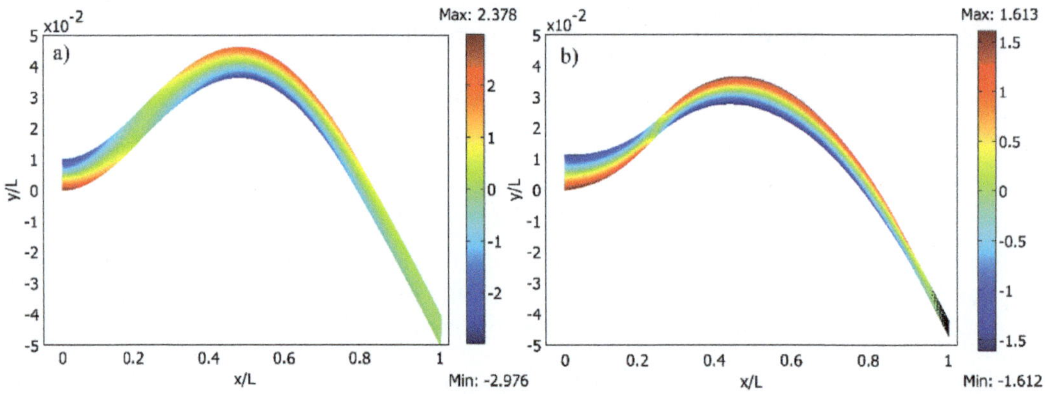

Figure 4. The second transverse vibration mode of the cantilever and the field of normal strain distribution: (**a**) cantilever of a constant cross-section; (**b**) optimally-shaped cantilever.

Strain distribution over the length of the beam changes sign, and collecting the induced piezoelectric charge with continuous electrodes results in an electrical cancellation effect accompanied by a reduction in harvested energy.

2.5. Determination of Strain Node Position for the Second Vibration Mode via Transient Analysis

The transient analysis of the cantilever was conducted to verify the position of the strain node in the cantilever during its base excitation with time varying force *f(t)* acting on it. The corresponding equation of motion is given in Equation (13).

$$M\ddot{u}(t) + C\dot{u}(t) + Ku(t) = f(t) \tag{13}$$

Here, C is the damping matrix. The time-dependent force f(t) is described as cantilever body load in the vertical direction and is defined as force/volume using the thickness.

$$f(t) = am\, sin\, \omega_n t \tag{14}$$

where a is acceleration, m is the mass of the system and ω_n is the excitation frequency. The equation of motion controls the linear dynamic behavior, and the dynamic response can be found by solving this equation of motion.

The transverse displacement of the free end of the cantilever for a given interval of time can be seen in Figure 5a. Figure 5b indicates that the period selected for further investigation is at the region of steady-state vibrations. Half of the period is selected for the investigation of the normal strain distribution along the face of the cantilever, as illustrated in Figure 5c.

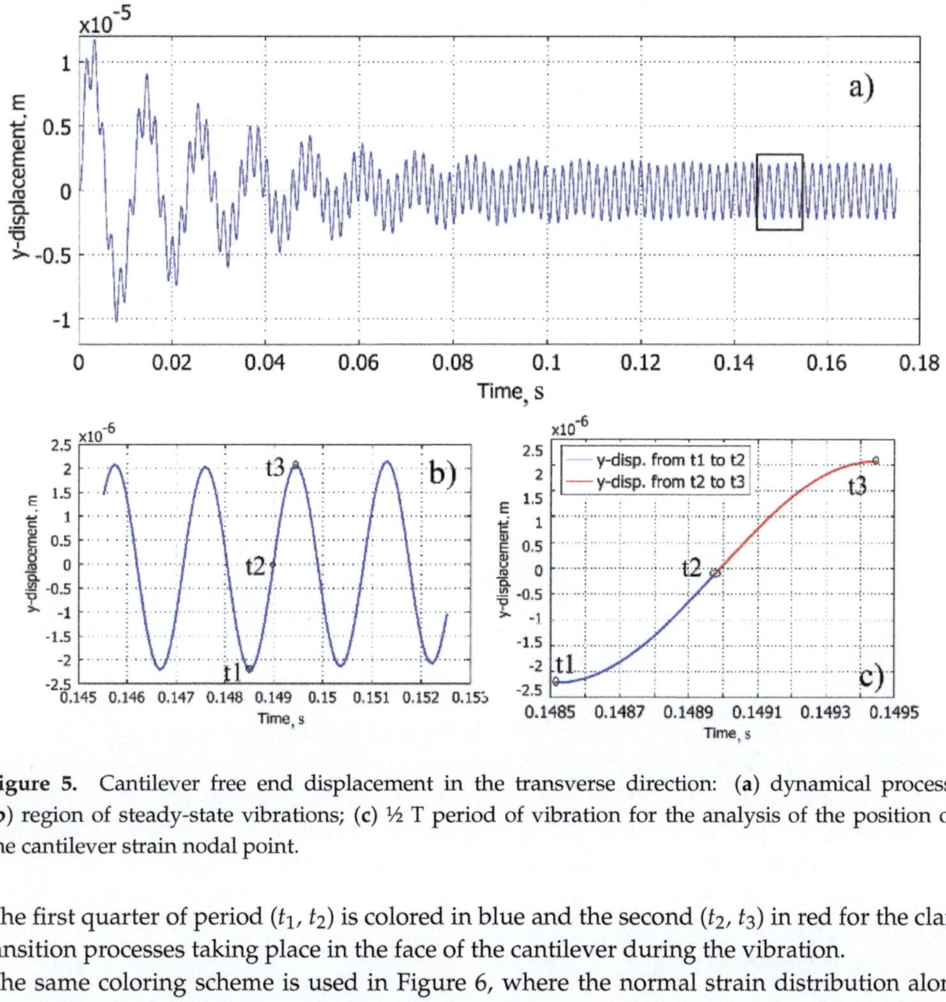

Figure 5. Cantilever free end displacement in the transverse direction: (**a**) dynamical process; (**b**) region of steady-state vibrations; (**c**) ½ T period of vibration for the analysis of the position of the cantilever strain nodal point.

The first quarter of period (t_1, t_2) is colored in blue and the second (t_2, t_3) in red for the clarity of the transition processes taking place in the face of the cantilever during the vibration.

The same coloring scheme is used in Figure 6, where the normal strain distribution along the upper face of both cantilever setups is shown per ½ T of the cantilevers' transverse vibration. The number of curves in Figure 6 represent the number of interpolated time step values Δt_i between t_1 and t_3. Figures 6 and 7 clearly reveal that, overall, normal strain output amplitudes are higher in the case of the cantilever of a constant cross-section. Meanwhile, the normal strain mode shape of the optimally-shaped cantilever forms a steeper angle at the cantilever end and is close to the predicted strain node, thus a higher strain density can be predicted. This forms strain mode curves that are stable at high strain output zones. In Figure 8, strain distribution along the face of the cantilever (Figure 8a: cantilever of a constant cross-section; Figure 8b: optimally-shaped cantilever) *versus* time can be observed.

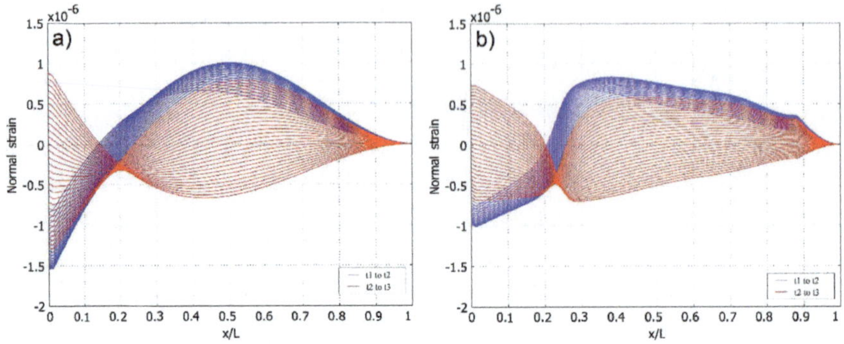

Figure 6. Normal strain distribution *versus* the length of the cantilever per ½ T obtained from transient analysis: (**a**) cantilever of a constant cross-section; (**b**) optimally-shaped cantilever.

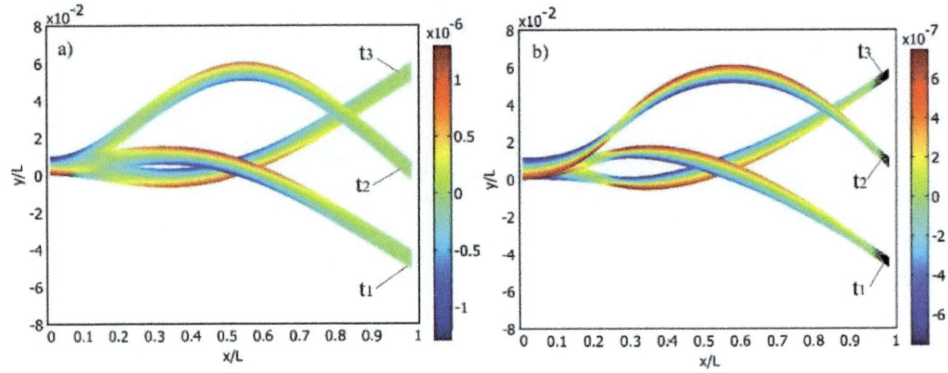

Figure 7. Mode shape and total normal strain distribution of the constant cross-section area cantilever at t_1, t_2 and t_3: (**a**) cantilever of a constant cross-section; (**b**) optimally-shaped cantilever.

The horizontal axis of Figure 8 represents a normal strain distribution along the upper face of the cantilever of a constant cross-section at time instant t_i, while the vertical axis represents strain change at any given point per time interval $[t_1, t_3]$. It can be observed that the cantilever of a constant cross-section produces higher strain amplitudes though the optimally-shaped cantilever exhibits larger average strain output with higher strain mode slopes.

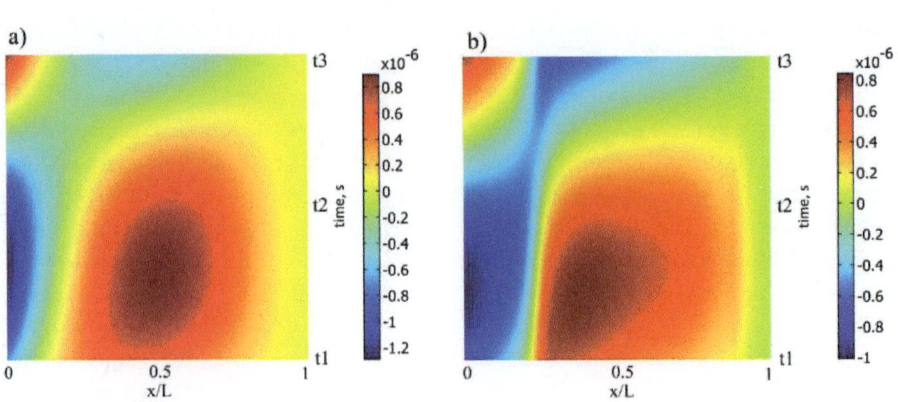

Figure 8. Normal strain distribution on the upper face of the cantilever in ½ T: (**a**) cantilever of a constant cross-section; (**b**) optimally-shaped cantilever.

Numerical analysis was performed in MATLAB in order to post-process results obtained from the COMSOL environment. Figure 6a,b demonstrates that the strain node is not as exactly defined; thus, a methodology has been developed to determine the strain node from transient analysis and to compare the amount of normal strain output of the cantilever for both. As the cantilever is vibrating, at any interpolated time step Δt_i, the cantilever is experiencing both compression and tension at the same face; this produces a drifting strain node where the normal strain is equal to zero. Δt_i is calculated as shown in Equation (15), where N is the number of integration steps.

$$\Delta t = \frac{t_3 - t_1}{N} \tag{15}$$

The aim of this optimization problem is to determine the exact strain node position along the cantilever by using transient analysis, where the average strain output per ½ T is equal to zero. The strain node from the transient analysis is estimated by integrating the area bounded by each interpolated time step (or curve) over an increment of the length of the cantilever L. The Simpson method was used for integration; the approximation is given in Equation (16).

$$\int_0^L \frac{du}{dx}dx \approx \frac{h}{3}\left[\frac{d(u_0)}{dx} + 2\sum_{j=1}^{\frac{N}{2}-1}\frac{d(u_{2j})}{dx} + 4\sum_{j=1}^{N/2}\frac{d(u_{2j-1})}{dx} + \frac{d(u_N)}{dx}\right] \tag{16}$$

where u is displacement in the axial direction along the cantilever. Numerical results obtained from COMSOL simulations were exported to MATLAB for further post-processing. To calculate the amount of normal strain at each time step Δt_i, the normal strain curve integration along the length of the cantilever was performed from both ends of the cantilever, as depicted in Figure 9a,b for a cantilever of a constant cross-section and an optimally-shaped cantilever, respectively. The blue curve represents the integration from the fixed end to the free end, red the integration of the opposite direction.

The curves represent the accumulation of the normal strain over an increment of length L of the cantilevers. Total strain per ½ T is negative at the fixed end and increases until it reaches the maximum at 0.216 L. The square marker in Figure 9a denotes the calculated maximum of the curve with the strain node obtained from the transient analysis, while the round marker denotes the normal strain output at the strain node obtained from the modal analysis (from Figure 3b).

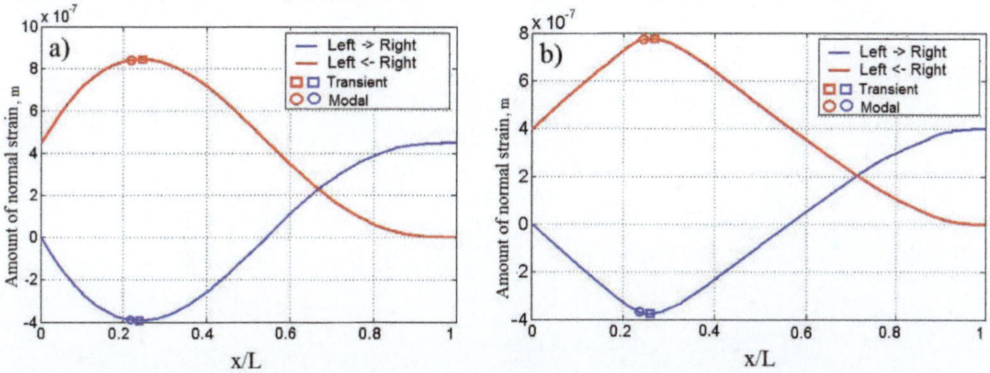

Figure 9. Accumulation of normal strain over length x/L per ½ T: (**a**) cantilever of a constant cross-section; (**b**) optimally-shaped cantilever. The blue curve represents the integration from the fixed end to the free end of the cantilever, red the integration of the opposite direction.

Table 3. Comparison of normal strain amount.

	Strain Node	Amount of Normal Strain-Left (Dimensionless)	Amount of Normal Strain-Right (Dimensionless)	Gain, %
		Cantilever of a constant cross-section		
Modal solution	0.216	6.53×10^{-10}	-9.7×10^{-10}	
Transient solution	0.238	6.73×10^{-10}	-9.9×10^{-10}	+5.5%
		Optimally-shaped cantilever		
Modal solution	0.239	6.3×10^{-10}	-9.6×10^{-10}	
Transient solution	0.259	6.5×10^{-10}	-9.85×10^{-10}	+5.2%

Table 3 summarizes the comparison of the effective total normal strain output integral for cantilevers segmented in strain node, determined via modal analysis, and a strain node obtained from the transient analysis. The total effective strain output consists of two components: strain produced by the region to the left of the strain node and the region to the right of the strain node, as illustrated in Figure 9a for the cantilever of a constant cross-section and Figure 9b for the optimally-shaped cantilever. Table 3 indicates that per ½ *T*, the left side of the cantilever of a constant cross-section (representing the first segment of the harvester) segmented at the strain node obtained from the modal solution undergoes strain equal to 6.53×10^{-10}, while the cantilever segmented at the strain node obtained from the transient solution underwent 6.73×10^{-10}, which is equal to a 3% increase in strain output. The right-hand side (representing the second segment) underwent 2.5% more strain; thus, the total gain of the transient *versus* modal solution was 5.5% more effective strain per ½ *T* of the second transverse vibration mode. Accordingly, the optimally-shaped cantilever segmented at the strain node obtained from the transient solution produced 5.2% more strain if compared to the strain output from a cantilever segmented at the strain node obtained from modal analysis.

Figure 10 illustrates the importance of the correct selection of the strain node for the segmentation of a piezoelectric cantilever subjected to ambient harmonic excitation that matches the second natural frequency of transverse vibrations of the cantilever. The red and black vertical lines in Figure 10 represent the location of segmentation lines at strain nodes obtained from modal and transient analysis, respectively. Table 3 indicates that the difference between two lines is 2.2×10^{-2} L for the cantilever of a constant cross-section and 2×10^{-2} L—for the optimally-shaped cantilever. Transferal of the segmentation line to the strain node obtained from transient analysis produced 5.5% and 5.2% more strain from the cantilever of a constant cross-section and the optimally-shaped cantilever, respectively, if compared to the modal solution. The period of vibration was chosen as ½ *T* because full transition from maximum negative to maximum positive strain output values can be observed.

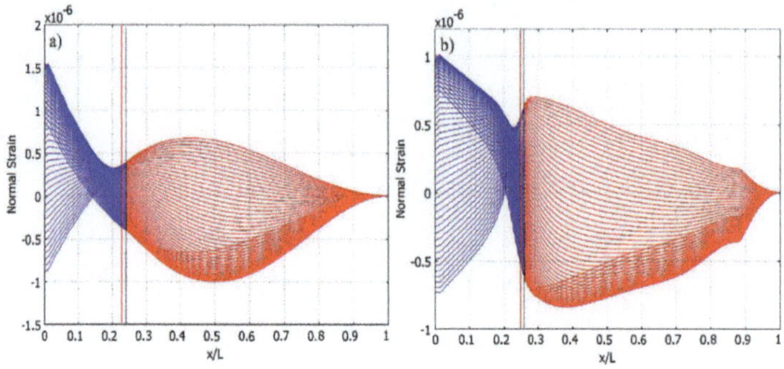

Figure 10. Normal strain distribution *vs.* the length of the cantilever per ½ T from transient analysis and the segmentation nodal point: (**a**) cantilever of a constant cross-section; (**b**) optimally-shaped cantilever. Vertical red line: nodal point position determined from modal analysis; black: from transient analysis.

3. Theoretical and Experimental Results: Open Circuit Voltage Outputs

Further, open-circuit voltage output generated by the cantilever of a constant cross-section and the optimally-shaped cantilever was compared to the experimentally-obtained open-circuit voltage values. The experimental setup and scheme are illustrated in Figures 11 and 12.

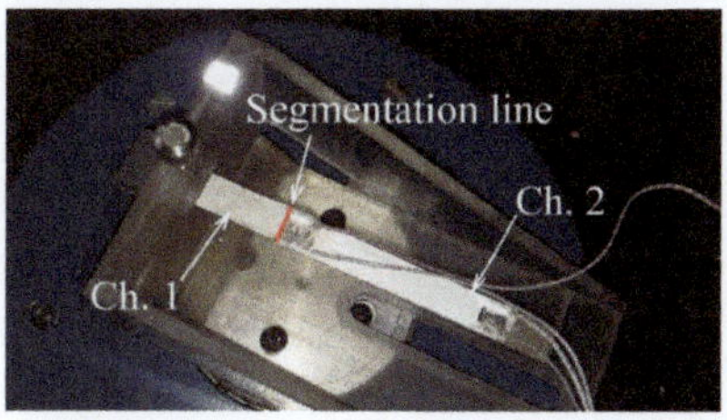

Figure 11. Experimental setup of the tested piezoelectric vibration energy harvester.

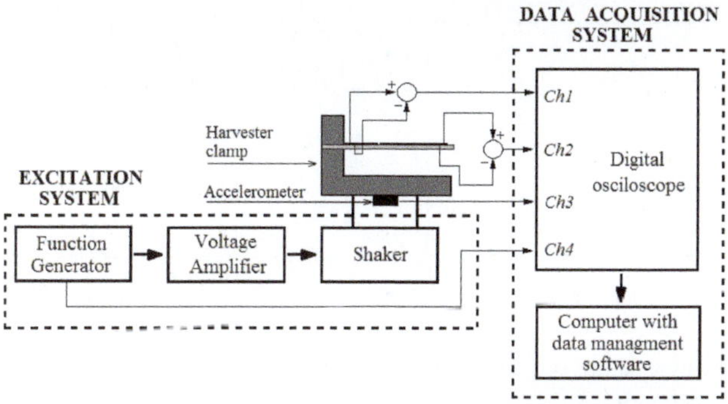

Figure 12. Scheme of the experimental setup.

The experimental setup shown in Figure 12 consists of a piezoelectric vibration energy harvester and two systems connected to it, the excitation system and the data acquisition system. The harvester clamp is mechanically mounted onto an electromagnetic shaker, which excites the PVEH. The substrate layer of the PVEH (with dimensions of 100 × 10 × 1 mm) is made of structural steel (Table 1), and two PVDF layers (transducers DT1-028K by Measurement Specialties Inc., Hampton, VA, United States) are mounted on top. The substrate layer was fabricated from structural steel by using water jet cutting. Function generator AGILENT 33220A is used to control the harmonic excitation signal transmitted to the electromagnetic shaker. Single-axis miniature piezoelectric charge-mode accelerometer METRA KS-93 (with a sensitivity of k = 5 mV/(m/s^2)) is attached at the bottom of the electromagnetic shaker for acceleration measurements. The experiments were performed with constant 1.3 g acceleration. Figure 13 provides the measured open-circuit voltage outputs for the cantilever of a constant cross-section that was excited at the second natural frequency of 551 Hz.

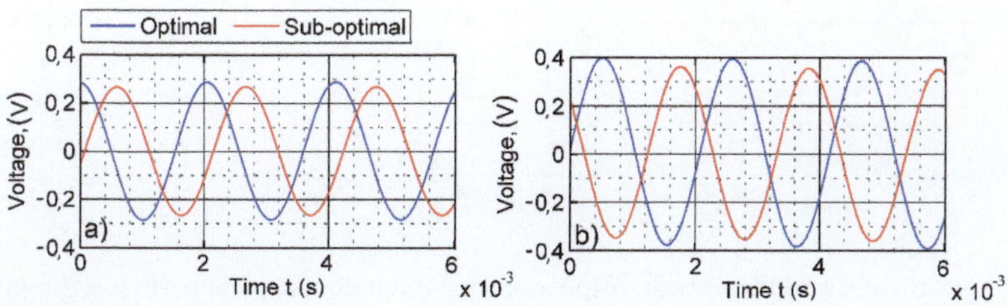

Figure 13. Open circuit voltage output for the cantilever of a constant cross-section (ω_2 = 551 Hz): (**a**) first segment; (**b**) second segment.

Figure 14 presents the open-circuit output of the left-hand side electrode (Channel 1 from Figure 11) and the right-hand side electrode (Channel 2 from Figure 11). The blue and red curves represent the voltage output of the optimal and suboptimal setups, respectively. The average voltage output at Channel 1 is 0.287 V for optimal segmentation and 0.264 V for the suboptimal segmentation (8.7% difference). Channel 2 produced 0.389 V for the optimal segmentation and 0.351 V for the suboptimal segmentation (10.2% difference).

Figure 14 provides the corresponding results for the optimally-shaped cantilever excited at its second natural frequency of 545 Hz. The average voltage output at Channel 1 is 0.275 V for the optimal segmentation and 0.261 V for the suboptimal segmentation (5.1% difference). Channel 2 produced 0.369 V for the optimal segmentation and 0.353 V for the suboptimal segmentation (4.6% difference). It can be observed that the optimally-shaped cantilever produces slightly lower voltage output than the cantilever of a constant cross-section, and the optimized segmentation produces higher voltage output in all cases.

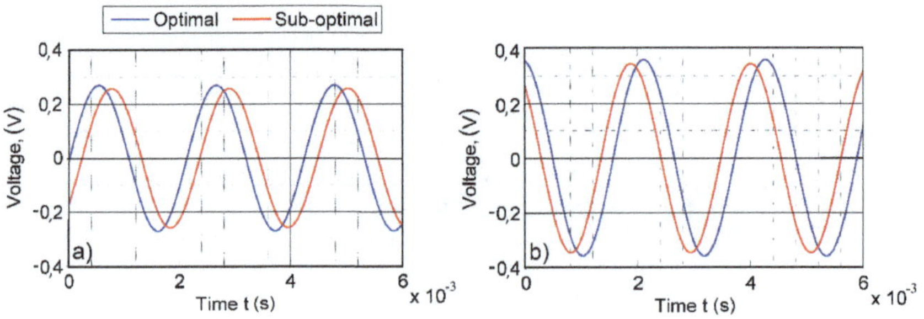

Figure 14. Open-circuit voltage output of the optimally-shaped cantilever at ω_2 = 545 Hz: (**a**) first segment; (**b**) second segment.

These results comply with the theoretical calculations of the normal strain distribution presented in the previous section and total voltage output plots. Figure 15 provides a comparison of the theoretical and experimental total voltage outputs for the cantilever of a constant cross-section and the optimally-shaped cantilever. In every case, the theoretical predictions for the voltage output has good agreement with the experimentally-obtained voltage values. The results demonstrate that the largest voltage output was reached by the optimal setup of the cantilever of a constant cross-section (segmentation line at 0.0238 m): 0.676 V. The suboptimal setup (segmentation line at 0.022 m) of the same cantilever produced 0.615 V. The optimally-shaped cantilever with optimal segmentation (segmentation line at 0.026 m) produced 0.536 V, while the suboptimal segmentation (segmentation line at 0.024 m) 0.511 V.

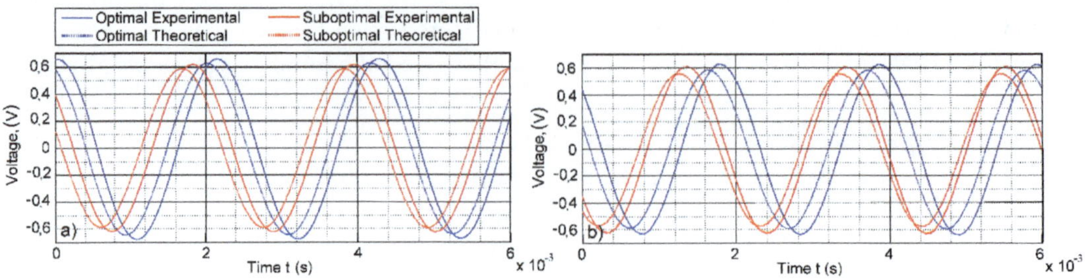

Figure 15. Experimental and theoretical outputs of total open-circuit voltage (first segment + second segment): (**a**) cantilever of a constant cross-section; (**b**) optimally-shaped cantilever.

4. Conclusions

For vibro-shock energy harvesters vibrating in higher modes, the segmentation of piezo-active layers is required due to the presence of strain nodes. Segmentation of the piezoelectric layers at the strain node obtained from modal analysis does not necessary guarantee the highest energy output. The paper presented the methodology for determining the optimal segmentation points of the piezoelectric layer vibrating in its second transverse mode.

1. The optimal segmentation point was determined by means of transient analysis for the two setups of the harvesters, the cantilever of a constant cross-section and the optimally-shaped, cantilever by integrating the strain distribution along the face of the cantilever during the ½ T period and then comparing to the total normal strain amount obtained from the cantilevers segmented at the strain node obtained from modal analysis.
2. Theoretically-obtained results indicating the superiority of the transient analysis *versus* modal analysis for finding the strain node for segmentation were verified experimentally.
3. The adjusted segmentation line increased the generated open-circuit voltage output per period of vibration by 7.2% for the optimally-shaped cantilever and 6% for the cantilever of a constant cross-section.

Acknowledgments: This work has been supported through EC funding for the FP7 project M-Future2013 (319179) and was funded by a grant (No. SEN-10/15) from the Research Council of Lithuania. Project acronym: "CaSpine" (Capillary spine).

Author Contributions: Darius Zizys wrote the paper, carried out the numerical simulation of the harvester dynamics and analysed the data; Vytautas Ostasevicius conceived and designed the experiments; Rimvydas Gaidys created the mathematical model and carried out the experiments; Rolanas Dauksevicius carried out the literature review; Vytautas Daniulaitis wrote MATLAB data analysis code.

Conflicts of Interest: The authors declare no conflict of interest.

References

1. Le, C.P.; Halvorsen, E.; Søråsen, O.; Yeatman, E.M. Wideband excitation of an electrostatic vibration energy harvester with power-extracting endstops. *Smart Mater. Struct.* **2013**, *22*, 50–62. [CrossRef]
2. Roundy, S.; Wright, P.K. A piezoelectric vibration based generator for wireless electronics. *Smart Mater. Struct.* **2004**, *13*, 1131–1142. [CrossRef]
3. Beeby, S.P.; Tudor, M.J.; White, N.M. Energy harvesting vibration sources for microsystems applications. *Meas. Sci. Technol.* **2006**, *17*, 175–195. [CrossRef]
4. Sirohi, J.; Chopra, I. Fundamental Understanding of Piezoelectric Strain Sensors. *J. Intell. Mater. Syst. Struct.* **2000**, *11*, 246–257. [CrossRef]
5. Sodano, H.A.; Park, G.; Inman, D.J. Estimation of Electric Charge Output for Piezoelectric Energy Harvesting. *Integr. Ferroelectr. Int. J.* **2004**, *40*, 49–58. [CrossRef]
6. Erturk, A.; Inman, D.J. Electromechanical Modeling of Cantilevered Piezoelectric Energy Harvesters. In *Energy Harvesting Technologies*; Springer: Berlin, Germany, 2009; pp. 41–77.
7. Mello, L.A.M.; Kiyono, C.Y.; Nakasone, P.H.; Silva, E.C.N. Design of quasi-static piezoelectric plate based transducer by using topology optimization. *Smart Mater. Struct.* **2014**, *23*. [CrossRef]
8. Nakasone, P.H.; Silva, N. Dynamic design of piezoelectric laminated sensors and actuators using topology optimization. *J. Intell. Mater. Syst. Struct.* **2010**, *21*, 1627–1652. [CrossRef]
9. Lefeuvre, E.; Badel, A.; Richard, C.; Petit, L.; Guyomar, D. Optimization of Piezoelectric Electrical Generators Powered by Random Vibrations. Available online: http://arxiv.org/abs/0711.3309 (accessed on 10 December 2015).
10. Ashraf, K.; Khir, M.H.M.; Dennis, J.O.; Baharudin, Z. A wideband, frequency up-converting bounded vibration energy harvester for a low-frequency environment. *J. Smart Mater. Struct.* **2013**, *22*. [CrossRef]
11. Li, W.G.; He, S.; Yu, S. Improving Power Density of a Cantilever Piezoelectric Power Harvester Through a Curved L-Shaped Proof Mass. *IEEE Trans. Ind. Electron.* **2010**, *57*, 868–876. [CrossRef]

12. El-Hebeary, M.M.R.; Arafa, M.H.; Megahed, S.M. Modeling and experimental verification of multi-modal vibration energy harvesting from plate structures. *Sens. Actuators* **2013**, *193*, 35–47. [CrossRef]
13. Bai, X.; Wen, Y.; Li, P.; Yang, J.; Peng, X.; Yue, X. Multi-modal vibration energy harvesting utilizing spiral cantilever with magnetic coupling. *Sens. Actuators* **2014**, *2009*, 78–86. [CrossRef]
14. Rezaeisaray, M.; El Gowini, M.; Sameoto, D.; Raboud, D.; Moussa, W. Low frequency piezoelectric energy harvesting at multi vibration mode shapes. *Sens. Actuators* **2015**, *228*, 104–111. [CrossRef]
15. Bucciarelli, L.L. Chapter 8. In *Engineering Mechanics for Structures*; Dover Publications: Mineola, NY, USA, 2009; p. 306.
16. Erturk, A.; Inman, D.J. On Mechanical Modeling of Cantilevered Piezoelectric Vibration Energy Harvesters. *J. Intel. Mater. Syst. Struct.* **2008**, *19*, 1311–1325. [CrossRef]
17. Erturk, A.; Tarazaga, P.A.; Farmer, J.R.; Inman, D.J. Effect of Strain Nodes and Electrode Configuration on Piezoelectric Energy Harvesting from Cantilevered Beams. *J. Vib. Acoust.* **2009**, *131*, 011010-1–011010-11. [CrossRef]
18. Kang, Z.; Tong, L. Integrated optimization of material layout and control voltage for piezoelectric laminated plates. *J. Intell. Mater. Syst. Struct.* **2008**. [CrossRef]
19. Cady, W.G. *Piezoelectricity: An Introduction to the Theory and Applications of Electromechanical Phenomena in Crystals*; McGraw-Hill: New York, NY, USA, 1946.
20. Djuguma, R.; Trivailo, P.; Graves, K. A study of energy harvesting from piezoelectrics using impact forces. *Eur. Phys. J. Appl. Phys.* **2009**, *48*. [CrossRef]
21. Renaud, M.; Fiorini, P.; van Hoof, C. Optimization of a piezoelectric unimorph for shock and impact energy harvesting. *Smart Mater. Struct.* **2007**, *16*, 1125–1135. [CrossRef]
22. Gilardi, G.; Sharf, I. Literature survey of contact dynamics modelling. *Mech. Mach. Theory* **2002**, *37*, 1213–1239. [CrossRef]
23. Impact test on cantilever Modal analysis. Available online: http://coep.vlab.co.in/?sub=34&brch=101&sim=957&cnt=2 (accessed on 23 September 2015).
24. Ostaševičius, V.; Janušas, G. Pecularities of The Third Natural Frequency Vibrations of a Cantilever for the Improvement of Energy Harvesting. *Sensors* **2015**, *15*, 12594–12612. [CrossRef] [PubMed]
25. Ferrari, M.; Cerini, F.; Ferrari, V. Autonomous Sensor Module Powered By Impact-Enchanced Energy Harvester From Broadband Low-Frequency Vibrations. In Proceedings of the 17th International Conference on Solid-State Sensors, Actuators and Microsystems (TRANSDUCERS & EUROSENSORS XXVII), Barcelona, Spain, 16–20 June 2013; pp. 2249–2252.
26. Edwards, B.; Xie, M.; Aw, K.C.; Hu, A.P.; Gao, W. An Impact Based Frequency Up-Conversion Mechanism for Low Frequency Vibration Energy. In Proceedings of the 17th International Conference on Solid-State Sensors, Actuators and Microsystems (TRANSDUCERS & EUROSENSORS XXVII), Barcelona, Spain, 16–20 June 2013; pp. 1344–1347.
27. Zhang, X.; Kang, Z. Dynamic topology optimization of piezoelectric structures with active control for reducing transient response. *Comput. Methods Appl. Mech. Eng.* **2014**, *281*, 200–219. [CrossRef]

12

Damage Detection in Flexible Plates through Reduced-Order Modeling and Hybrid Particle-Kalman Filtering

Giovanni Capellari [1], Saeed Eftekhar Azam [2,*] and Stefano Mariani [1]

Academic Editor: Vittorio M. N. Passaro

[1] Politecnico di Milano, Dipartimento di Ingegneria Civile e Ambientale, Piazza L. da Vinci 32, 20133 Milano, Italy; giovanni.capellari@polimi.it (G.C.); stefano.mariani@polimi.it (S.M.)

[2] University of Thessaly, Department of Mechanical Engineering, Leoforos Athinon, Pedion Areos, 38334 Volos, Greece

* Correspondence: eftekhar@uth.gr

Abstract: Health monitoring of lightweight structures, like thin flexible plates, is of interest in several engineering fields. In this paper, a recursive Bayesian procedure is proposed to monitor the health of such structures through data collected by a network of optimally placed inertial sensors. As a main drawback of standard monitoring procedures is linked to the computational costs, two remedies are jointly considered: first, an order-reduction of the numerical model used to track the structural dynamics, enforced with proper orthogonal decomposition; and, second, an improved particle filter, which features an extended Kalman updating of each evolving particle before the resampling stage. The former remedy can reduce the number of effective degrees-of-freedom of the structural model to a few only (depending on the excitation), whereas the latter one allows to track the evolution of damage and to locate it thanks to an intricate formulation. To assess the effectiveness of the proposed procedure, the case of a plate subject to bending is investigated; it is shown that, when the procedure is appropriately fed by measurements, damage is efficiently and accurately estimated.

Keywords: structural health monitoring; reduced-order modeling; proper orthogonal decomposition; particle-Kalman filtering; inertial sensors

---

1. Introduction

The health monitoring of aging structures and infrastructures [1–4] is nowadays becoming more and more important, and can exploit devices and methodologies developed within the field of embedded or inclusive smart technologies [5–7]. The envisioned structural health monitoring (SHM) systems have to sense in real-time the changing environment, so as to send out early warnings if dangerous situations are approached. Such feature would also be appealing if structures have to be monitored in regions where environmental and geological risks are of concern [8].

To provide some data highlighting the timeliness of smart SHM technologies, it is worth noting that a high percentage of civil structures and infrastructures in the developed and industrialized nations was built in the first half of the twentieth century: over 50% of the bridges in the USA were built prior to 1940 [9]; furthermore, over 42% of all the aforementioned bridges are structurally deficient, as reported in [10]. In Canada, over 40% of the bridges were built earlier than 1970, and a majority of them demands prompt rehabilitation, strengthening or replacement [11].

The aim of this paper is the development of an online damage identification method, or SHM strategy based on recursive Bayesian filters [12]. Such filters have been already successfully applied to the (grey-box) identification of shear-type buildings: in [13], unscented Kalman and particle filters were adopted for the online parametric identification of nonlinear hysteretic models with uncertainties; in [14], an efficient extended Kalman-particle filter scheme was adopted for the online identification of a structural model featuring a softening inter-story behavior. Recently, the attention has also been drawn to input-state estimation so that the obtained structural response can be used for the fatigue damage assessment, via vibration measurements, of structures subject to cyclic loadings [15,16].

The focus of the present study is on the design of a SHM strategy featuring fast, robust and unbiased estimations of damage indexes, obtained thanks to partial observations of the structural response. Within this frame, the aforementioned local damage indexes quantify the reduction of the stiffness properties of the structure. A similar approach can be adopted to also estimate the local reduction of the materials strength properties, although this topic is out of the scope of the current work and will be investigated in future activities. We therefore assume that the structure behaves piece-wise linearly, within any time window between subsequent observations of the structural response; accordingly, only the time-varying elastic properties of the system need to be identified. The goal of fast and accurate monitoring of these systems is here attained by introducing an effective model order reduction technique, to guarantee low computational costs, and by optimizing the deployment of a limited number of (inertial) sensors, collecting the observations of the system response.

The model order reduction is achieved through a Galerkin projection of the original full structural model onto a sub-space spanned by the so-called proper orthogonal modes (POMs), computed via proper orthogonal decomposition (POD) in its snapshot version [17–19]. The simultaneous, dual estimation of the reduced-order structural state and of the mentioned damage indexes is obtained with a particle filter, enhanced through the use of an extended Kalman filter before the resampling stage [14,19]. The analysis of such dual estimation procedure, featuring an augmented state vector that gathers both the state of the system and the parameters to be tuned, can be traced back to [20] and, somehow, to the seminal paper [21]. In recent years, nonlinear dual estimation problems related to structural dynamics have been tackled with the use of the extended Kalman filter (see [22,23] among others), of the unscented Kalman filter [24], of particle filters [25,26], and of hybrid particle-Kalman filters [19,27]. Hence, in this work we do not focus on the development of a new filtering technique but we instead show how a hybrid particle-Kalman filter can be adopted to identify on-the-fly the time-dependent properties of a reduced-order model of the structural system.

It is known that, as number of parameters to be identified increases, the accuracy of identification tends to decrease: this issue can be linked to the so-called curse of dimensionality. When dealing with the identification of structural systems featuring a large number of degrees-of-freedom (DOFs), methods like component mode synthesis were developed for reducing the number of unknown system parameters in procedures for model updating; with this approach, modal properties are identified offline. In this study, the goal is the online and real-time estimation of the damage parameters of the system, so the POMs are directly obtained from the response of the system in the initial training phase. As the inception and growth of damage modify the structural properties and, thereby, the relevant response to the external actions, the sub-space spanned by the POMs needs to be continuously updated during the filtering process; such update is here obtained thanks to a further Kalman filter. The resulting intricate formulation allows tracking the time evolution of the damage parameters as well as of the state of the partially observed system.

To assess the capability of the proposed SHM procedure, a thin plate subject to bending-dominated deformations and (possibly) time evolving damage is considered. It is shown that the filter is able to identify the spatial distribution of damage even if an extremely low number of POMs (on the order of 1 to 4) is adopted. This strongly reduced-order of the model handled by the filter is not strictly related to the kind of excitation considered; it takes instead advantage of the filter adopted to track the evolution of the POMs, which beneficially affects the overall accuracy of the procedure. Concerning observations, it has been assumed that a network of inertial sensors is surface-mounted on the plate, according to what proposed in [28,29]. To investigate the computational advantage of the proposed method, the computational complexity and the CPU time have been considered: it is demonstrated that the proposed method can speed-up the analyses up to hundreds of times in comparison to the full finite element model.

Methods similar to the one adopted herein for dual estimation and sub-space update, were recently developed and successfully applied to shear-type buildings in [18,19]. The current study represents an evolution of previous works, as it proves to be able to locate and estimate a structural damage via a few vibration measurements only.

The remainder of this paper is organized as follows. In Section 2 the proposed SHM strategy is described: first, the model order reduction approach is detailed; next, the intricate filtering strategy, developed to simultaneously track the damage evolution and the response of the structure in the reduced-order space, is reported. In Section 3, results are shown for the SHM of a thin square plate displaying a reduction of its stiffness properties. Some closing remarks and suggestions for future developments are gathered in Section 4. Finally, Appendix A collects some major algorithmic details of the proposed procedure.

2. Methodology

As already discussed, the health monitoring of real-life structures with complex geometry and/or boundary and loading conditions can become extremely time consuming, as an accurate model to interpret the collected measurements has to be provided.

In order to send out warnings as soon as dangerous conditions are approached, the SHM procedure cannot be run offline; accordingly, a reduced-order model of the system has to be set with a high level of fidelity; this issue is here tackled through POD. Since reduced-order modeling introduces additional uncertainties into the structural model, the Bayesian procedure used to provide the estimations of damage amount (if any) and location can become inaccurate. Such problem is next handled through a fully automated filtering methodology able by itself to enhance the accuracy of the reduced-order model, adapt it to possible structural changes and simultaneously estimate the local damage.

2.1. Model Order Reduction

The dynamic behavior of any space-discretized (e.g., through finite elements) mechanical system can be described through the following vector-valued equation of motion [19]:

$$\boldsymbol{M}\ddot{\boldsymbol{u}}\,(t) + \boldsymbol{D}\dot{\boldsymbol{u}}\,(t) + \boldsymbol{K}\boldsymbol{u}\,(t) = \boldsymbol{F}\,(t) \tag{1}$$

where t is time; $\boldsymbol{u} \in \mathbb{R}^n$ is the vector that collects the kinematic quantities describing the structural response, either displacements or rotations for the considered plates; as a superposed dot stands for time derivative, $\dot{\boldsymbol{u}}$ and $\ddot{\boldsymbol{u}}$, respectively, gather the relevant velocities and accelerations; $\boldsymbol{M}$ is the mass matrix; $\boldsymbol{D}$ is the viscous damping matrix; $\boldsymbol{K}$ is the stiffness matrix; and $\boldsymbol{F}$ is the vector of the external forces.

The model order reduction technique is set to decrease the dimension of the vectors gathering the kinematic quantities and their time derivatives and, therefore, of the matrices in Equation (1). The reduction is obtained via a projection of system dynamics onto a low order sub-space, leading to:

$$\boldsymbol{\mathcal{M}}\ddot{\boldsymbol{\alpha}}\,(t) + \boldsymbol{\mathcal{D}}\dot{\boldsymbol{\alpha}}\,(t) + \boldsymbol{\mathcal{K}}\boldsymbol{\alpha}\,(t) = \boldsymbol{\mathcal{F}}\,(t) \tag{2}$$

where $\boldsymbol{\alpha} \in \mathbb{R}^l$, with $l \ll n$, and [18,19]:

$$\begin{gathered}
\boldsymbol{\mathcal{M}} = \boldsymbol{\Phi}_l^{\mathrm{T}}\boldsymbol{M}\boldsymbol{\Phi}_l \\
\boldsymbol{\mathcal{D}} = \boldsymbol{\Phi}_l^{\mathrm{T}}\boldsymbol{D}\boldsymbol{\Phi}_l \\
\boldsymbol{\mathcal{K}} = \boldsymbol{\Phi}_l^{\mathrm{T}}\boldsymbol{K}\boldsymbol{\Phi}_l \\
\boldsymbol{\mathcal{F}} = \boldsymbol{\Phi}_l^{\mathrm{T}}\boldsymbol{F}
\end{gathered} \tag{3}$$

where, above, $\mathbf{\Phi}_l \in \mathbb{R}^{n \times l}$ is a matrix wherein POMs are arranged according to:

$$\mathbf{\Phi}_l = \left[\mathbf{\Phi}^1 \cdots \mathbf{\Phi}^l\right]$$

This projection matrix allows linking the kinematic variable vectors in Equations (1) and (2) through $\boldsymbol{u}(t) \cong \mathbf{\Phi}_l \boldsymbol{\alpha}(t)$, and gathers the required number l of POMs to attain an *a-priori* defined degree of (energy-like) accuracy. Details relevant to this topic can be found in [18,19,30,31] and are not further discussed here. POMs are obtained through a principal component analysis of a matrix built around the so-called structural snapshots, which provide the evolution in time of the system response to the external loads (see [17] and [32]). The snapshots are collected in the initial training stage of the reduced-order modeling procedure. The computational burden of this stage is basically linked to the number of snapshots that guarantee the convergence towards a steady-state solution $\mathbf{\Phi}_l$; adaptive procedures are available to reduce this to a minimum, see e.g., [32]. As already reported in [18], for systems like shear buildings subject to earthquake excitations, the duration of the training stage can be simply based on the fundamental period of vibration of the structure, and a collection of around $30 - 50$ snapshots proves sufficient to approach the mentioned steady-state.

POD has been devised for linear systems; to cope with an evolving damage state, leading to a time-dependent stiffness matrix $\boldsymbol{K}$, the POD is here coupled with a Kalman filter that continuously adapts the sub-space onto which the model evolution is projected. Such simultaneous use of a filter has been devised for systems whose nonlinearity slowly evolves in time; if abrupt changes of the system properties occur, the proposed procedure requires some time for the reduced-order model to adapt to the new structural health, and so some delay in the tracked damage conditions can emerge.

2.2. Damage Detection and Localization

To estimate the possible local damage, the SHM procedure is driven by the reduced-order model only. The problem is framed as a dual estimation one, within which both the dynamic evolution of the structure and its local integrity are retained in a state vector that reads, at time t_k:

$$\boldsymbol{x}_k = \left\{ \begin{array}{c} \boldsymbol{\alpha}_k \\ \dot{\boldsymbol{\alpha}}_k \\ \ddot{\boldsymbol{\alpha}}_k \\ \boldsymbol{d}_k \end{array} \right\} \tag{4}$$

where vector $\boldsymbol{\alpha}_k$ collects the current values of the generalized coordinates associated with the reduced-order model; vector $\boldsymbol{d}_k$ collects instead the local damage indexes $d_{i,k}$ (for $i = 1, \ldots, N_p$) of all the N_p zones in which the structure is conceptually decomposed. Hence, it turns out that $\boldsymbol{x}_k \in \mathbb{R}^{3l+N_p}$. As reported in Equation (4), not only the generalized coordinates $\boldsymbol{\alpha}_k$ but also their time derivatives $\dot{\boldsymbol{\alpha}}_k$ and $\ddot{\boldsymbol{\alpha}}_k$ are retained in $\boldsymbol{x}_k$. Such increase of the number of the state components has been already targeted in [24,33,34] as a necessary condition to track the nonlinear evolution of the statistics of the system within a stochastic framework. In fact, in the presence of dissipative phenomena like the considered degradation of the mechanical properties of the structure, activation conditions are typically adopted to model the evolution of internal state variables (in our case, of the damage indexes), and so complementarity conditions arise. The Jacobian of the state evolution equations, that will be shown to have a role in the filtering procedure, is linked to such activation conditions; the evolution of the state variables thus becomes path-dependent, and requires all the kinematic variables to be gathered in the augmented state vector $\boldsymbol{x}_k$.

Considering all the equations governing the evolution of system and damage indexes in a stochastic environment, the dynamics of the augmented state vector in the generic time interval $[t_{k-1}\ t_k]$ reads:

$$\boldsymbol{x}_k = \boldsymbol{f}_k(\boldsymbol{x}_{k-1}) + \boldsymbol{w}_k \tag{5}$$

where w_k is a zero mean, additive white Gaussian noise with covariance W, that represents the uncertainties related to the mathematical/numerical model of the system and, further to any standard full-order formulation, the errors introduced by the reduced-order modeling; the nonlinear operator f_k depends on the adopted time integration algorithm and on the mechanical characteristics of the structure, *i.e.*, on the matrices $\mathcal{M}, \mathcal{D}$ and $\mathcal{K}$ showing up in the reduced-order model. A thorough discussion on this topic is provided in Appendix A, see also [14].

To model the structural effects of damage, we now follow a rather standard approach in damage mechanics, see e.g., [34]: the mass and damping matrices are not affected by the damage state; the stiffness matrix is instead assumed to scale proportionally to the damage. To account for multiple damage scenarios with different patterns, the reduced-order stiffness matrix $\mathcal{K}$ is represented as a linear combination of N_p sub-matrices, each one relevant to a region featuring a uniform damage:

$$\mathcal{K}_k = \sum_{i=1}^{N_p} (1 - d_{i,k})\, \mathcal{K}^i_{und} \tag{6}$$

where the index k is now also adopted for $\mathcal{K}$ to denote that its value is damage-dependent, and therefore time-dependent. The summation in Equation (6) has to be correctly interpreted as the assemblage [35] of the undamaged stiffness contributions $\mathcal{K}^i_{und}$, all scaled by the corresponding values of the damage indexes. Equation (6) shows that damage indexes are dimensionless variables. Here, we do not aim to model the evolution of damage induced by the current loading; we therefore assume that indexes $d_{i,k}$ are material parameters, which might change in time (according to thermodynamic requirements, they can only monotonically increase in time in the range [0–1)) and are homogeneous inside each region (so, independent of the adopted space discretization inside the region itself).

The structural response thus becomes nonlinear whenever damage grows. The linear structural dynamics described by Equation (1) or (2) does not provide any means to explicitly model such evolving state of the system; by assuming in the filtering procedure that $\mathcal{K}$ can evolve in time in a step-wise fashion, such issue gets negligible if the time scales corresponding to the damage evolution and to the monitored structural response are well separated.

To estimate the state of the system and the hidden (*i.e.*, not explicitly observable) damage parameters, observations of the system look necessary. Allowing for the formulated reduced-order model of the structure, the stochastic measurement equation reads

$$y_k = H L_k x_k + v_k \tag{7}$$

and provides a link of the N_{obs} observations collected in the vector y_k with the reduced-order state vector x_k. Here, H is an appropriate Boolean matrix that allows to pick up the components of u, $\dot{u}$ and $\ddot{u}$ actually observed. Since the type of sensors and their positions are not supposed to change in time, H turns out to be time-independent. In Equation (7), v_k is a zero mean, additive white Gaussian noise featuring covariance V, that represents the uncertainties due to measurements, typically related to the accuracy of the sensors deployed over the structure. As observations in y_k are defined in the full-order space while x_k lies in the reduced sub-space, the transformation matrix L_k is required for the state vector to be recovered in the original space. According to the definition (4) of x_k, we therefore have:

$$L_k = \begin{bmatrix} \Phi_{l,k} & & & \\ & \Phi_{l,k} & & \\ & & \Phi_{l,k} & \\ & & & \mathbf{0} \end{bmatrix} \tag{8}$$

where $\Phi_{l,k}$ is the projection matrix estimated at time t_k, and all the off-diagonal matrices not reported are null. Matrix L_k obviously features null diagonal block corresponding to the vector d_k of the non-observable damage variables, as they do not have a role in the sub-space projection.

Damage evolution in the structure depends on the loading conditions; hence, the evolution in time of the POMs linked to the changes of the structural health, cannot be *a-priori* assigned. In our approach, we assume that damage evolution is smooth and slow enough, so that POMs can be considered constant within each time step. Since in a Kalman filtering implementation state variables are all gathered in vector form, POMs of the reduced model are rearranged now inside vector $\boldsymbol{\varphi}_{l,k} \in \mathbb{R}^{nl}$ according to:

$$\boldsymbol{\varphi}_{l,k} = \begin{Bmatrix} \boldsymbol{\Phi}_k^1 \\ \vdots \\ \boldsymbol{\Phi}_k^l \end{Bmatrix}$$

Within a stochastic frame, we can therefore model the evolution of the projection sub-space as:

$$\boldsymbol{\varphi}_{l,k} = \boldsymbol{\varphi}_{l,k-1} + \mathbf{w}_k^{ss} \tag{9}$$

where $\boldsymbol{w}_k^{ss}$ is a fictitious zero mean Gaussian noise, whose covariance $\boldsymbol{W}^{ss}$ is to be tuned in order to guarantee stability and accuracy of the estimations. The measurement equation relevant to the sub-space tracking reads:

$$\boldsymbol{y}_k = \boldsymbol{H}_k^{ss}\, \boldsymbol{\varphi}_{l,k} + \boldsymbol{v}_k \tag{10}$$

where $\boldsymbol{H}_k^{ss}$ is an appropriate matrix that links the observations $\boldsymbol{y}_k$ and the sub-space vector $\boldsymbol{\varphi}_{l,k}$. It must be noted that Equations (7) and (10) do represent the same link between observables $\boldsymbol{y}_k$, state variables $\boldsymbol{x}_k$ and POMs $\boldsymbol{\Phi}_k^i$, with $i = 1, \ldots, l$; they are only written in somehow different forms to explicitly state what is identified in the two stages of the proposed procedure, namely $\boldsymbol{x}_k$ first and $\boldsymbol{\varphi}_{l,k}$ next.

Sensor inaccuracies can potentially affect sub-space and damage estimates. As far as the sub-space estimation is concerned, it should be noted that the POMs are the set of basis vectors which optimally capture the maximum variations in the structural response. Therefore, a constant unexpected offset in the sensor measurement is not supposed to alter the estimates of POMs; conversely, in case of a time varying offset, POMs may be affected. To alleviate issues raised by an inaccurate calibration of noises, some methodologies were proposed for their online and real-time, yet optimal estimates through Kalman-type filters, see e.g., [36,37]. While the focus of these works was on developing Bayesian probabilistic methods for noise covariance estimation, it was shown in [37] that they could be generalized to also account for a time variability of the noise mean value.

To cope with all the details reported here above, the adopted recursive Bayesian method has been developed starting from what proposed in [19]: a particle filter is exploited first to estimate the state vector $\boldsymbol{x}_k$, given a certain set of data $\boldsymbol{y}_k$ acquired from the network of sensors deployed over the structure; next a Kalman filter is adopted to update the POMs, still on the basis of observations $\boldsymbol{y}_k$. The latter stage is obviously necessary whenever the damage state changes, and so a different structural response arises; thanks to the final update stage of the Kalman filter, biases in the observables can be exploited to achieve model updating. Both filtering procedures deliver the estimate of damage or the model updating via two stages: first, a prediction is achieved through the evolution or state equations; second, the prediction is updated by taking advantage of observations. Prediction is usually pursued through Chapman-Kolmogorov integrals, whereas the update through the Bayes rule of conditional probability. A closed form solution to the aforementioned procedure can be found only for linear systems with white Gaussian uncertainties. In case of a nonlinear state-space model featuring weak nonlinearities, the solution is obtained by linearizing the state-space equations; if instead the probability distribution of the state variable is non-Gaussian or if severe nonlinearities prevail, the Chapman-Kolmogorov integrals can be computed with numerical quadrature rules. A well-known method to deal with general nonlinear, non-Gaussian systems is the adopted particle filter [38], which is based on a Monte Carlo sampling performed according to the posterior probability distribution of the state variables. As such distribution is not available while sampling, a major aim of the filtering method is to approximate it. Alternatively, sequential importance sampling [38] usually makes recourse, in the absence of the optimal distribution, to the prior probability density function of the state variables. Hence, with this latter approach samples drawn do not contain information brought by the latest observations.

According to what discussed, particle filtering has been preferred to other methods, e.g., the extended or unscented Kalman filtering [14,19,33,34], in order to better match the real non-Gaussian statistics of the state of the nonlinearly evolving system. As this approach would require a large set of particles to get an accurate estimation of damage, an extended Kalman filter is adopted to push the samples towards zones of higher probability and therefore reduce the overall number of particles to be adopted [19].

To perform the dual estimation, a linearization of the evolution Equation (5) is required. In Algorithm 1, $\boldsymbol{F}_k$ denotes the Jacobian of function $\boldsymbol{f}_k$ computed with the currently available estimation $\boldsymbol{x}_{k-1}$ of the state of the system; further details are provided in Appendix A. The whole procedure, as detailed in Algorithm 1, consists of an initialization step, and of a recursive update of the estimates and their covariances. As already described, estimations are provided for the whole state vector $\boldsymbol{x}_k$, including damage indexes $\boldsymbol{d}_k$, and for the POMs in $\boldsymbol{\varphi}_{l,k}$.

In the initialization step, the initial guess (in terms of $\hat{\boldsymbol{x}}_0$ and $\hat{\boldsymbol{\varphi}}_{l,0}$ and relevant covariance matrices) is first set, then all the filter particles ${}^j\boldsymbol{x}_0$, with $j = 1, \ldots, N_s$ being the particle or sample index, are consistently selected along with the corresponding weights ${}^j\omega_0$.

The state vector $\hat{\boldsymbol{x}}_0$ can be easily determined according to initial conditions at time t_0. For instance, if the structural system is initially at rest, all the entries of vectors $\boldsymbol{u}$ and $\dot{\boldsymbol{u}}$ are null, whereas entries of $\ddot{\boldsymbol{u}}$ could be computed by solving Equation (1) at instant t_0; components of $\hat{\boldsymbol{x}}_0$ can next be obtained once the POMs are initialized too. According to real situations encountered whenever a structural system (like a building) has to be monitored, we assume that at time t_0 a preliminary stage of training of the reduced-order model has been actually carried out. This means that, besides possible random fluctuations due to noises, the initial estimation of $\hat{\boldsymbol{\varphi}}_{l,0}$ is indeed accurate. Vector $\hat{\boldsymbol{\varphi}}_{l,k}$ will be subsequently updated to cope with a changing damage state leading to a variation in the structural responses to loading, avoiding the need of a time consuming retraining stage; for further details readers are referred to [14,19].

Within the subsequent time steps, particles are first drawn on the basis of the current expected value and covariance of the estimates, then allowed to (nonlinearly) evolve in time and, eventually, the estimations are updated by averaging what brought by the particles themselves. As proved in [39], the variance of the weights increases over time and, therefore, after few steps only one weight will be not negligible, while all the other ones will converge to zero. In order to avoid the computational costs associated with negligible terms, a resampling stage is adopted to select only the most likely particles; accordingly, ${}^ju$ is a random value drawn at this stage from the uniform probability distribution $\mathfrak{U}[0,1]$. Finally, the tracking of the POMs of the reduced-order model is obtained with Kalman filter; hence, no particles have been defined for this step of the proposed SHM strategy. A subtle difference between the approach discussed above and what typically reported in the literature is that the estimation of the augmented state and POMs is pursued separately. Linearity of POMs evolution and the related observation equation, then allows computing a closed-form solution to their estimation through a Kalman filter. When POMs are updated, the orthogonality condition among them is destroyed and so they must be re-orthogonolized right after.

Algorithm 1: Scheme of the proposed damage identification and state tracking method

- Initialization at t_0:

$$\hat{x}_0 = L_0{}^{\mathrm{T}}\mathbb{E}\left[x_0\right]$$
$$P_0 = L_0{}^{\mathrm{T}}\mathbb{E}\left[(x_0 - \mathbb{E}\left[x_0\right])\left(x_0 - \mathbb{E}\left[x_0\right]\right)^{\mathrm{T}}\right] L_0$$

$$^{j}x_0 = \hat{x}_0$$
$$^{j}\omega_0 = p(y_0|^{j}x_0)\ j = 1, \ldots, N_s$$
$$\hat{\varphi}_{l,0} = \mathbb{E}\left[\varphi_{l,0}\right]$$
$$P_0^{ss} = \mathbb{E}[(\varphi_{l,0} - \hat{\varphi}_{l,0})\left(\varphi_{l,0} - \hat{\varphi}_{l,0}\right)^{\mathrm{T}}]$$

- Recursive computation at $t_k = t_1, \ldots, T_{end}; j = 1, \ldots, N_s$

Prediction stage:

1. Draw samples
   $^{j}x_k^- \sim p(x_k|^{j}x_{k-1})$
2. Push particles toward the region of high probability through an EKF

$$^{j}P_k^- = F_k{}^{j}P_{k-1}F_k^T + W$$
$$^{j}G_k = {}^{j}P_k^- H^T\left(H\,{}^{j}P_k^- H^T + V\right)^{-1}$$
$$^{j}x_k = {}^{j}x_k^- + {}^{j}G_k^-\left(y_k - H^T\,{}^{j}x_k^-\right)$$
$$^{j}P_k = {}^{j}P_k^- - {}^{j}P_k H\,{}^{j}P_k^-$$

- Update stage:

1. Evolve weights

$$^{j}\omega_k = {}^{j}\omega_{k-1}\, p(y_k|^{j}x_k)$$

2. Resampling

$$^{j}u \sim \mathfrak{U}[0, 1]$$
$$\text{find } m \text{ s.t. } \sum_{i=1}^{m-1}{}^{i}\omega_k < {}^{j}u < \sum_{i=1}^{m}{}^{i}\omega_k$$
$$^{j}x_k = {}^{m}x_k$$
$$^{j}\omega_k^* = \frac{1}{N_s}$$

3. Compute expected value

$$\hat{x}_k = \sum_{j=1}^{N_s}{}^{j}\omega_k^*\,{}^{j}x_k$$

4. Predict sub-space and associated covariance

$$\varphi_{l,k}^- = \hat{\varphi}_{l,k-1}$$
$$P_k^{ss-} = P_{k-1}^{ss} + W^{ss}$$

5. Calculate Kalman filter gain for updating sub-space

$$G_k^{ss} = P_k^{ss-} H_k^{ss\mathrm{T}}\left(H_k^{ss} P_k^{i-} H_k^{ss\mathrm{T}} + V\right)^{-1}$$

6. Update sub-space and associated covariance

$$\hat{\varphi}_{l,k} = \varphi_{l,k}^- + G_k^{ss}\left(y_k - H_k^{ss}\varphi_{l,k}^-\right)$$
$$P_k^{ss} = P_k^{ss-} - G_k^{ss} H_k^{ss} P_k^{ss-}$$

To investigate the computational aspects of incorporating a reduced-order model within the proposed hybrid particle-Kalman filtering scheme, the computational complexity of the procedure is considered in terms of the order of the required floating-point operations, and compared with that of the full order model. Concerning reduced-order modeling, the complexity associated with obtaining the reduced bases from the snapshot matrix is $O\left(n\ N_{snap}^2 + l\ n\ N_{obs} + N_{snap}^3\right)$, where we recall that n is the number of DOFs of the full-order model, l is the number of POMs retained in the reduced-order one, and N_{snap} stands for the number of snapshots collected, see [40]. Table 1 provides instead the computational complexity of all the stages of the proposed hybrid particle-Kalman filtering, as listed in Algorithm 1. In the prediction phase, the drawing of samples from a multivariate probability distribution scales linearly with the size $3l + N_p$ of the state vector, and the number N_s of samples; next, by adopting an Extended Kalman Filter (EKF) to improve the quality of the particle ensemble, a square $n \times n$ matrix must be inverted and the related computational complexity is $O\left(n^3\right)$ according to a Gauss-Jordan procedure [41]. In the subsequent update stage, the complexity is dominated by the calculation of the determinant of the covariance matrix $\boldsymbol{P}_k$ of the state vector; resampling is endowed with a $O\left(N_s\right)$ complexity, according to [42]. When dealing with the reduced-order modeling, an additional computational burden is associated to the updating of the sub-space, as reported at Stages 4–6 of the Algorithm. Overall, the computational complexity of the whole procedure is:

$$O\left(n^2 N_{obs} + N_s\left(\left(N_p + l\right)^3 + N_{obs}^3 + N_{obs}\left(N_p + l\right)^2 + N_{obs}^2\left(N_p + l\right)\right)\right) \tag{11}$$

In the case that the full-order model is used for hybrid particle-Kalman filtering, the relevant complexity instead reads:

$$O\left(N_s\left(\left(N_p + n\right)^3 + N_{obs}^3 + N_{obs}\left(N_p + n\right)^2 + N_{obs}^2\left(N_p + n\right)\right)\right) \tag{12}$$

By making recourse to the above defined relations for complexities associated with the use of reduced- and full-order models in the hybrid particle filtering procedure, an estimate of the analysis speed-up can be computed as the ratio between the number of operations required to run the SHM with the full-order model and that required instead to run the SHM with the reduced-order model. It can be observed in Equations (11) and (12) that the complexity features at most a term $O\left(n^2\right)$ in the first case, and $O\left(n^3\right)$ in the second. The obtained complexities for the two procedures refer to asymptotic scenarios; the real CPU times and so the corresponding speed-up could be different from the estimates obtained via complexity due to the effects of, e.g., processor architecture, hardware settings, and structure of matrices.

Table 1. Computational complexity of the stages of the proposed algorithm.

		Computational Complexity (Flops)
prediction	1	$O\left(N_s\left(l + N_p\right)\right)$
	2	$O\left(N_s\left(\left(l + N_p\right)^3 + N_{obs}\left(l + N_p\right)^2 + N_{obs}^2\left(l + N_p\right) + N_{obs}^3\right)\right)$
update	1	$O\left(N_s\left(l + N_p\right)^3\right)$
	2	$O\left(N_s\right)$
	3	$O\left(N_s\left(l + N_p\right)\right)$
	4	$O\left(n^2\right)$
	5	$O\left(n^2 N_{obs} + n\ N_{obs}^2 + N_{obs}^3\right)$
	6	$O\left(n N_{obs} + n\ N_{obs}^2\right)$

3. Results and Discussion

We consider in this Section a benchmark test represented by a thin plate subject to bending. We provide first the geometrical features of the plate and the mechanical properties of the material considered, along with loading/boundary conditions. We next discuss possible effects on the health monitoring capabilities of the major parameters of the proposed approach, namely: the number l of the POMs retained in the reduced-order model; the process noise; the measurement noise. We provide outcomes at varying damage level in one region of plate only, and also at a varying spreading of the damage over the plate. Finally, data are reported considering the speed-up provided by the reduced-order modeling, at varying space discretization adopted for the initial full-order finite element model.

As shown in Figure 1, the structure to be monitored is a thin square plate, with side length 200 mm and thickness 5 mm. The plate is assumed to be made of aluminum (6061-T6), whose relevant mechanical properties (in the virgin state) are: Young's modulus $E = 68.9$ GPa, Poisson's ratio $\nu = 0.3$, and density $\rho = 2500\ \mathrm{kg/m^3}$.

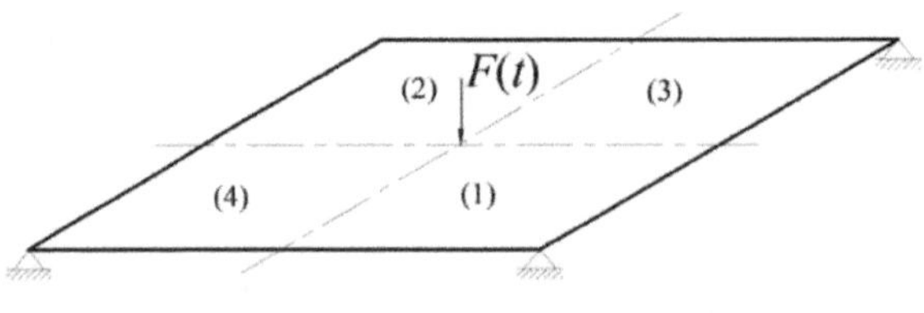

Figure 1. Benchmark plate test: boundary and loading conditions, and zone numbering.

The plate has been modeled with the commercial finite element code Abaqus, using the S4R general-purpose, conventional shell elements which take into account also transverse shear deformations [43]. The kinematics of this element type is fully characterized by six DOFs per node, that are the three displacements along the axes of any orthonormal reference frame (typically with two of them belonging to the mid-plane of the plate and one pointing perpendicularly to such mid-plane) and the three rotations about the same axes.

The plate is simply supported at the four corners, and subject to a sinusoidally varying force $F(t) = F^M \sin(2\pi f t)$ at its center (Figure 1). The maximum value of the force $F^M = 100$ N has been set to avoid damage development in the plate caused by it; the plate is therefore considered already damaged during the test, and to possibly suffer damage growth due to external causes. The frequency of the load has been instead set to $f = 80$ Hz, which is smaller than the fundamental frequency of vibrations of the aluminum plate, which amounts to $f' = 315$ Hz for the undamaged case, and $f'' = 280$ Hz for the reference damage case described in what follows.

The structure is decomposed into four regions, each of them associated with the relevant target damage index d_i, with $i = 1, \ldots, 4$. If not stated otherwise, in the examples to follow the structure is supposed to be damaged only in zone 2 (see Figure 1), where the Young's modulus is reduced to one half of its virgin value: therefore, $d_2 = 0.5$ is the target damage value to be estimated.

In a pseudo-experimental test frame, finite element analyses have been adopted to first obtain the mass matrix $\boldsymbol{M}$ of the whole system and the (undamaged) stiffness submatrices $\boldsymbol{K}^i_{und}$ (see Equation (6)) of the mentioned $N_p = 4$ regions. Next, for each damage scenario here considered, the same code has been used to run the full-order analysis (although it can be run independently, once the mass and stiffness matrices have been obtained); accordingly, the snapshots are collected for the considered loading/boundary conditions, and POMs in $\hat{\boldsymbol{\varphi}}_{l,0}$ are obtained. As for measurements, results of the simulations have been corrupted with a Gaussian noise of known variance according to Equation (7). For the plate bending model considered in this paper, rotations about the in-plane reference axes, at the mid-point of each plate edge are handled as observations; this is in line with the optimal spatial distribution of measurements gathered through a network of surface-mounted inertial sensors, proposed in [28,29] to achieve sensitivity to damage independently of its location and amount.

The uncertainty and noise levels in the model are quantified for the structural system by the time-independent matrices $\boldsymbol{W}$ and $\boldsymbol{V}$ (Section 2.2). As for the process noise, which takes into account the uncertainties introduced by the model (due to the spatial and temporal discretization and to the order reduction procedure), its covariance W is scaled by a dimensionless factor σ_w, having the role of a standard deviation for the DOFs in the reduced-order model, according to:

$$W = \sigma_w^2 I_w$$

where $I_w \in \mathbb{R}^{3l+N_p}$ is an identity matrix of appropriate size.

The measurement noise takes instead into account what related to the physical medium of communication between the sensors and the acquisition system, the random electrical noise due to the sensors and the electrical circuits, the quality of the measurement devices and also the environmental noise. Similar to the process noise, the measurement covariance V is scaled by a dimensionless factor σ_v, playing the role of the standard deviation of the collected rotation angles, according to:

$$V = \sigma_v^2 I_v$$

where now $I_v \in \mathbb{R}^{N_{obs}}$. So, all the measurements are assumed uncorrelated. It is well-known that the performance of Bayesian filters rests on a proper choice of the above covariance matrices. By tuning the values of the process and measurement noise variances, the level of confidence in the model and measurements equations can be set; for instance, in case the measurements are deemed more reliable than the model, a low measurement noise variance σ_v^2 should be assigned, when compared with the process one σ_w^2. In practice, measurement noise variance is known for the used sensors; tuning of the process noise variance is in general more intricate, and is normally achieved by a trial and error procedures offline. The optimal estimation of noise parameters in an online fashion has recently gained attention; reader are referred to [36,37,44] for further details.

In this study, the number of samples in particle filtering has been always assumed $N_s = 10$. Moreover, to build the initial POMs, a number of snapshots $N_{snap} = 50$ always led to convergence of those retained in the analyses.

Moving to the results, let us first investigate the effects of the number l of POMs taken in the reduced-order model, on the accuracy of the estimations obtained. In Figure 2, it is possible to see that such accuracy grows as the number of POMs gets higher. If the plate is coarsely meshed using only one finite element for each single region, the resulting structural DOFs turn out to amount to 50, once the constraints at plate corners are taken into account. Although both the geometry and the loading condition are two-fold symmetric, such symmetry has not been exploited in the analyses, as damage in only one region breaks it. A remarkable result reported in Figure 2d is the proof that, for this benchmark, it is possible to reduce the number of DOFs to only 4 and get an estimation error smaller than 10%.

To assess the influence on the results of the initialization of the filter in terms of the damage state d_0, Figure 3 shows for $l = 3$ that estimates are stable whenever the initial guess is close enough to the target one. Such performance test guarantees a wide range of applicability of the method, since the undamaged state $d_0 = \mathbf{0}$ and also an underestimation by 50% of the initial elastic moduli allow to assure stability. The other way around, an initial guess too far from the target solution (Figure 3c) is shown to insert instabilities or biases in the estimations.

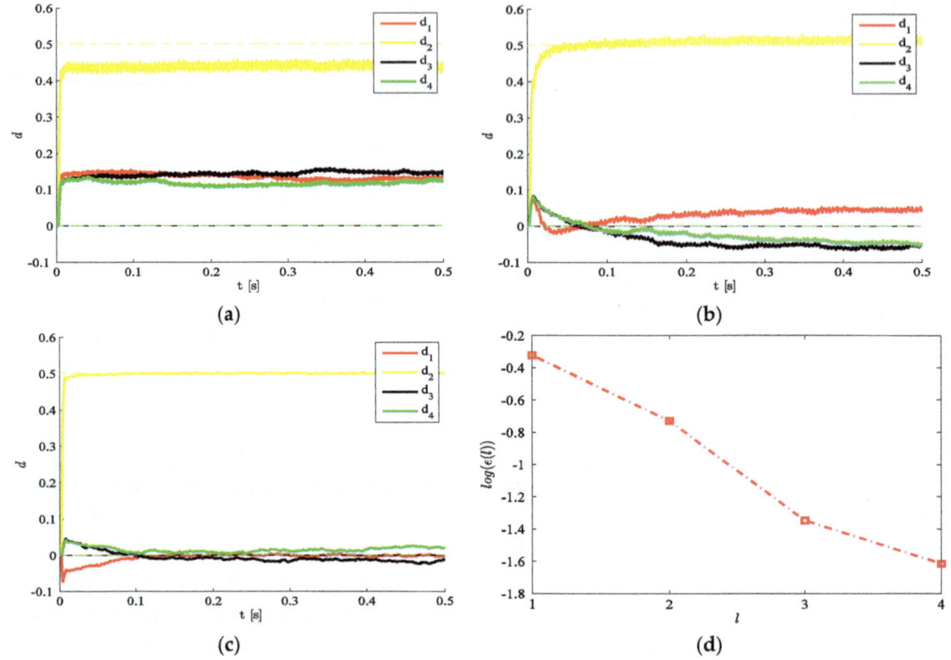

Figure 2. Time-invariant target damage state, $\sigma_v = 10^{-5}$, $\sigma_w = 10^{-5}$, $\boldsymbol{d}_0 = 0$: time evolution of estimations of the damage indexes d_i, $i = 1, \ldots, 4$ identified with (**a**) $l = 1$; (**b**) $l = 2$; (**c**) $l = 3$; and (**d**) relative error $\epsilon(l) = \dfrac{||d_l - d||_{L^2}}{||d||_{L^2}}$ in the damage indexes, at varying order l of the reduced-order model.

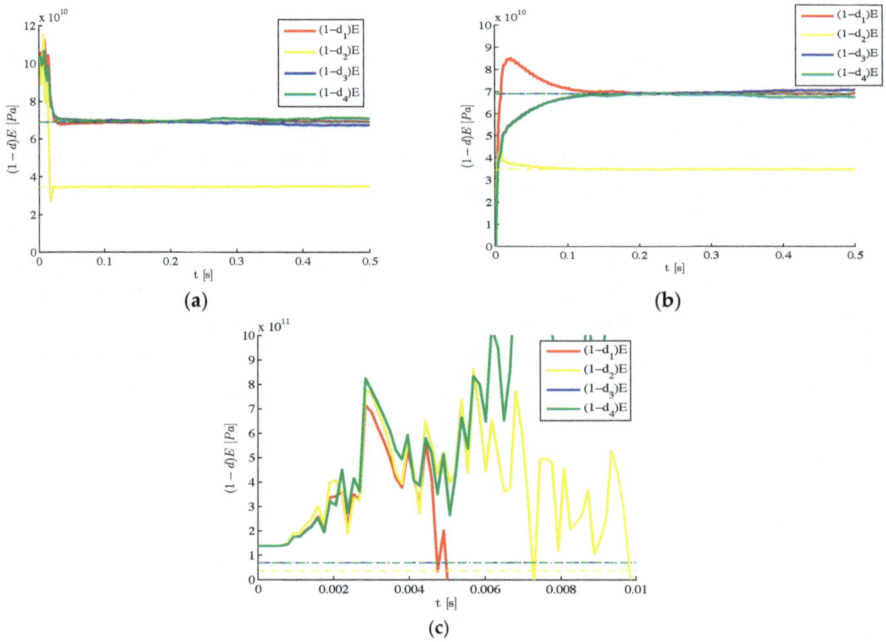

Figure 3. Time-invariant target damage state, $l = 3$, $\sigma_v = 10^{-5}$, $\sigma_w = 10^{-5}$: time evolution of estimations of the scaled elastic moduli (and, therefore, of the damage indexes d_i), identified with initial values (**a**) $E_{i,0} = \dfrac{3}{2}E$, $i = 1, \ldots, 4$; (**b**) $E_{i,0} = 0$; and (**c**) $E_{i,0} = 2E$.

If we consider the effects of the process noise, some exemplary results are gathered in Figure 4. As expected, a high noise level amounting to $\sigma_w = 10^{-2}$ (Figure 4a), can incept large variations of the estimates over each time interval, and therefore gives rise to possible biases or even divergence in the final solution. On the other hand, if the process noise is lowered to $\sigma_w = 10^{-5}$ (Figure 4b), wild oscillations do not show up any longer and the estimation evolutions become smooth.

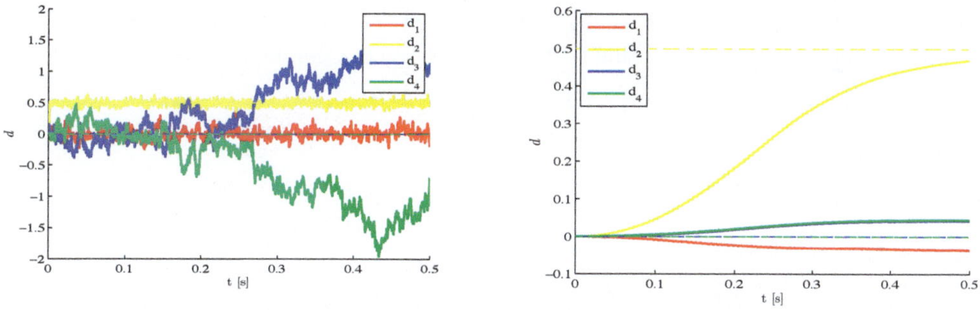

(a) (b)

Figure 4. Time-invariant target damage state, $l = 3$, $\sigma_v = 10^{-5}$, $\boldsymbol{d}_0 = \boldsymbol{0}$: time evolution of estimations of the damage indexes d_i, identified with a process noise (**a**) $\sigma_w = 10^{-2}$ and (**b**) $\sigma_w = 10^{-5}$.

In the analyses so far, the measurement noise level was kept low through $\sigma_v = 10^{-5}$, *i.e.*, (even if not explicitly shown) smaller than 10% of the maximum amplitude of the structural response under the considered loading. Figure 5 testifies that, as the measurement noise gets bigger (up to $\sigma_v = 10^{-3}$), the ability of the procedure to estimate the damage indexes drops; as results are here shown for $l = 3$, they can be compared to the plots already reported in Figure 2a for $\sigma_v = 10^{-5}$. The trend shown by the accuracy of the estimates and by the readiness to approach the target values is basically due to the fact that the structural response gets hidden by the noise if σ_v becomes bigger, and the filter is not able to extract significant information from measurements. Overall, by increasing σ_v the transitory stage in the time evolution of the damage indexes lasts more and more (Figure 5a), or the SHM procedure can even lose the ability to estimate the damage in a region (Figure 5b). This response has been checked to vary monotonically and smoothly when σ_v changes; moreover, plots in Figure 5 are reported for a null initialization of damage parameters, namely for $\boldsymbol{d}_0 = \boldsymbol{0}$, but similar results can be obtained for any other initial guess guaranteeing the stability of estimations.

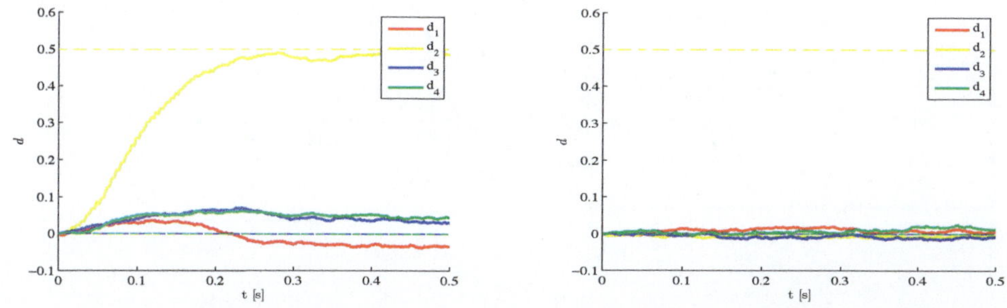

(a) (b)

Figure 5. Time-invariant target damage state, $l = 3$, $\sigma_w = 10^{-5}$, $\boldsymbol{d}_0 = \boldsymbol{0}$: time evolution of estimations of the damage indexes d_i, identified with a measurement noise featuring (**a**) $\sigma_v = 1.5 \times 10^{-5}$ and (**b**) $\sigma_v = 10^{-3}$.

We next investigate the method performance at a varying value of the target damage index d_2, with $l = 3$ and still keeping the noise factors $\sigma_v = \sigma_w = 10^{-5}$ adopted in Figure 3. Figure 6 shows the results in terms of time evolutions of the estimate of d_2 for the four target cases $d_2 = 0.1,\ 0.25,\ 0.5,\ 0.75$. Such evolutions are obtained along with those relevant to the other damage indexes (all zero valued); since the trend shown by these latter ones is basically the same reported in Figure 3b, we focus on the initial transient stage of the estimations of d_2 starting from the initial guess $d_2 = 1$. Such transient stage is reported to last less than 0.01 s; after that, estimates are affected by small fluctuations only. It also appears that target values are promptly matched if damage values are large. This is somehow expected, as small target values of d_2 provide small drifts from the response of the healthy structure, and so the sensitivity to damage of the monitoring system in the noisy environment gets detrimentally affected.

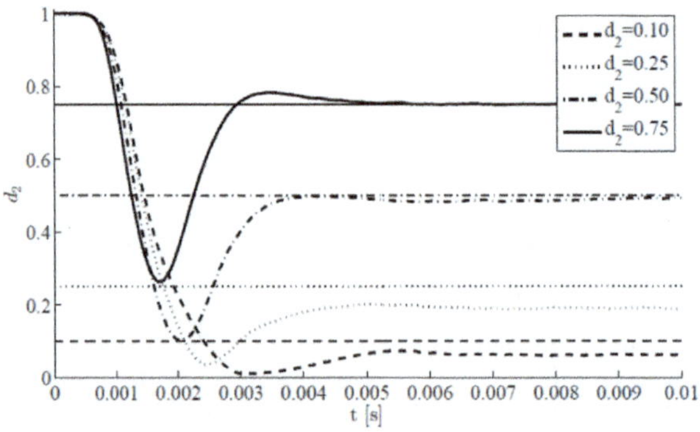

Figure 6. Time-invariant target damage state, $l = 3,\ \sigma_v = 10^{-5},\ \sigma_w = 10^{-5}, \boldsymbol{d}_0 = \mathbf{1}$: time evolution of estimations of the damage index d_2 at varying target value.

The proposed methodology also features the same accuracy level if all the plate regions are damaged. Figure 7 provides the time evolutions of the estimates, in the exemplary case characterized by target values $d_1 = 0.75$, $d_2 = 0.5$, $d_3 = 0.9$ and $d_4 = 0.25$. For two different initialization sets, it is shown that estimates converge fast towards the target values, and small oscillations around them are next linked to the hybrid filtering methodology adopted. Comparing these results with those reported in Figure 6, it can be seen once again that smaller values of the damage indexes (and, so bigger values of the residual stiffness reported in the graphs) are connected to a delayed convergence of the filter estimates. Indeed, the provided steady-state solutions always perfectly match the target one.

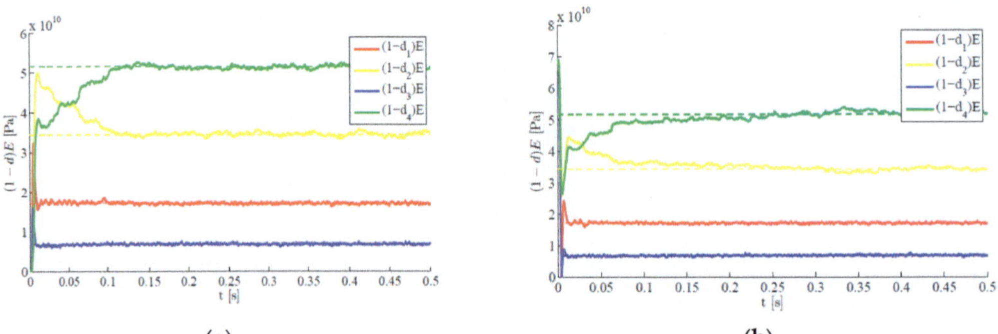

Figure 7. Time-invariant target damage state, four plate regions damaged, $l = 3\ \sigma_v = 10^{-5}, \sigma_w = 10^{-5}$: time evolution of estimations of the scaled elastic moduli, identified with initial values (**a**) $\boldsymbol{d}_0 = \mathbf{1}$ and (**b**) $\boldsymbol{d}_0 = \mathbf{0}$.

We move now to a case characterized by a time evolving damage state. The structure is still assumed to be initially damaged with $d_2 = 0.5$, but such damage is suddenly increased to $d_2 = 0.7$

around $t = 0.25$ s due to a further event, possibly linked in real-life situations to unexpected or extreme loadings. In Figure 8 results are provided concerning the estimated time evolution of the damage indexes, by either disregarding (Figure 8a) or allowing for (Figure 8b) the sub-space update. Such update, according to the setting defined in Section 2, is automatically driven by the filtering procedure as soon as the structural observations display a drift away from the response expected on the basis of the current damage state. POMs update is obviously compulsory to keep a similar degree of accuracy of the reduced-order model when the damage state is altered. If such update is not performed, the reduced-order model (that has been trained with the initial damage state) does not match the current structural health, and the damage estimates diverge or are affected by unacceptable biases.

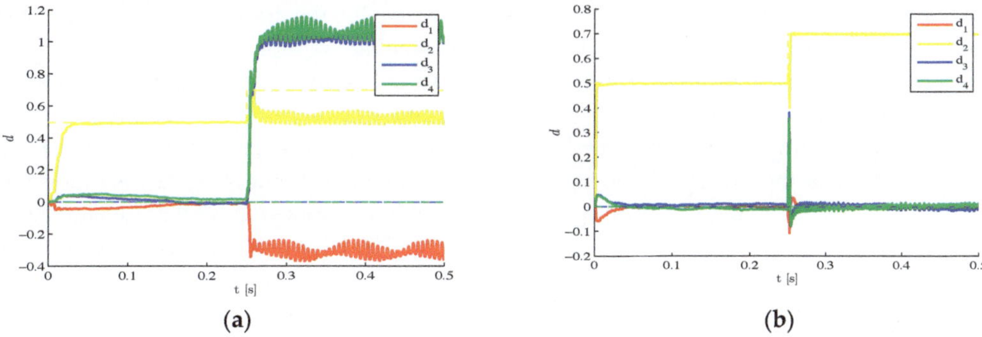

Figure 8. Time-varying target damage state, $l = 3$, $\sigma_v = 10^{-5}$, $\sigma_w = 10^{-5}, \boldsymbol{d}_0 = \mathbf{0}$: time evolution of estimations of the damage indexes d_i, identified (**a**) without update of POMs and (**b**) with POMs update.

Some additional results are provided in Figure 9 in relation to a finer mesh adopted to build the full-order model; in this case, the free DOFs of the initial finite element model amount to 722. Compared to the former case linked to the coarsest possible mesh, a slightly higher number of POMs would be necessary to attain a high level of accuracy of the reduced-order model, if the proposed filtering procedure were not adopted. However, as stated in Section 2, the intricate formulation consisting in three filters processing the data, allows also to increase the accuracy of the reduced-order model, so that only two POMs prove enough to achieve an unbiased estimation of the damage state in a short time, see Figure 9b.

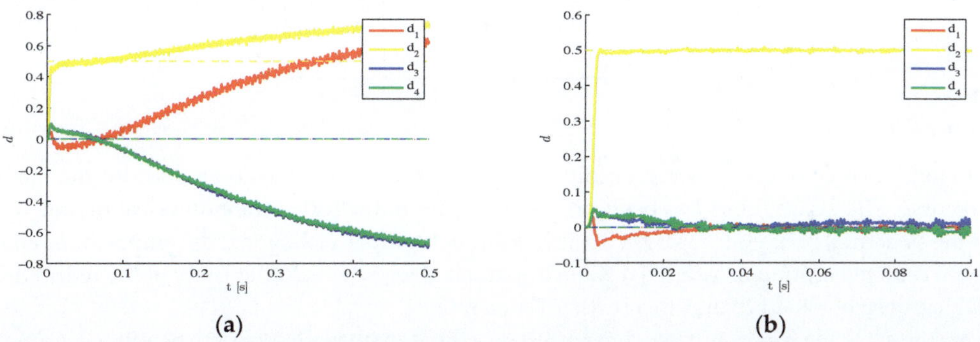

Figure 9. Time-invariant target damage state, fine space discretization, $\sigma_v = 10^{-5}$, $\sigma_w = 10^{-5}$, $\boldsymbol{d}_0 = \mathbf{0}$: time evolution of estimations of the damage indexes d_i, identified with (**a**) $l = 1$ and (**b**) $l = 2$.

Results in the figures have shown that each damage index, although theoretically constrained to belong to the [0–1) interval, may be estimated out of the bounds on such interval. If damage becomes negative, then the stiffness properties of the structure results to be enhanced, hence greater than in the virgin state; that is obviously of no practical interest, and should be considered as a wrong outcome of filtering procedure. Anyhow, this kind of results does not violate the thermodynamic requirement to always have a positive reduced Young's modulus (see [34]). The other way around, damage can become greater than one, and so the reduced stiffness turns out to be negative. Although negative tangent stiffness values can be encountered at the structural level when local failure phenomena take place, in our case those values are to be considered wrong outcomes of the filtering procedure. Nonetheless, values of the damage indexes have not been constrained in our procedure, so that a standard particle-Kalman filter can be adopted. Alternatively, implementations centered on the so-called constrained Kalman filters were offered [45–47]; such possible alternative implementations have not been considered in our work since, as clearly depicted in the graphs of this Section, estimates of the damage indexes largely exceed the bounds only when the solution becomes unstable. In all the analyses providing reliable results, it usually happens that only some indexes slightly move below the zero value threshold.

Finally, concerning the analysis speed-up, relevant outcomes are reported in Table 2 at varying mesh and order l of the model, in terms of both CPU time and floating-point operation metrics. These data have been obtained by running the procedure implemented in Matlab (release 2014a) on a personal computer featuring an Intel Xeon E3-1270 V2 @ 3.50 GHz processor, with 8.00 Gb of RAM and Windows 7 64-bit as OS. It can be seen that the finer the full-order space discretization, the higher the speed-up; on the basis of a minimum reported speed-up value exceeding 30 for the coarse mesh, this joint use of intricate filtering schemes and reduced-order modeling can prove successful in the design of real-time SHM procedures.

Table 2 also shows that the speed-up computed through the algorithmic complexity discussed in Section 2.2, actually represents an upper bound on the real one given by the CPU time. It decreases at a smaller rate than that linked to CPU time when the number l of retained POMs is increased, but it converges to unitary values when l tends to the number n of total DOFs of the full-order model (actually it converges to values slightly less than one, as POM update is not carried out with the full-order model).

Table 2. Analysis speed-up provided by CPU time and flops, at varying mesh and order l of the reduced-order model.

l	Coarse Mesh		Fine Mesh	
	CPU Time	**Flops**	**CPU Time**	**Flops**
1	41.8	100	1319.5	1812
2	37.2	89	701.7	1810
3	33.3	79	611.7	1809

4. Conclusions

In this work, an online structural health monitoring procedure based on reduced-order modeling and recursive Bayesian filters has been proposed. Order reduction of the numerical model, used to track the structural behavior, has been achieved through proper orthogonal decomposition. Filtering has been centered around an extended Kalman-particle filter, whereas adaptivity of the reduced-order model has been obtained thanks to a further Kalman filter.

According to the results of a benchmark test on a thin square plate subject to bending deformations, the main outcomes shown here can be outlined as follows:

- the model order reduction method allows decreasing dramatically the number of system degrees-of-freedom, without losing accuracy as for the estimation of damage;
- The initial conditions for the damage state can be easily set, so that the estimation procedure keeps stability;

- Measurement and process noises should be small in comparison with the amplitude of measurements, to avoid biases in the estimates.

In future investigations, the proposed approach will be further extended to attack real-life, more complex cases in terms of structural geometry, material behavior and loading/boundary conditions. It is in fact well known that by increasing the number of parameters to be estimated, *i.e.*, by increasing the number of regions wherein damage is assumed to be homogeneous, problems connected to the curse of dimensionality arise. This generally means that by increasing the number of parameters to estimate, a lower accuracy of the filtering procedure is expected.

Acknowledgments: Giovanni Capellari and Stefano Mariani wish to acknowledge financial support by Fondazione Cariplo through project Safer Helmets.

Author Contributions: Authors contributed equally to this work.

Conflicts of Interest: The authors declare no conflict of interest.

Appendix A. Algorithmic Details

In this Appendix, we go through the main algorithmic details of the proposed procedure to handle the dual estimation of the continuously updated, reduced-order model of the structural system.

We start by showing how Equation (5) can be obtained from the equation of motion (Equation (2)) of the reduced-order model. Within the considered time step $[t_{k-1}\ t_k]$, the solution is advanced through the following three-stage scheme:

a. Predictor stage

$$\widetilde{\boldsymbol{\alpha}}_k = \boldsymbol{\alpha}_{k-1} + \Delta t \dot{\boldsymbol{\alpha}}_{k-1} + \Delta t^2 \left(\frac{1}{2} - \beta\right) \ddot{\boldsymbol{\alpha}}_{k-1}$$

$$\widetilde{\dot{\boldsymbol{\alpha}}}_k = \dot{\boldsymbol{\alpha}}_{k-1} + \Delta t \left(1 - \gamma\right) \ddot{\boldsymbol{\alpha}}_{k-1}$$

b. Explicit integrator stage

$$\ddot{\boldsymbol{\alpha}}_k = \boldsymbol{\mathcal{M}}_{k-1}^{-1} \left(\boldsymbol{\mathcal{F}}_k - \boldsymbol{\mathcal{D}}_{k-1} \widetilde{\dot{\boldsymbol{\alpha}}}_k - \boldsymbol{\mathcal{K}}_{k-1} \widetilde{\boldsymbol{\alpha}}_k\right)$$

c. Corrector stage

$$\boldsymbol{\alpha}_k = \widetilde{\boldsymbol{\alpha}}_k + \Delta t^2 \beta \ddot{\boldsymbol{\alpha}}_k$$

$$\dot{\boldsymbol{\alpha}}_k = \widetilde{\dot{\boldsymbol{\alpha}}}_k + \Delta t\ \gamma \ddot{\boldsymbol{\alpha}}_k$$

where: β and γ are the Newmark's parameters (see [35]); and matrices $\boldsymbol{\mathcal{M}}_{k-1}$, $\boldsymbol{\mathcal{D}}_{k-1}$ and $\boldsymbol{\mathcal{K}}_{k-1}$ are time-dependent, since the sub-space model can be continuously updated at the end of each time step.

To simplify the equations, we assume in what follows that system damping is null. It was already discussed in [23] that, whenever the system is continuously excited, the filter results will not be affected by the damping properties of the structure. By considering the components of the state vector $\boldsymbol{x}_k$ related to the structural state only, arranged according to:

$$\boldsymbol{z}_k = \begin{Bmatrix} \boldsymbol{\alpha}_k \\ \dot{\boldsymbol{\alpha}}_k \\ \ddot{\boldsymbol{\alpha}}_k \end{Bmatrix}$$

the relevant evolution equation can be written, as provided by the time integration scheme described above:

$$\boldsymbol{z}_k = \boldsymbol{A}_{k-1} \boldsymbol{z}_{k-1} + \boldsymbol{b}_{k-1}$$

where:

$$A_{k-1} = \begin{bmatrix} I - \beta\Delta t^2 \mathcal{M}_{k-1}^{-1}\mathcal{K}_{k-1} & \Delta t I - \beta\Delta t^3 \mathcal{M}_{k-1}^{-1}\mathcal{K}_{k-1} & \Delta t^2\left(\frac{1}{2}-\beta\right)\left[I - \beta\Delta t^2 \mathcal{M}_{k-1}^{-1}\mathcal{K}_{k-1}\right] \\ -\Delta t\gamma \mathcal{M}_{k-1}^{-1}\mathcal{K}_{k-1} & I - \Delta t^2\gamma \mathcal{M}_{k-1}^{-1}\mathcal{K}_{k-1} & (1-\gamma)\, dt I - \Delta t^3\gamma\left(\frac{1}{2}-\beta\right)\mathcal{M}_{k-1}^{-1}\mathcal{K}_{k-1} \\ -\mathcal{M}_{k-1}^{-1}\mathcal{K}_{k-1} & -\Delta t \mathcal{M}_{k-1}^{-1}\mathcal{K}_{k-1} & -\Delta t^2\left(\frac{1}{2}-\beta\right)\mathcal{M}_{k-1}^{-1}\mathcal{K}_{k-1} \end{bmatrix}$$

$$b_{k-1} = \begin{Bmatrix} \Delta t^2 \beta \mathcal{M}_{k-1}^{-1}\mathcal{F}_k \\ \Delta t\, \gamma \mathcal{M}_{k-1}^{-1}\mathcal{F}_k \\ \mathcal{M}_{k-1}^{-1}\mathcal{F}_k \end{Bmatrix}$$

Entries of A_{k-1} and b_{k-1} vary in time not only because POMs are updated, but also because damage indexes d_{k-1} affecting the stiffness matrix are tracked in their time evolution.

In a noisy environment, the whole process model thus becomes:

$$x_k = \begin{Bmatrix} z_k \\ d_k \end{Bmatrix} = \begin{Bmatrix} A_{k-1} z_{k-1} + b_{k-1} \\ d_{k-1} \end{Bmatrix} + w_k = f_k\left(x_{k-1}\right) + w_k$$

whose Jacobian matrix F_k reads:

$$F_k = \begin{bmatrix} A_{k-1} & \dfrac{\partial\left(A_{k-1} z_{k-1}\right)}{\partial d_{k-1}} \\ 0 & I_d \end{bmatrix}$$

where $I_d \in \mathbb{R}^{N_p}$ is an identity matrix, and the upper-right term of the matrix is made of terms like:

$$\frac{\partial\left(A_{k-1} z_{k-1}\right)}{\partial d_{k-1}^i} = \begin{bmatrix} \Delta t^2 \beta \mathcal{M}_{k-1}^{-1}\mathcal{K}_{und}^i & \Delta t^3 \beta \mathcal{M}_{k-1}^{-1}\mathcal{K}_{und}^i & \Delta t^2\left(\frac{1}{2}-\beta\right)\beta \mathcal{M}_{k-1}^{-1}\mathcal{K}_{und}^i \\ \Delta t\gamma \mathcal{M}_{k-1}^{-1}\mathcal{K}_{k-1} & \Delta t^2\gamma \mathcal{M}_{k-1}^{-1}\mathcal{K}_{und}^i & \Delta t^3\gamma\left(\frac{1}{2}-\beta\right)\mathcal{M}_{k-1}^{-1}\mathcal{K}_{und}^i \\ \mathcal{M}_{k-1}^{-1}\mathcal{K}_{und}^i & \Delta t \mathcal{M}_{k-1}^{-1}\mathcal{K}_{und}^i & \Delta t^2\left(\frac{1}{2}-\beta\right)\mathcal{M}_{k-1}^{-1}\mathcal{K}_{und}^i \end{bmatrix}$$

$\mathcal{K}_{\mathrm{und}}^{\mathrm{i}}$ terms being already defined in Equation (6).

References

1. Achenbach, J.D. Structural health monitoring—What is the prescription? *Mech. Res. Commun.* **2009**, *36*, 137–142. [CrossRef]
2. Glaser, S.D.; Tolman, A. Sense of sensing. *J. Infrastruct. Syst. ACSE* **2008**, *14*, 4–14. [CrossRef]
3. Hong, A.; Ubertini, F.; Betti, R. New stochastic subspace approach for system identification and its application to long-span bridges. *J. Eng. Mech.* **2013**, *139*, 724–736. [CrossRef]
4. Bao, Y.; Xia, Y.; Li, H.; Xu, Y.; Zhang, P. Data fusion-based structural damage detection under varying temperature conditions. *Int. J. Struct. Stab. Dyn.* **2012**, *12*, 1–27. [CrossRef]
5. Chae, M.J.; Yoo, H.S.; Kim, J.Y.; Cho, M.Y. Development of a wireless sensor network system for suspension bridge health monitoring. *Autom. Constr.* **2012**, *21*, 237–252. [CrossRef]
6. Jo, H.; Sim, S.; Nagayama, T.; Spencer, B.J. Development and application of high-sensitivity wireless smart sensors for decentralized stochastic modal identification. *J. Eng. Mech.* **2012**, *138*, 683–694. [CrossRef]
7. Ratcliffe, C.; Heider, D.; Crane, R. Investigation into the use of low cost MEMS accelerometers for vibration based damage detection. *Compos. Struct.* **2008**, *82*, 61–70. [CrossRef]
8. Yang, H.; Shan, J.; Randall, C.; Hansma, P.; Shi, W. Integration of health monitoring and control of building structures during earthquakes. *J. Eng. Mech.* **2014**, *140*, 04014013. [CrossRef]

9. Klaiber, F.W.; Dunker, K.F.; Wipf, T.J.; Sanders, W.W. Methods of strengthening existing highway bridges. *Transp. Res. Rec.* **1987**, *1380*, 1–6.
10. Stallings, J.M.; Tedesco, J.W.; El-Mihilmy, M.; McCauley, M. Field performance of FRP bridge repairs. *J. Bridge Eng.* **2000**, *5*, 107–113. [CrossRef]
11. Canada, I. Reinforcing Concrete Structures with Fibre Reinforced Polymers. Available online: http://www.vrodaustralia.com.au/wp-content/themes/vrod/pdf/engineering/codes_and_guides/01.pdf (accessed on 12 October 2015).
12. Ljung, L. *System Identification. Theory for the User*; Prentice Hall: Englewood Cliffs, NJ, USA, 1999.
13. Chatzi, E.N.; Smyth, A.W.; Masri, S.F. Experimental application of on-line parametric identification for nonlinear hysteretic systems with model uncertainty. *Struct. Saf.* **2010**, *32*, 326–337. [CrossRef]
14. Eftekhar Azam, S.; Mariani, S. Dual estimation of partially observed nonlinear structural systems: A particle filter approach. *Mech. Res. Commun.* **2012**, *46*, 54–61. [CrossRef]
15. Eftekhar Azam, S.; Chatzi, E.; Papadimitriou, C. A dual Kalman filter approach for state estimation via output-only acceleration measurements. *Mech. Syst. Signal Process.* **2015**, *60*, 866–886. [CrossRef]
16. Lourens, E.; Reynders, E.; de Roeck, G.; Degrande, G.; Lombaert, G. An augmented Kalman filter for force identification in structural dynamics. *Mech. Syst. Signal Process.* **2012**, *27*, 446–460. [CrossRef]
17. Sirovich, L. Turbulence and the dynamics of coherent structures. *Q. Appl. Math.* **1987**, *45*, 573–590.
18. Eftekhar Azam, S.; Mariani, S. Investigation of computational and accuracy issues in POD-based reduced order modeling of dynamic structural systems. *Eng. Struct.* **2013**, *54*, 150–167. [CrossRef]
19. Eftekhar Azam, S. *Online Damage Detection in Structural Systems*; PoliMI SpringerBriefs, Springer International Publishing: Cham, Switzerland, 2014.
20. Ljung, L. Asymptotic behavior of the extended Kalman filter as a parameter estimator for linear systems. *IEEE Trans. Autom. Control* **1979**, *24*, 36–50. [CrossRef]
21. Kalman, R.E.; Bucy, R.S. New results in linear filtering and prediction theory. *ASME J. Basic Eng.* **1961**, *83*, 95–108. [CrossRef]
22. Bittanti, S.; Maier, G.; Nappi, A. Inverse problems in structural elastoplasticity: A Kalman filter approach. *Plast. Today* **1984**, 311–329.
23. Corigliano, A.; Mariani, S. Parameter identification in explicit structural dynamics: Performance of the extended Kalman filter. *Comput. Methods Appl. Mech. Eng.* **2004**, *193*, 3807–3830. [CrossRef]
24. Mariani, S.; Ghisi, A. Unscented Kalman filtering for nonlinear structural dynamics. *Nonlinear Dyn.* **2007**, *49*, 131–150. [CrossRef]
25. Ching, J.; Beck, J.L.; Porter, K.A. Bayesian state and parameter estimation of uncertain dynamical systems. *Probab. Eng. Mech.* **2006**, *21*, 81–96. [CrossRef]
26. Khalil, M.; Sarkar, A.; Adhikari, S. Nonlinear filters for chaotic oscillatory systems. *Nonlinear Dyn.* **2009**, *55*, 113–137. [CrossRef]
27. Zghal, M.; Mevel, L.; Del Moral, P. Modal parameter estimation using interacting Kalman filter. *Mech. Syst. Signal Process.* **2014**, *47*, 139–150. [CrossRef]
28. Mariani, S.; Bruggi, M.; Caimmi, F.; Bendiscioli, P.; de Fazio, M. Sensor deployment over damage-containing plates: A topology optimization approach. *J. Intell. Mater. Syst. Struct.* **2013**, *24*, 1105–1122. [CrossRef]
29. Caimmi, F.; Mariani, S.; De Fazio, M.; Bendiscioli, P. Investigation of the effectiveness and robustness of a MEMS-based structural health monitoring system for composite laminates. *IEEE Sens. J.* **2014**, *14*, 2208–2215. [CrossRef]
30. Kerschen, G.; Golinval, G.C. Physical interpretation of the proper orthogonal modes using the singular value decomposition. *J. Sound Vib.* **2002**, *249*, 849–865. [CrossRef]
31. Schilders, W.H.; van der Vorst, H.A.; Rommes, J. *Model Order Reduction: Theory, Research Aspects and Applications*; Springer-Verlag: Berlin, Germany, 2008.
32. Corigliano, A.; Dossi, M.; Mariani, S. Domain decomposition and model order reduction methods applied to the simulation of multiphysics problems in MEMS. *Comput. Struct.* **2013**, *122*, 113–127. [CrossRef]
33. Eftekhar Azam, S.; Ghisi, A.; Mariani, S. Parallelized sigma-point Kalman filtering for structural dynamics. *Comput. Struct.* **2012**, *92–93*, 193–205. [CrossRef]
34. Krajcinovic, D. Damage mechanics. *Mech. Mater.* **1989**, *8*, 117–197. [CrossRef]
35. Hughes, T.J.R. *The Finite Element Method: Linear Static and Dynamic Finite Element Analysis*; Dover: New York, NY, USA, 2000.

36. Yuen, K.-V.; Kuok, S.-C. Online updating and uncertainty quantification using nonstationary output-only measurement. *Mech. Syst. Signal Process.* **2016**, *66–67*, 62–77. [CrossRef]
37. Kontoroupi, K.; Smyth, A.W. Online noise identification for joint state and parameter estimation of nonlinear systems. *ASCE-ASME J. Risk Uncertain. Eng. Syst.* **2015**. [CrossRef]
38. Gordon, N.J.; Salmond, D.J.; Smith, A.F.M. Novel approach to nonlinear/non-gaussian bayesian state estimation. *IEE Proc. F Radar Signal Process.* **1993**, *140*, 107–113. [CrossRef]
39. Doucet, A. On sequential Monte Carlo methods for bayesian filtering. *Stat. Comput.* **2000**, *10*, 197–208. [CrossRef]
40. Wang, Z.; McBee, B.; Iliescu, T. Approximate partitioned method of snapshots for pod. *J. Comput. Appl. Math.* **2015**, in press. [CrossRef]
41. Sharma, G.; Agarwala, A.; Bhattacharya, B. A fast parallel Gauss Jordan algorithm for matrix inversion using CUDA. *Comput. Struct.* **2013**, *128*, 31–37. [CrossRef]
42. Hol, J.D.; Schon, T.B.; Gustafsson, F. *In On Resampling Algorithms for Particle Filters*; Nonlinear Statistical Signal Processing Workshop: Cambridge, UK, 2006.
43. Simulia, D.S. *Abaqus 6.13 Theory Manual*; Dassault Systèmes: Vélizy-Villacoublay, France, 2013.
44. Yuen, K.V.; Liang, P.F.; Kuok, S.C. Online estimation of noise parameters for Kalman filter. *Struct. Eng. Mech.* **2013**, *47*, 361–381. [CrossRef]
45. Chen, Y.-H.; Chiang, C.-T. Adaptive beamforming using the constrained Kalman filter. *IEEE Trans. Antennas Propag.* **1993**, *41*, 1576–1580. [CrossRef]
46. De Geeter, J.; Van Brussel, H.; de Schutter, J. A smoothly constrained Kalman filter. *IEEE Trans. Pattern Anal. Mach. Intell.* **1997**, *19*, 1171–1177. [CrossRef]
47. Liu, G. Active fault tolerant control with actuation reconfiguration. *IEEE Trans. Aerosp. Electron. Syst.* **2004**, *40*, 1110–1117.

13

A Microwave Ring-Resonator Sensor for Non-Invasive Assessment of Meat Aging

Muhammad Taha Jilani [1], Wong Peng Wen [1,*], Lee Yen Cheong [2] and Muhammad Zaka ur Rehman [1]

Academic Editor: Vittorio M. N. Passaro

1 Department of Electrical and Electronics Engineering, University Technology PETRONAS, Bandar Seri Iskandar 32610, Malaysia; mtaha.jilani@gmail.com (M.T.J.); zakarahman@gmail.com (M.Z.R.)

2 Department of Fundamental and Applied Science, University Technology PETRONAS, Bandar Seri Iskandar 32610, Malaysia; lee_yencheong@petronas.com.my

* Correspondence: wong_pengwen@petronas.com.my

Abstract: The assessment of moisture loss from meat during the aging period is a critical issue for the meat industry. In this article, a non-invasive microwave ring-resonator sensor is presented to evaluate the moisture content, or more precisely water holding capacity (WHC) of broiler meat over a four-week period. The developed sensor has shown significant changes in its resonance frequency and return loss due to reduction in WHC in the studied duration. The obtained results are also confirmed by physical measurements. Further, these results are evaluated using the Fricke model, which provides a good fit for electric circuit components in biological tissue. Significant changes were observed in membrane integrity, where the corresponding capacitance decreases 30% in the early aging (0D-7D) period. Similarly, the losses associated with intracellular and extracellular fluids exhibit changed up to 42% and 53%, respectively. Ultimately, empirical polynomial models are developed to predict the electrical component values for a better understanding of aging effects. The measured and calculated values are found to be in good agreement.

Keywords: dielectric properties; complex permittivity; dielectric spectroscopy; meat quality; microwave sensor; microstrip ring-resonator; biosensor

---

1. Introduction

The electrical characteristics of food materials have been analyzed for many years to determine their physical, chemical and physiological states. Particularly for meat products, the non-invasive and online assessment of these properties enables the industry to evaluate meat quality effectively. Meat quality is usually evaluated based on physical attributes including pH, color, water holding capacity (WHC) and cook loss [1]. The water holding capacity (WHC) is considered the most important attribute in meat quality, since water is the main constituent, holding about 75% of the total body mass. As a matter of fact, the strong correlation between moisture content and dielectric permittivity allows researchers to evaluate the state of meat usinga microwave technique [2]. Several research studies have reported the determination of these attributes through examination of dielectric properties [3–6]. Discrimination of low quality classes of meat, such as pale, soft and exudative (PSE) and dark, firm and dry (DFD) is carried out at 500 MHz to 20 GHz frequency using an open-ended coaxial probe [7]. In the early period of animal slaughter, the muscle glycolysis procedure triggers rigor mortis (stiffness) in the animal carcass, where the pH level approaches the myofibrillar protein isoelectric point (pH near to 5.4) which results in high ionic mobility in muscle fluids [8]. This ultimately increases the ionic conductivity, and apparently changes the dielectric properties significantly. Results obtained from [7] suggest that determination between PSE and DFD grade meat is possible in early postmortem period, and the changes in water holding capacity (WHC) are more viable at two punctual frequencies (500 MHz and 10 GHz). Similarly, structural changes produced during the aging have been identified in the 200 MHz to 20 GHz frequency range [9]. It is observed that during the aging period, the dielectric constant decreases at all frequencies while the loss-factor increases at frequencies below 4 GHz and beyond that it remains constant. This effect on dielectric properties is mainly contributed by the water loss during storage [9]. The build-up of lactic acid during the aging period is known to be a main reason for reduction in water holding capacity (WHC). Therefore, different concentrations of lactic acid solution are evaluated at microwave frequencies ranging from 500 MHz to 20 GHz [10]. Good correlations are found among lactic acid, ATP and IMP content with the loss factor at 500 MHz, 915 MHz and 1 GHz frequencies. The aging period induces structural change, particularly in membrane integrity, insulating properties of cellular membrane, and intracellular fluid (ICF) and extracellular fluid (ECF) composition [11]. These structural changes can also be produced by freezing. The permeability of cell membranes produces a mix of extracellular and intracellular fluids, and eventually loss of moisture (as can be seen in Figure 1). This increases the conductivity of muscle and therefore produces substantial changes in dielectric properties.

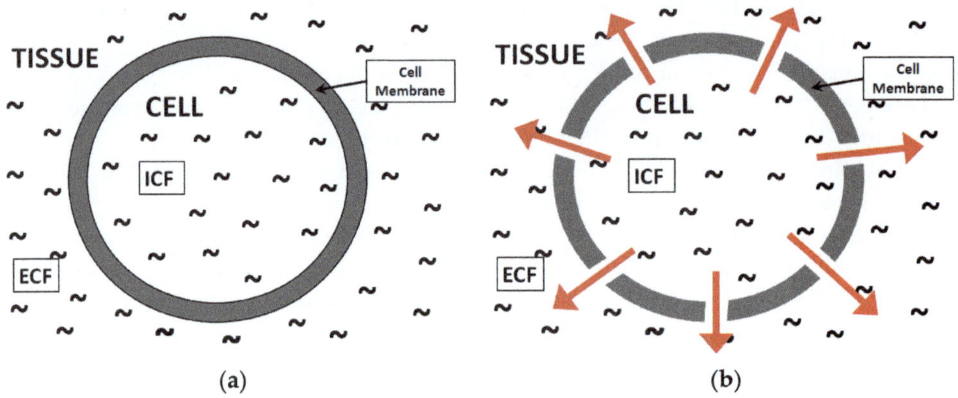

Figure 1. Fluids within a muscle, intracellular (ICF) and extracellular (ECF) compartment fluids. (**a**) Fresh muscle with intact cell membrane; (**b**) muscle with permeable cell membrane.

The ability to measure the dielectric properties of meat, and to correlate them with the amount of moisture content, offers new opportunities to identify moisture lost during the aging period. In order to determine the moisture lost, several microwave methods are being reported. Among them, the coaxial dielectric probe method is widely used to evaluate permittivity measurements. It has high accuracy but is still subject to errors. The main source of error includes radiation effects [12], the air-gap problem and cable movement during measurements [13]. Further, while evaluating the physical properties it would be useful to measure the average permittivity of a material. Considering these aspects, recently, a highly sensitive microwave resonator sensor has been proposed to evaluate the moisture amount [14]. The ring resonator is based on the resonance mechanism, which offers not just high accuracy but also simple, cost-effective and fast measurements [15]. In this method sample loading and unloading is easier, thus enabling the evaluation of large and irregularly shaped samples. The ring-resonator sensor has shown commendable accuracy compared to the well-established coaxial probe technique [16]. In addition to this, the ring resonator technique is also evident in other biological applications, such as for hyperthermia [17] and liquid solvent sensing [18].

In this study a microwave ring-resonator sensor has been used for non-invasive assessment of meat aging. For this purpose, broiler meat is investigated, since it is considered the second highest source of meat after pork, globally. The physical and the electrical measurements are carried out over a four week period, and their results are investigated to assess the moisture lost during the aging period. These results are then used to analyze the impact of moisture lost on the electrical properties of the muscle. Further, the obtained data is evaluated using the Fricke model, which provides a good fit for electric parameters. This will help to identify the relationship between electrical and physical characteristics.

2. Microwave Ring-Resonator Sensor: Design and Theoretical Analysis

2.1. Design and Development

A single-port microstrip ring-resonator with a resonant frequency of 1 GHz is designed on Ansys HFSS software. The Roger RT/Duriod 5880 substrate with its dielectric constant of 2.2 and a height of 787 μm is used for the sensor structure, as reported in [14]. All design parameters of the microstrip ring-resonator sensor are tabulated in Table 1. The sensor (Figure 2) is then fabricated by using the standard lithography process. The frequency responses from simulation and measurement are then compared, over a frequency ranging from 200 MHz to 10 GHz. However, only the fundamental frequency response is considered for further evaluation, due to its greater sensitivity than harmonic frequencies. Simulation and measurement results over the frequency range 800 MHz to 2.3 GHz are presented in Figure 3. The resonator shows periodic resonances over the regular interval of 1 GHz frequency, in the absence of any overlay sample. The simulated and measured responses of the resonance frequency are in good agreement; however, some discrepancy is observed for the return-loss level. This is mainly due to limitations on accuracy, which are introduced during the fabrication process. Furthermore, the surrounding environment (such as temperature or humidity) can also affect losses.

Table 1. Design parameters of a highly sensitive microwave ring-resonator sensor.

Design Parameter	Value
Substrate dielectric constant (ε_r)	2.2
Dissipation factor ($tan\ \delta$)	0.0009
Ring outer-radius (r_a)	34.87 mm
Ring inner-radius (r_b)	34.27 mm
Feed width (w_l)	35 mm
Ring width (w_r)	2.4 mm
Coupling gap (s)	600 μm
Conductor strip (t)	17.5 μm
Substrate height (h)	787 μm

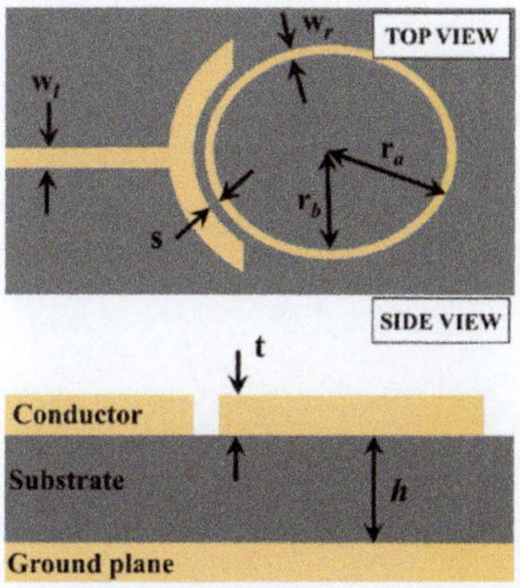

(a)

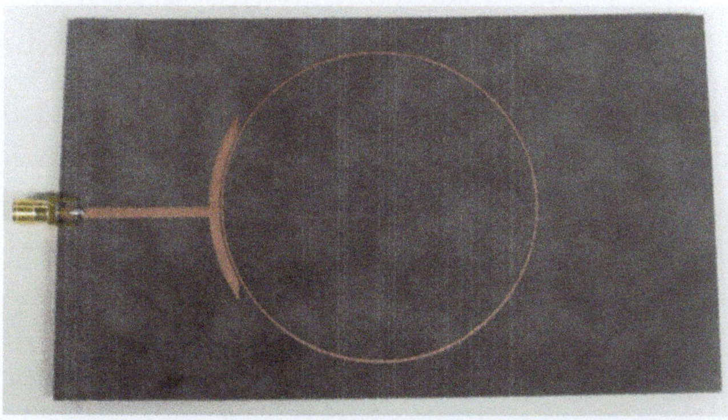

(b)

Figure 2. Highly sensitive microwave ring-resonator sensor. (**a**) Illustration; (**b**) Fabricated prototype.

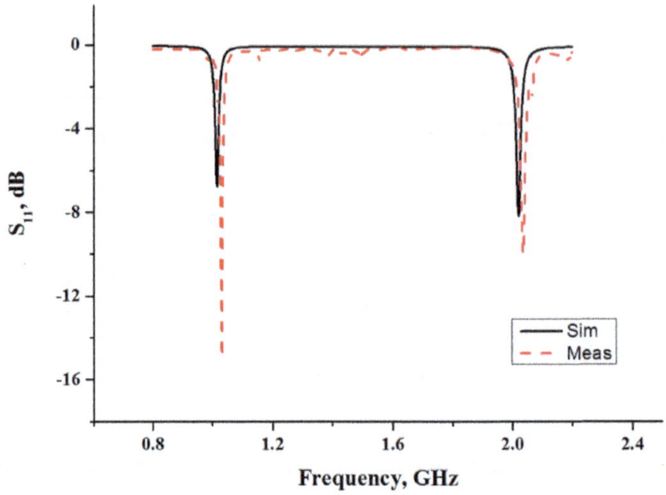

Figure 3. Simulated and measured resonance frequency response of a microwave ring-resonator sensor (without sample).

2.2. Theoretical Analysis

The most straightforward way to analyze the ring structure is an equivalent RLC circuit. As shown in Figure 4a, the ring-resonator can be presented as a simple parallel RLC circuit [19]. The equivalent circuit consists of the ring's capacitance (C_r), inductance (L_r) and resistance (R_r), and also includes the capacitive coupling (C_g) and feed line losses (R_f). To determine these equivalent lumped components the standard transmission line theory can be used, where the ring RLC elements are described as [20].

$$R_r = \frac{1}{\left(\frac{1}{R_{rad}} + \frac{1}{R_{cond}} + \frac{1}{R_{dielec}}\right)} (\Omega) \tag{1}$$

$$C_r = \frac{\pi}{Z_0 \omega_0} \text{ (F)} \tag{2}$$

and

$$L_r = 1/\omega_0^2 C_r \text{ (H)} \tag{3}$$

where R_{rad}, R_{cond} and R_{dielec} are the losses associated with radiation, conductor and dielectric material, as reported elsewhere [21]. Further, Z_0 and ω_0 are the characteristic impedance and angular frequency, respectively. The coupling-gap capacitance (C_g), which coupled the power to the resonator through feed lines, is extensively analyzed [22]. The closed-form expression to cater its effect is defined by

$$C_g = \frac{\left(C_{odd} - \frac{C_{even}}{2}\right)}{2} \tag{4}$$

$$\frac{C_{odd}}{w} = \left(\frac{s}{w}\right)^{m_0} e^{k_0} \text{pF/m}$$
$$\frac{C_{even}}{w} = 12\left(\frac{s}{w}\right)^{m_e} e^{k_e} \text{pF/m} \tag{5}$$

where s is the gap size, w is the strip width, m_e and m_o are the even and odd mode values, and k_e and k_e are the even and odd constants [22]. The losses occurring due to the feed line (R_f) can be obtained by using a computer program to fit its value to simulation data.

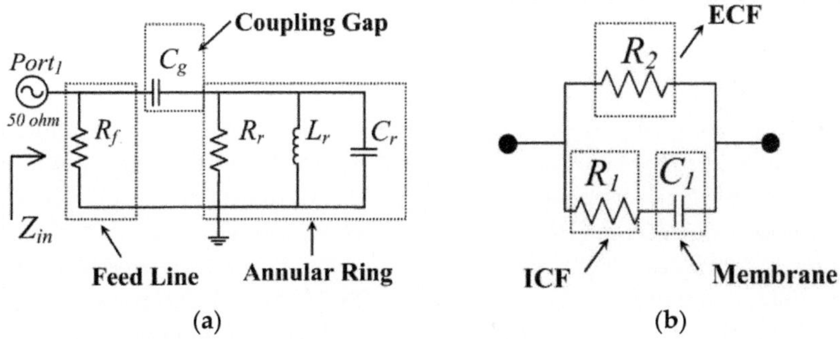

Figure 4. Equivalent RLC circuit of (**a**) Capacitively coupled ring resonator; (**b**) Fricke model representing the biological tissue.

Figure 4b depicts a Fricke model [2], which provides a good fit to electric circuit components in the biological tissue. The tissue contains extracellular fluid (ECF), which behaves as an electrolyte which mainly comprises Na^+ and Cl^- ions [11]. The resistance R_2 represents the losses associated with the ECF. Similarly, the resistance R_1 represents the associated losses produced by another fluid, *i.e.*, intra-cellular fluid (ICF). The cellular membrane, which has insulating properties, behaves likes a capacitor between two electrolytes and is represented as C_1 capacitance.

The dielectric material loaded ring-resonator can be analyzed by considering the parallel connection between the tissue and ring circuits, as reported earlier [19]. Initially, the individual transfer matrix of each component of tissue is determined and their corresponding Y-parameters are obtained:

$$[Y]_{R1} = \begin{bmatrix} \frac{1}{R_2} & -\frac{1}{R_2} \\ -\frac{1}{R_2} & \frac{1}{R_2} \end{bmatrix} \tag{6}$$

$$[Y]_{C1} = \begin{bmatrix} \frac{1}{\frac{-j}{\omega C_1} + R_1} & -\frac{1}{\frac{-j}{\omega C_1} + R_1} \\ -\frac{1}{\frac{-j}{\omega C_1} + R_1} & \frac{1}{\frac{-j}{\omega C_1} + R_1} \end{bmatrix} \tag{7}$$

and

$$[Y]_{Cg} = \begin{bmatrix} \frac{1}{\frac{-j}{\omega C_g}} & -\frac{1}{\frac{-j}{\omega C_g}} \\ -\frac{1}{\frac{-j}{\omega C_g}} & \frac{1}{\frac{-j}{\omega C_g}} \end{bmatrix} \tag{8}$$

Adding $[Y]_{R1} + [Y]_{C1} + [Y]_{Cg}$ will give the overall Y-parameter, which is then transform back into the ABCD parameter [19]. This ABCD matrix can be used to evaluate the input impedance (Z_{in}) of the whole circuit, using:

$$Z_{in} = \frac{AZ_L + B}{CZ_L + D} \tag{9}$$

Eventually, the return loss (S_{11}) of the sample loaded ring-resonator sensor can be obtained by:

$$S_{11} = \frac{Z_{in} - Z_0}{Z_{in} + Z_0} \tag{10}$$

3. Experimental

3.1. Sample Handling and Preparation

Fresh broiler carcasses were collected from the local supplier at 45 min postmortem time and then immediately transported to the laboratory in an ice-filled box. Breast fillets (pectoralis major) from the right half of the carcass were excised and their weight was measured with weighing scale (Ohaus, Valor Precision 3000, New York, NY, USA). The left half of the breast remained intact with the carcass for the physical analysis. All samples were then vacuum packed by using a vacuum sealer (Fresheild, Elite III, Malaysia) and then stored at a chilling temperature at 4 °C (Figure 5a). Subsequent measurements were performed after 0, 1, 2, 7, 14, 21 and 28 days. On each measurement day, samples were taken out of the refrigerator and allowed to equilibrate at room temperature for 3 to 4 h, as reported elsewhere [9]. The cubic samples (30 × 35 × 15 mm, approximately) were obtained by trimming the fillet by using a sharp cutter. Special care was taken to extract samples of similar size, considering the material density influence on measurements [23]; however, due to its malleable nature a minor variation was found in size, which was negligible [24]. Since ageing can affect the anisotropy of muscles [25], dielectric measurements were taken both parallel and perpendicular to the fiber direction.

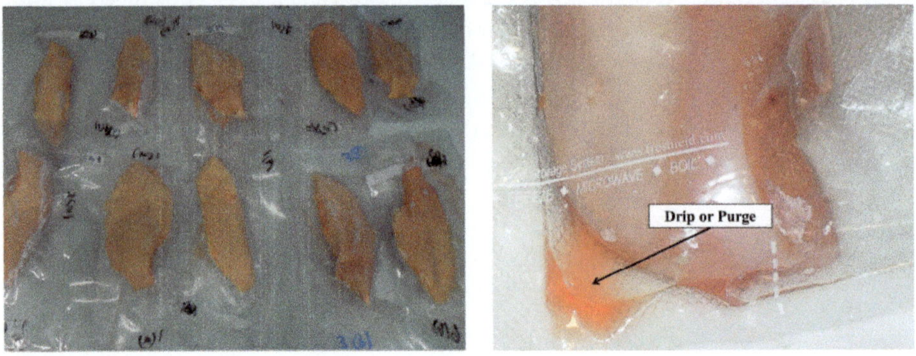

(a) (b)

Figure 5. Studied broiler breast fillets. (**a**) Samples stored in freezer; (**b**) Equilibrated sample with collected drip.

3.2. Physical Measurements—pH, Color and Temperature

The relationship between meat quality, muscle pH, and color is well established [26,27]. Therefore, to identify that relationship, mean values of pH and CIE (Color L*, a*, and b*) were determined. The pH measurements were taken with a portable pH meter (Luton, PH-220S, China) by inserting the electrode approximately 10 mm into a sample. Initially it was measured at 8 h and 24 h postmortem time, and then it was recorded for 2D, 3D, 7D, 14D, 21D and 28D periods. Over the whole aging period, the meter was calibrated every time using standard calibration liquids provided with the product. For the refrigerated samples, once they were equilibrated at room temperature, the pH measurements were carried out before deboning of samples.

Meat color (CIE L*, a*, and b* values) was measured on the dorsal surface (bone side) using a Chroma-meter (Konica Minolta, CR-300, Osaka, Japan) at 1.5 h postmortem time. Fresh broiler fillets were selected for the color measurements which were free from any obvious color defects, such as blood spots or bruises. The Chroma-meter was used to measure the reflectance between 400 and 700 nm, while the CIE values were calculated on a D65 illuminant, as reported elsewhere [24]. Prior to measurements white and black color tiles were used for the calibration. Three measurements were taken for each sample, and their average value was further considered. For the aging and refrigerated samples measurements, moisture at the surface of samples was gently removed using absorbent tissue prior to color analyses. Particularly in the aging study, once the samples were measured the same areas of the samples were marked, such that the same surfaces of the samples could be used for repeated measurements.

Muscle temperature has a significant impact on meat tenderness during aging and accelerates the rate of pH decline [27]. Therefore, temperature of fresh fillets was measured with a probe-thermometer (Testo, Mini Thermometer 560), starting at 1.5 h postmortem time with regular intervals of 30 min until 24 h postmortem time. The temperature probe was inserted into the sample and left until it showed a constant value for temperature. During the study, the temperature of each fillet was measured before deboning the sample for the measurements. This ensured the equilibrium state of sample temperature with room temperature.

3.3. Electrical Measurements

The enhanced coupled high impedance microstrip-ring resonator was connected to an Agilent E8363C vector network analyzer and prior to measurements, it was calibrated with an open circuit, short circuit and load circuit using an Agilent standard calibration-kit. The VNA was configured to provide the S_{11} parameter for a range of 500 MHz to 20 GHz with 1980 frequency points. Initially, measurements were taken with unloaded resonator, once the unloaded resonance frequency was determined. The resonator was loaded with cubic samples (with a size of 30 × 35 × 15 mm, approximately), which were partially overlaid on the coupling region of the ring-resonator. Triplicate measurements were performed for the dielectric measurements of each sample. Samples were placed in a manner such that the orientation of fiber was perpendicular to the direction of electric field. Measurements were conducted at a controlled temperature of 25 ± 2 °C and humidity within the 55 RH% ± 5 range. The experimental design for both physical and electrical measurements is shown in Figure 6.

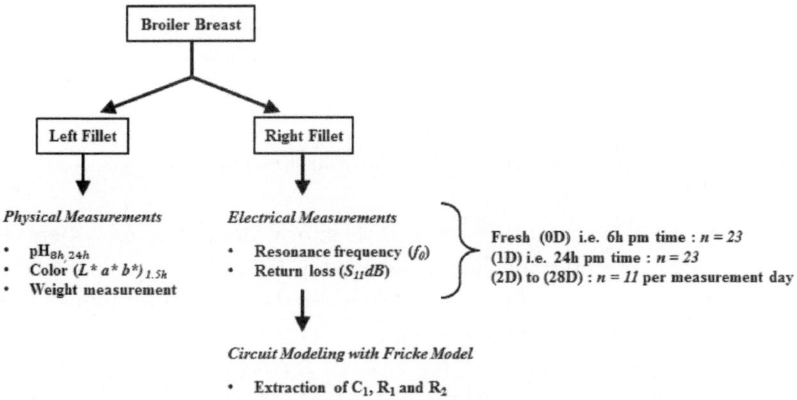

Figure 6. Experimental design for physical and electrical measurements.

4. Results and Discussion

The effect of meat aging was determined over four weeks post-mortem. The physical properties, including color and pH mean values, are presented in Table 2. The measured color (L*, a* and b*) values are close to those observed for the low quality meat, such as pale, soft and exudate (PSE) meat. Similarly, the pH values obtained at 8 h and 24 h postmortem time are related to those of low quality grade meat. The values are close to those published in the literature [28,29]. For the electrical measurements, overall seven observations were taken during the studied period. On each measurement day, the behavior of the resonance frequency as observed by using a microwave resonator sensor. Figure 7 shows the response of the resonance frequency for fresh (6 h after slaughter) to aged (28 days after slaughter) pectoralis major muscles from a broiler chicken (samples with similar physical characteristics are used for this purpose). It can be seen from the plot (Figure 7) that the fresh samples (0D) have the highest frequency shift of all the samples. However, over the storage period, the measured resonance frequency of the other samples is shifted toward the ring's fundamental-frequency. This behavior is expected, since a reduction in the water holding capacity of the samples over the aging period causes an increase in drip-loss, thus reducing the amount of resonance shift.

Table 2. Measured physical properties of fresh broiler meat.

Physical Properties		Measured Value
Color [1]	L*	57.61
	a*	1.77
	b*	5.19
pH [2]	8 h	5.67
	24 h (postmortem time)	5.44

[1] At 1.5 h postmortem time; [2] $n = 23$ per mean with a variance of 0.0109, SD = 0.1044.

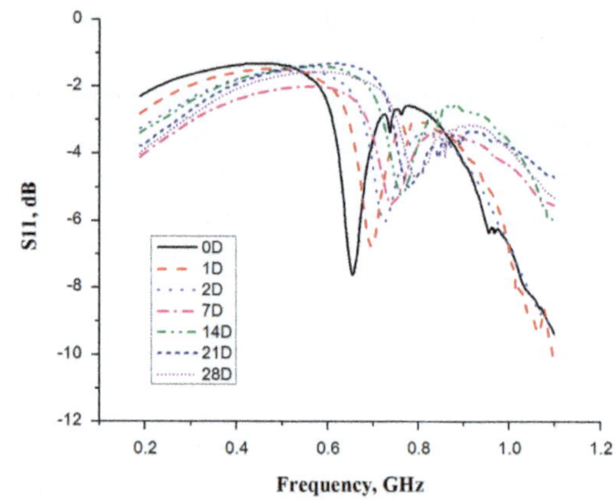

Figure 7. Resonance frequency observed for the meat stored up to four weeks (D = days).

The main reason for this drip-loss is due to structural degradations, where the changes in intracellular and extracellular fluid composition took place [11,30]. The reduction in WHC changes the permittivity of meat samples, where the dielectric constant decreases continuously over the studied period. The change in the dielectric constant is significant for fresh (0D) sample, *i.e.*, 6 h postmortem time, but as the time progresses (0D to 28D) the amount of resonance shift is reduced. In this study, the shift observed with respect to unloaded resonance for the first day (0D *i.e.*, 6 h postmortem time) was 345 MHz, whereas in fourth week of aging (21D) this difference was reduced to 206 MHz. The change, or more particularly, amount of resonance shift was significant for the fresh (0D) sample measurement, while with passage of time, that is until 21D, the amount of resonance shift was reduced. These substantial changes in resonance frequency are mainly caused by the reduction in WHC during the storage period. It was determined that during the storage period the myofibrillar proteins (*i.e.*, actin and myosin) undergo a progressive denaturation, which even involves the bond breaking [31]. Since most of the muscle water is present within the myofibrils, a general hypothesis for explaining drip loss is that it originates from shrinkage of myofibrils after death, causing the water to be expelled into the extracellular space [32]. The shrinkage of fiber expels the water from intracellular spaces to the extracellular spaces, and from where this water ultimately exudates in the form of drip. Since, myosin makes up approximately 50% of myofibrillar proteins, it causes a considerable amount of purge-loss [33]. Additionally, with conventional freezing the formation of large ice-crystals causes expansion and even rupture of the cell membrane [34]. Upon thawing of such muscles, the water is purged from the damaged cellular structure. This also increases the drip-loss in addition to the low pH level.

Other than change in resonance frequency, it is observed that the increase in bandwidth of the notches was also apparent in the obtained results. As described earlier, the loss-factor of the samples increases with increasing storage time, whereas this increase causes noticeable changes in the return-loss and the bandwidth of the resonance. However, as time progresses the resolution of this change in amplitude and bandwidth decreases. The increase of loss factor with storage time is due to the increase in ionic liquid volume. The intracellular fluid (ICF) mixes into extracellular fluids (ECF), increases the amount of ionic liquid, and apparently, increases conductivity. The abundant availability of free ions in these electrolytes provides transportation for electric charges [4]. As described earlier, the main contributing ions to support the current passage are Na^+, K^+ and Cl^-, where the release of these ions is not only pH-dependent but also due to structural degradation during the storage period [35]. In addition Mg^{2+} and Ca^{2+} strongly bind to myosin proteins; however, when the pH reached the isoelectric point, the cancellation of charges releases these two cations.

Anisotropy of the meat sample was also analyzed; the measurements were taken in parallel and perpendicular to the fiber directions at 6 h and 1D postmortem time. The change in resonance frequency due to anisotropic variation is minimal, in both measurement periods. Therefore, further study on meat anisotropy was not conducted. However, due to reater dielectric sensitivity of samples with direction placement perpendicular to fiber [35], in further studies perpendicular orientation was used for the measurements.

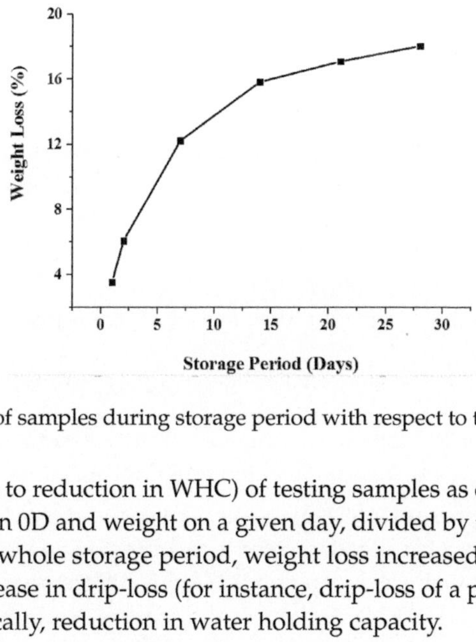

Figure 8. Weight loss of samples during storage period with respect to their fresh (0D) weight.

The loss of weight (due to reduction in WHC) of testing samples as determined by the difference between weight measured on 0D and weight on a given day, divided by the 0D weight. It can be seen from Figure 8 that over the whole storage period, weight loss increased with the storage days. This weight loss is due to an increase in drip-loss (for instance, drip-loss of a particular sample is shown in Figure 5b) and more specifically, reduction in water holding capacity.

Equivalent Circuit Modeling for Storage Effect

Meat equivalent circuit parameters are calculated from fittings of the resonance frequency and return loss plots to the Fricke model. The cellular membrane, which is characterized as capacitance (C_1), deteriorated over the time period. This can be observed in Figure 9, which shows a decrease in capacitance relative to the storage time period [4]. The insulating properties of the cell-membrane decrease with the aging period, due to oxidation of phospholipid membrane layers, which makes it permeable [11]. Moreover, the capacitance of membrane was decreases rapidly, about a 30% change in the early aging (0D–7D) period, which shows the significant structural changes in that period.

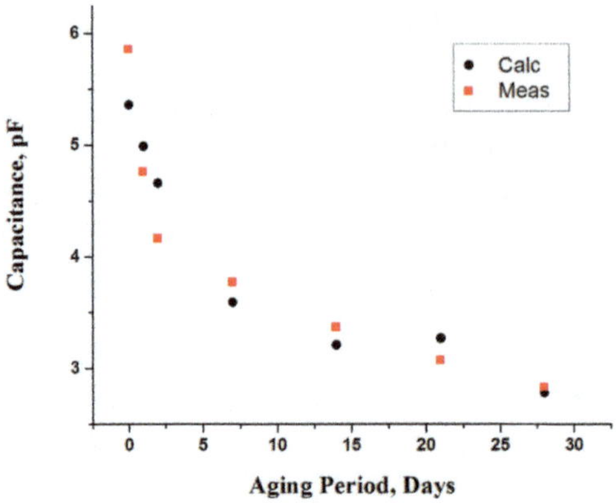

Figure 9. Capacitance (C_1) of cell membrane as a function of aging period.

In contrast, the change in capacitance for the rest of period was not more than 15%, which indicates that cellular membrane became porous [4]. The first observation (0D) was made at 6h postmortem time, and beyond that there were minimal changes in structure once it achieved the ultimate pH level. The obtained data is fitted using a polynomial model, which is given as:

$$C_1 = 5.36377 - 0.40039 \times \text{Aging} + 0.0242434 \times \text{Aging}^2 - 0.000473023 \times \text{Aging}^3 \text{ (pF)} \quad (11)$$

As shown in Figure 9, this expression provides a good approximation of the values found in the measured data, where the overall discrepancy is minimal. Further, as depicted in Figure 10, the resistance of an intercellular fluid (R_1) decreases during the storage period. This behavior can be explained as reducing the water binding ability of muscle over the storage period. The myofibrillar shrinkage (between thick and thin filaments) reduces the space for intracellular water. Since these myofibrils account for the major portion (about 2/3rd) of muscle water, upon their release and flow between the myofibrils, the conductivity increases. Thus, resistivity decreases with fluid movement as time progresses; however, significant change (42%) is observed in first week of the aging period. The measured data is then used to obtain an empirical model, which is given as:

$$R_1 = 17.2538 - 1.38201 \times \text{Aging} + 0.0844434 \times \text{Aging}^2 - 0.00165975 \times \text{Aging}^3 \text{ } (\Omega) \quad (12)$$

The calculated values of R_1, which are obtained from expression Equation (12) are in good agreement with the measured data, where the variation in calculated values is negligible (Figure 10).

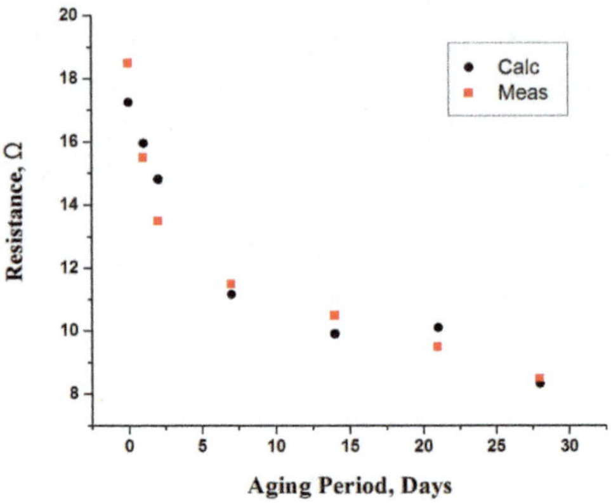

Figure 10. Resistance (R_1) of an intracellular fluid (ICF) as a function of aging period.

Similarly, the behavior of an extracellular fluid is modeled as resistance (R_2), where its value progressively increases (Figure 11) during the storage period. The relation of this resistance seems to be related with the drip-loss (comparing with weight-loss in Figure 8). Since, the amount of drip-loss is rapidly increased in the first week of storage, the resistance steadily increases (about 53%), due to a decrease in conductivity. In this period, the loss of water fluid is at the maximum; however, as time progresses the loss tends to decrease, thus changing the resistivity marginally. The empirical model based on polynomials for the extracellular resistance (R_2) is given by:

$$R_2 = 568.811 + 68.7006 \times \text{Aging} - 4.19321 \times \text{Aging}^2 + 0.0910305 \times \text{Aging}^3\ (\Omega) \quad (13)$$

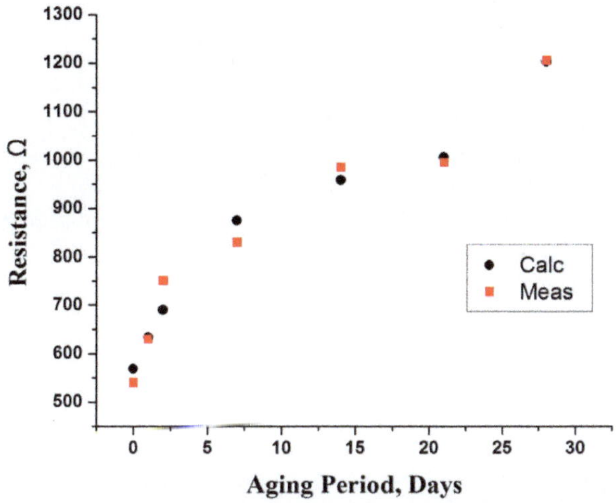

Figure 11. Resistance (R_2) of an extracellular fluid (ECF) as a function of aging period.

As depicted in Figure 11, the calculated values for the extracellular resistance (R_2) using Equation (13) are in good agreement with measured values, where the overall difference is minimal. Finally, it is deduced that the obtained empirical models for the C_1, R_1 and R_2 can be used to define the relationship between aging and electrical parameters for the aging period (0D–28D) with other considered parameters.

5. Conclusions

The effects of aging on broiler meat were studied over a four-week period in which significant changes were observed in resonance frequency and return loss. These changes are even affirmed with weight loss measurements. The anisotropy of muscles was also analyzed, but no significant changes were observed using the proposed resonator. Data from electrical measurements was then evaluated with the Fricke model, which provides a good fit for electric circuit components in biological tissue. Based on the measurement data, empirical models for the circuit parameters were then developed. The measured and calculated values were found to be in good agreement. The results obtained from this research will act as a step towards the development of an effective and reliable microwave sensing device which can be used in meat quality evaluation. Also, the established relationship of equivalent circuit parameters with the physical properties is useful to define electrical characteristics. This will eventually provide an in-depth understanding of dielectric properties associated with the aging period.

Acknowledgments: This study was supported by the Centre of Graduate Studies (CGS) from Universiti Teknologi PETRONAS, Malaysia.

Author Contributions: The work in this paper was completed by Muhammad Taha Jilani under the guidance of Wong Peng Wen and Lee Yen Cheong. Muhammad Zaka ur Rehman assisted in performing the experiments.

Conflicts of Interest: The authors declare no conflict of interest.

References

1. Trabelsi, S. Investigating Effects of Aging on Radiofrequency Dielectric Properties of Chicken Meat. In Proceedings of the IEEE International Instrumentation and Measurement Technology Conference (I2MTC), Minneapolis, MN, USA, 6–9 May 2013; pp. 1685–1688.
2. Damez, J.-L.; Clerjon, S.; Abouelkaram, S.; Lepetit, J. Beef meat electrical impedance spectroscopy and anisotropy sensing for non-invasive early assessment of meat ageing. *J. Food Engin.* **2008**, *85*, 116–122. [CrossRef]
3. You, K.Y.; Mun, H.K.; You, L.L.; Salleh, J.; Abbas, Z. A small and slim coaxial probe for single rice grain moisture sensing. *Sensors* **2013**, *13*, 3652–3663. [CrossRef] [PubMed]
4. Guermazi, M.; Kanoun, O.; Derbel, N. Investigation of long time beef and veal meat behavior by bioimpedance spectroscopy for meat monitoring. *IEEE Sens. J.* **2014**, *14*, 3624–3630. [CrossRef]
5. Damez, J.-L.; Clerjon, S. Quantifying and predicting meat and meat products quality attributes using electromagnetic waves: An overview. *Meat Sci.* **2013**, *95*, 879–896. [CrossRef] [PubMed]
6. Siddiqui, O.; Kashanianfard, M.; Ramahi, O. Dielectric sensors based on electromagnetic energy tunneling. *Sensors* **2015**, *15*, 7844–7856. [CrossRef] [PubMed]
7. Castro-Giráldez, M.; Aristoy, M.; Toldrá, F.; Fito, P. Microwave dielectric spectroscopy for the determination of pork meat quality. *Food Res. Int.* **2010**, *43*, 2369–2377. [CrossRef]
8. Huff-Lonergan, E.; Lonergan, S.M. Mechanisms of water-holding capacity of meat: The role of postmortem biochemical and structural changes. *Meat Sci.* **2005**, *71*, 194–204. [CrossRef] [PubMed]
9. Trabelsi, S.; Roelvink, J. Investigating the influence of aging on radiofrequency dielectric properties of chicken meat. *J. Microw. Power Electromagn. Energy* **2014**, *48*, 215–220.
10. Castro-Giráldez, M.; Dols, L.; Toldrá, F.; Fito, P. Development of a dielectric spectroscopy technique for the determination of key biochemical markers of meat quality. *Food Chem.* **2011**, *127*, 228–233. [CrossRef]
11. Damez, J.-L.; Clerjon, S.; Abouelkaram, S.; Lepetit, J. Dielectric behavior of beef meat in the 1–1500 kHz range: Simulation with the fricke/cole–cole model. *Meat Sci.* **2007**, *77*, 512–519. [CrossRef] [PubMed]
12. Roelvink, J.; Trabelsi, S. Measuring the Complex Permittivity of Poultry Meat with a Planar Transmission-Line Sensor. In Proceedings of the IEEE International Instrumentation and Measurement Technology Conference (I2MTC), Minneapolis, MN, USA, 6–9 May 2013; pp. 1689–1693.
13. Note, Application. In *Basics of Measuring the Dielectric properties of Material*; Agilent Technologies: Santa Clara, CA, USA, 2006.
14. Jilani, M.T.; Wen, W.P.; Zakariya, M.A.; Cheong, L.Y.; Rehman, M.Z.U. An Improved Design of Microwave Biosensor for Measurement of Tissues Moisture. In Proceedings of the IEEE MTT-S International Microwave Workshop Series on RF and Wireless Technologies for Biomedical and Healthcare Applications (IMWS-Bio), London, UK, 8–10 December 2014; pp. 1–3.
15. Bernard, P.A.; Gautray, J.M. Measurement of dielectric constant using a microstrip ring resonator. *IEEE Trans. Microw. Theory Tech.* **1991**, *39*, 592–595. [CrossRef]
16. Jilani, M.T.; Wen, W.P.; Rehman, M.Z.U.; Khan, A.M.; Cheong, L.Y. Microwave sensor for non-destructive dielectric characterization of biological systems. *Int. J. Appl. Electromagn. Mech.* **2016**. in press.
17. Bahl, I.; Stuchly, S.S.; Stuchly, M. A new microstrip radiator for medical applications. *EEE Trans. Microw. Theory Tech.* **1980**, *28*, 1464–1469. [CrossRef]
18. Saeed, K.; Guyette, A.C.; Hunter, I.C.; Pollard, R.D. Microstrip Resonator Technique for Measuring Dielectric Permittivity of Liquid Solvents and for Solution Sensing. In Proceedings of the IEEE/MTT-S International Microwave Symposium, Honolulu, HI, USA, 3–8 June 2007; pp. 1185–1188.
19. Jilani, M.T.; Wen, W.P.; Cheong, L.Y.; Zakariya, M.A.; Rehman, M.Z.U. Equivalent circuit modeling of the dielectric loaded microwave biosensor. *Radioengineering* **2014**, *23*, 1038–1047.
20. Hsieh, L.-H.; Chang, K. Equivalent lumped elements g, l, c, and unloaded q's of closed-and open-loop ring resonators. *IEEE Trans. Microw. Theory Tech.* **2002**, *50*, 453–460. [CrossRef]
21. Hopkins, R.; Free, C. Equivalent circuit for the microstrip ring resonator suitable for broadband materials characterisation. *Microw. Antennas Propag. IET* **2008**, *2*, 66–73. [CrossRef]
22. Garg, R.; Bahl, I. Microstrip discontinuities. *Int. J. Electron. Theor. Exp.* **1978**, *45*, 81–87. [CrossRef]

23. Jilani, M.T.; Wen, W.P.; Cheong, L.Y.; Rehman, M.Z.U.; Khan, M.T. Determination of size-independent effective permittivity of an overlay material using microstrip ring resonator. *Microw. Op. Technol. Lett.* **2016**, *58*, 4–9. [CrossRef]
24. Castro-Giráldez, M.; Botella, P.; Toldrá, F.; Fito, P. Low-frequency dielectric spectrum to determine pork meat quality. *Innov. Food Sci. Emerg. Technol.* **2010**, *11*, 376–386. [CrossRef]
25. Clerjon, S.; Damez, J.-L. Microwave sensing for an objective evaluation of meat ageing. *J. Food Engin.* **2009**, *94*, 379–389. [CrossRef]
26. Samuel, D.; Trabelsi, S. Measurement of dielectric properties of intact and ground broiler breast meat over the frequency range from 500 MHz to 50 GHz. *Int. J. Poult. Sci.* **2012**, *11*, 172–176. [CrossRef]
27. Byrne, C.; Troy, D.; Buckley, D. Postmortem changes in muscle electrical properties of bovine M. longissimus dorsi and their relationship to meat quality attributes and pH fall. *Meat Sci.* **2000**, *54*, 23–34. [CrossRef]
28. Lesiów, T.; Kijowski, J. Impact of PSE and DFD meat on poultry processing—A review. *Pol. J. Food Nutr. Sci.* **2003**, *12*, 3–8.
29. Adzitey, F.; Nurul, H. Pale soft exudative (PSE) and dark firm dry (DFD) meats: Causes and measures to reduce these incidences—A mini review. *Int. Food Res. J.* **2011**, *18*, 11–20.
30. Fennema, O. Comparative water holding properties of various muscle foods. *J. Muscle Foods* **1990**, *1*, 363–381. [CrossRef]
31. Wagner, J.R.; AÑOn, M.C. Effect of frozen storage on protein denaturation in bovine muscle 1. Myofibrillar ATPase activity and differential scanning calorimetric studies. *Int. J. Food Sci. Technol.* **1986**, *21*, 9–18. [CrossRef]
32. Den Hertog-Meischke, M.J.A.; van Laack, R.J.L.M.; Smulders, F.J.M. The water-holding capacity of fresh meat. *Vet. Q.* **1997**, *19*, 175–181. [CrossRef] [PubMed]
33. Huff-Lonergan, E. *Chemistry and Biochemistry of Meat*; Wiley-Blackwell: Oxford, UK, 2010.
34. Huff-Lonergan, E. Water-holding capacity of fresh meat. *Fact. Sheet* **2002**, *4669*.
35. Damez, J.-L.; Clerjon, S.; Abouelkaram, S.; Lepetit, J. Electrical impedance probing of the muscle food anisotropy for meat ageing control. *Food Control* **2008**, *19*, 931–939. [CrossRef]

14

Characterization of *N*-Acylhomoserine Lactones Produced by Bacteria Isolated from Industrial Cooling Water Systems

Noriya Okutsu [1], Tomohiro Morohoshi [1,*], Xiaonan Xie [2], Norihiro Kato [1] and Tsukasa Ikeda [1]

Academic Editor: Alexander Star

[1] Department of Material and Environmental Chemistry, Graduate School of Engineering, Utsunomiya University, 7-1-2 Yoto, Utsunomiya, Tochigi 321-8585, Japan; noriya.okutsu@kurita.co.jp (N.O.); katon@cc.utsunomiya-u.ac.jp (N.K.); tikeda@cc.utsunomiya-u.ac.jp (T.I.)

[2] Center for Bioscience Research and Education, Utsunomiya University, 350 Mine-machi, Utsunomiya, Tochigi 321-8505, Japan; xie@cc.utsunomiya-u.ac.jp

* Correspondence: morohosi@cc.utsunomiya-u.ac.jp

Abstract: The cooling water systems are used to remove heat generated in the various industries. Biofouling of the cooling water systems causes blocking of condenser pipes and the heat exchanger tubes. In many Gram-negative bacteria, *N*-acylhomoserine lactone (AHL) are used as quorum-sensing signal molecule and associated with biofilm formation. To investigate the relationship between quorum sensing and biofouling in the cooling water system, we isolated a total of 192 bacterial strains from the five cooling water systems, and screened for AHL production. Seven isolates stimulated AHL-mediated purple pigment production in AHL reporter strain *Chromobacterium violaceum* CV026 or VIR07. Based on their 16S rRNA gene sequences, AHL-producing isolates were assigned to *Aeromonas hydrophila*, *Lysobacter* sp., *Methylobacterium oryzae*, and *Bosea massiliensis*. To the best of our knowledge, *B. massiliensis* and *Lysobacter* sp. have not been reported as AHL-producing species in the previous researches. AHLs extracted from the culture supernatants of *B. massiliensis* and *Lysobacter* sp. were identified by liquid chromatography-mass spectrometry. AHLs produced by *B. massiliensis* were assigned as *N*-hexanoyl-L-homoserine lactone (C6-HSL), *N*-(3-oxohexanoyl)-L-homoserine lactone (3-oxo-C6-HSL), and *N*-(3-oxooctanoyl)-L-homoserine lactone (3-oxo-C8-HSL). AHLs produced by *Lysobacter* sp. were assigned as *N*-decanoyl-L-homoserine lactone (C10-HSL) and *N*-(3-oxodecanoyl)-L-homoserine lactone (3-oxo-C10-HSL). This is the first report of identification of AHLs produced by *B. massiliensis* and *Lysobacter* sp. isolated from the cooling water system.

Keywords: quorum sensing; *N*-acylhomoserine lactone; industrial cooling water system; *Bosea massiliensis*; *Lysobacter* sp.

---

1. Introduction

Quorum sensing is one of the cell-cell communication systems depending on cell population density [1]. Quorum-sensing signaling is mediated by an autoinducer, which is a self-produced signal molecule [1]. In many Gram-negative bacteria, several kinds of *N*-acylhomoserine lactone (AHL) have been identified as autoinducer signal compounds involved in this mechanism [2]. The substrates in AHL synthesis are *S*-adenosyl methionine (SAM) and acyl-acyl carrier proteins. AHLs are synthesized bacteria by LuxI family protein, and diffused outside and inside the cell. Some AHLs requires active transport through an efflux pump [3]. When intracellular concentration of AHLs increases, LuxR family protein binds AHL and regulates the expression of specific genes [2]. AHL-mediated quorum sensing regulates the various phenotypes such as bioluminescence, motility, pigment production, and biofilm formation [1,2]. In general, AHL-deficient mutants show defects in quorum-sensing regulated gene expression. Therefore, it is expected that disrupting or manipulating the AHL signals could inhibit the expression of quorum-sensing regulated phenotypes such as virulence and infection of host cells [4].

The cooling water systems are used to remove heat generated in the various industries such as chemical and petrochemical, electric power generating stations, and paper mills [5]. Biofouling, which is accumulation of microorganisms on the wet surfaces, is one of the serious problems in the cooling water systems. Biofouling of the cooling water systems causes blocking of condenser pipes and the heat exchanger tubes [5]. Chlorine is one of the most used chemicals for controlling biofouling in the industrial cooling water systems. However, to inhibit biofouling with effect, it is necessary for continuous or intermittent chlorination in the industrial cooling water systems. In the case of the membrane bioreactor (MBR) wastewater treatment systems, quorum sensing is associated with the formation of a biofouling layer on the membrane surface [6]. Quorum-quenching technique has been investigated as an approach to controlling biofouling in MBR wastewater treatment systems [6,7]. However, there is no research example that describes the relationship between biofouling of the cooling water systems and quorum sensing. In a previous study, the AHL synthase gene homolog were found within 26% of the 265 completed Proteobacterial genomes [8]. It was revealed that the majority of the bacterial community in an industrial cooling water system was Proteobacteria [5]. From these facts, there is a high possibility that AHL-mediated quorum sensing affects the development of biofouling in the cooling water system. To our knowledge, the AHL-producing bacterial community in the industrial cooling water system has not been examined. In this study, we screened for AHL-producing bacteria from the industrial cooling water systems, and identified AHL structures produced by them.

2. Experimental Section

2.1. Isolation of Bacteria from Industrial Cooling Water Systems

The cooling water samples were collected from the industrial cooling water systems of factory A, B, C, D, and E. The water samples were serially diluted in sterilized distilled water and spread on 1/5-diluted trypticase soy broth (TSB) agar plate (Becton, Dickinson and Company). The plates were incubated in the dark at 30 °C for 72 h. Thereafter, bacterial colonies were randomly selected from among those that grew on the 1/5-diluted TSB plate. Colonies from each sample were randomly selected and transferred onto a new 1/5-diluted TSB agar medium.

2.2. Screening of AHL-Producing Bacteria

AHLs were detected by two AHL biosensors, *Chromobacterium violaceum* CV026 and VIR07. Strain CV026 and VIR07 respond to short-chain and long-chain AHLs respectively, by producing the purple pigment violacein [9,10]. The bacterial isolates were inoculated onto a 1/5-diluted TSB agar medium prepared in the 96-well plates. Strain CV026 and VIR07 were inoculated onto the lower right or left positions of the same well. The bacterial colonies were incubated at 30 °C for 24 h. AHL-producing colonies, which induced the production of purple pigments of the AHL reporter strains, were selected.

2.3. Identification of AHL-Producing Bacteria

Chromosomal DNA of AHL-producing bacteria was extracted using DNeasy Blood and Tissue Kit (QIAGEN). To identify the bacterial species of AHL-producing bacteria, the 16S rRNA gene-encoding regions were amplified by PCR with Blend Taq plus DNA polymerase (Toyobo, Osaka, Japan) and the previously described universal primers, 27f (5′-AGA GTT TGA TCM TGG CTC AG-3′) and 1525r (5′-AGG AGG TGW TCC ARC C-3′) [11]. After electrophoretic separation, 16S rRNA fragments were purified from agarose by using MagExtractor PCR and Gel Clean up (Toyobo). Sequencing reactions were performed with the BigDye Terminator, ver. 3.1 and Applied Biosystems 3500 Series Genetic Analyzer (Applied Biosystems). Closest type-strain 16S rRNA gene relatives to each clone sequence were determined using the RDP II sequence match tool [12]. A phylogenetic tree by the neighbor-joining method was constructed by Molecular Evolutionary Genetic Analysis (MEGA) [13].

2.4. Extraction of AHLs

AHL-producing bacterial strains were inoculated into 4 mL of 1/5-diluted TSB medium and incubated at 30 °C for 18 h with shaking. Two milliliter of full-grown cultures of bacterial strains were transferred into 200 mL of fresh 1/5-diluted TSB medium and incubated at 30 °C for 24 h with shaking. After cultivation, cells were removed by centrifugation at 10,000 × g for 5 min. The culture supernatant was concentrated by evaporation using a rotary evaporator. The concentrated supernatant was extracted with three times its volume of ethyl acetate in a separatory funnel. The extract was evaporated to dryness using a rotary evaporator and was then dissolved in 500 µL of dimethylsulfoxide. To check the presence of AHL, full-grown culture of CV026 or VIR07 was mixed with LB agar medium. Eight-millimeter paper disks, which were applied AHL samples, were placed on LB agar medium containing CV026 or VIR07. After overnight incubation at 30 °C, the appearance of purple pigment was determined as the presence of AHLs.

2.5. AHL Identification by Liquid Chromatography-Mass Spectrometry (LC-MS)

Identification of AHL were performed using a triple quadrupole/linear ion trap instrument (LIT) (QTRAP5500; AB Sciex) with an electrospray ionization (ESI) source and coupled to an UHPLC system (Nexera X2; Shimadzu). Chromatographic separation was achieved on a C18 column (Kinetex F5, φ 2.1 × 150 mm, 2.6 µm; Phenomenex). For the analysis of AHL, the elution of the samples was carried out using acetonitrile (solvent A) and water (solvent B), both of which contained 0.1% (vol/vol) acetic acid the following gradient (vol/vol): initially 10% A, rising to 50% A at 1.5 min, followed by a 7 min gradient to 99% A. Finally, the column was equilibrated for 3 min, using this solvent composition. The column was operated at 30 °C with a flow rate of 0.2 mL/min. MS/MS spectra were recorded in product ion scan mode using LIT. Ion source was maintained at 400 °C with curtain gas at 20, collisional activated dissociation (CAD) gas at 7 psi (12 psi for LIT), ion source gas at 80 psi, and ion source gas2 at 70 psi. Ionspray voltage was set at 5500 V in positive ion mode. Declustering, entrance, and collision cell exit potentials were maintained at 15 V. Injection volume (2 µL) and the settings used was as described previously. Precursor ion scanning experiments were performed in positive ion mode for analysis, where Q1 was set to scan a mass range of *m/z* 80 to 500 Da and Q3 was monitor for the product ion that indicates presence of lactone ring at *m/z* 102. AHLs were analyzed by LC-MS/MS with MS/MS and precursor ion scanning experiments by identification from synthetic AHLs standards. AHL standards, *N*-hexanoyl-L-homoserine lactone (C6-HSL), *N*-(3-oxohexanoyl)-L-homoserine lactone (3-oxo-C6-HSL), *N*-(3-oxooctanoyl)-L-homoserine lactone (3-oxo-C8-HSL), *N*-decanoyl-L-homoserine lactone (C10-HSL), and *N*-(3-oxodecanoyl)-L-homoserine lactone (3-oxo-C10-HSL), were synthesized using a previously described method [14].

2.6. Nucleotide Sequence Accession Numbers

The nucleotide sequences of 16S rRNA were deposited in the DDBJ/ENA/GenBank database with the following accession numbers, LC102389-LC102396.

3. Results and Discussion

3.1. Isolation and Identification of AHL-Producing Bacteria

A total of 192 bacterial strains were isolated from 5 different factories and examined for the further study. As the results of screening of AHL-producing bacteria using *C. violaceum* CV026 and VIR07, 7 bacterial strains showed AHL-producing activity. 16S rRNA gene were amplified by PCR and sequenced for identification of bacterial species (Table 1). AHL-producing strain D10, which was isolated from factory B, was assigned to the *Aeromonas hydrophila*. Two AHL-producing strains F6 and

F13, which were isolated from factory C, were assigned to the *A. hydrophila* and *Lysobacter brunescens*, respectively. Four AHL-producing isolates, which were isolated from factory D, were assigned to the *Methylobacterium oryzae* (strain D7 and F9) and *Bosea massiliensis* (strain E9 and F12).

Table 1. Identification and characterization of *N*-acylhomoserine lactone (AHL)-producing strains on 96-well plates. Induction of AHL-responsible violacein production by CV026 or VIR07 is represented by the number of plus signs; +, low induction; ++, high induction; -, no induction.

Strain	Source	Related Type Strain	Identity	CV026	VIR07
D10	Factory B	*Aeromonas hydrophila*	100%	++	+
F6	Factory C	*Aeromonas hydrophila*	100%	++	+
F13	Factory C	*Lysobacter brunescens*	96.5%	-	++
D7	Factory E	*Methylobacterium oryzae*	99.5%	+	+
E9	Factory E	*Bosea massiliensis*	100%	++	+
F9	Factory E	*Methylobacterium oryzae*	99.5%	-	+
F12	Factory E	*Bosea massiliensis*	100%	++	+

3.2. Phylogenetic Analysis of AHL-Producing Isolates

Phylogenetic analysis was carried out to examine the relationships of AHL-producing isolates (Figure 1). *A. hydrophila*, which is close to strain D10 and F6, is found in freshwater environments and well-known AHL-producing bacteria [15]. It has been reported that AHL-producing *Aeromonas* strains were isolated from activated sludge [11,15]. It has been reported that *Aeromonas* strains mainly produced *N*-butyryl-L-homoserine lactone (C4-HSL) and *N*-hexanoyl-L-homoserine lactone (C6-HSL) [11]. From the results of AHL-production assay, it was assumed that the *A. hydrophila* D10 and F6 can produce the short chain AHLs such as C4-HSL and C6-HSL as well as reported AHL-producing *Aeromonas* species. *Methylobacterium* species were often isolated from the tap water and well-known AHL-producing bacteria [16,17]. *Methylobacterium* have been reported to exhibit higher stress-tolerance than the other species found in biofilms, particularly under conditions of extreme nutrient limitation [18]. On the other hand, *Lysobacter* sp. and *Bosea massiliensis* have not been reported as AHL-producing bacteria. This is the first report of an AHL-producing bacterial strain belonging to the genus *Lysobacter* and *Bosea*.

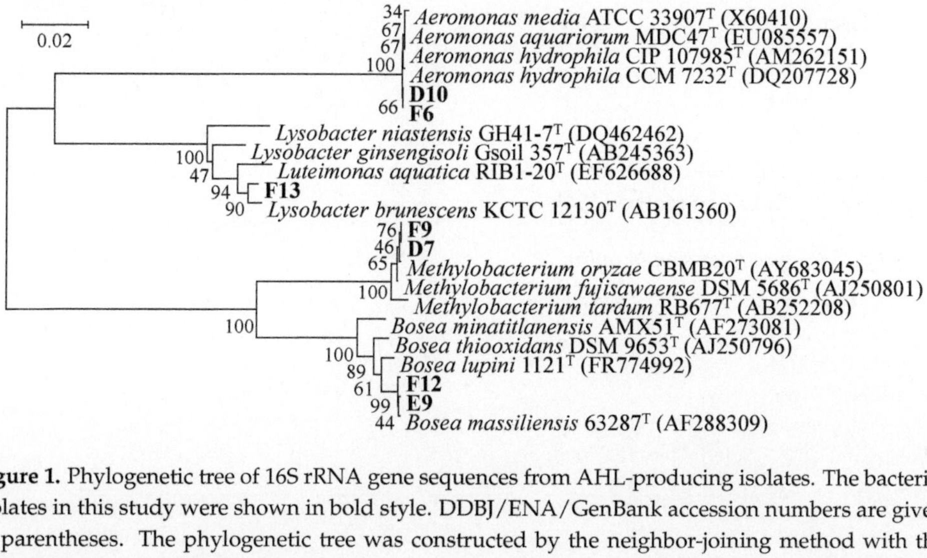

Figure 1. Phylogenetic tree of 16S rRNA gene sequences from AHL-producing isolates. The bacterial isolates in this study were shown in bold style. DDBJ/ENA/GenBank accession numbers are given in parentheses. The phylogenetic tree was constructed by the neighbor-joining method with the ClustalW program of MEGA. The percentage of replicate trees in which the associated taxa clustered together in the bootstrap test (1000 replicates) are shown next to the branches. The scale bar represents 0.02 substitutions per nucleotide position.

3.3. Analytical Identification of AHL Molecules

In this study, we identified the AHL profiles of *B. massiliensis* E9 and F12, and *Lysobacter* sp. F13, which have not been reported as AHL-producing species. To identify the structure of AHL produced by these strains, AHL samples were extracted from the culture supernatants of these strains. The presence of AHLs was checked by LB agar plates containing *C. violaceum* CV026 and VIR07. As the results, AHL extracts from *B. massiliensis* and *Lysobacter* sp. strongly stimulated *C. violaceum* CV026 and VIR07, respectively (Figure 2). These results corresponded to those of 96-well plate assay in Table 1. HPLC fractionation followed by MS analysis of the AHL extracts of these strains were performed to identify the types of AHLs. The presence of C6-HSL, 3-oxo-C6-HSL, and 3-oxo-C8-HSL was identified in the AHL extracts of *B. massiliensis* E9 (Figures 3–5). The profile of the AHL extracts of strain F12 completely corresponded to that of strain E9. The presence of C10-HSL and 3-oxo-C10-HSL was identified in the AHL extracts of *Lysobacter* sp. F13 (Figures 6 and 7). Two short-chain AHLs, C6-HSL and 3-oxo-C6-HSL, can stimulate the production of violacein by *C. violaceum* CV026 [9]. Three long chain AHLs, 3-oxo-C8-HSL, C10-HSL, and 3-oxo-C10-HSL, can stimulate the production of violacein by *C. violaceum* VIR07 [10]. AHL profiles of these three strains support the results of the AHL-production assay shown in Table 1 and Figure 2.

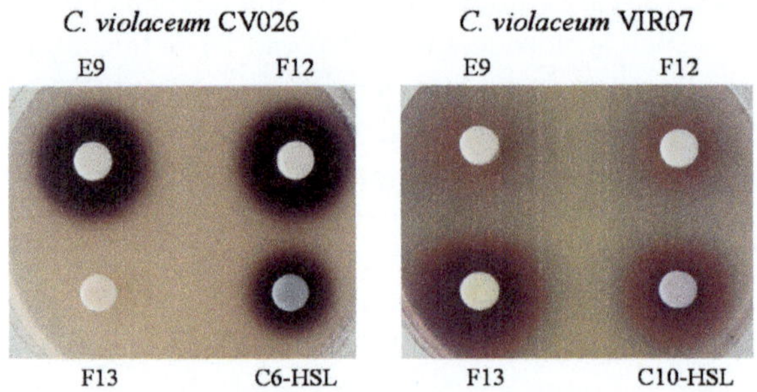

Figure 2. Identification of AHLs produced by *B. massiliensis* E9 and F12, and *Lysobacter* sp. F13. The presence of AHLs was visualized as pigments produced by *C. violaceum* CV026 and VIR07. *N*-hexanoyl-L-homoserine lactone (C6-HSL) and C10-HSL were used for positive control.

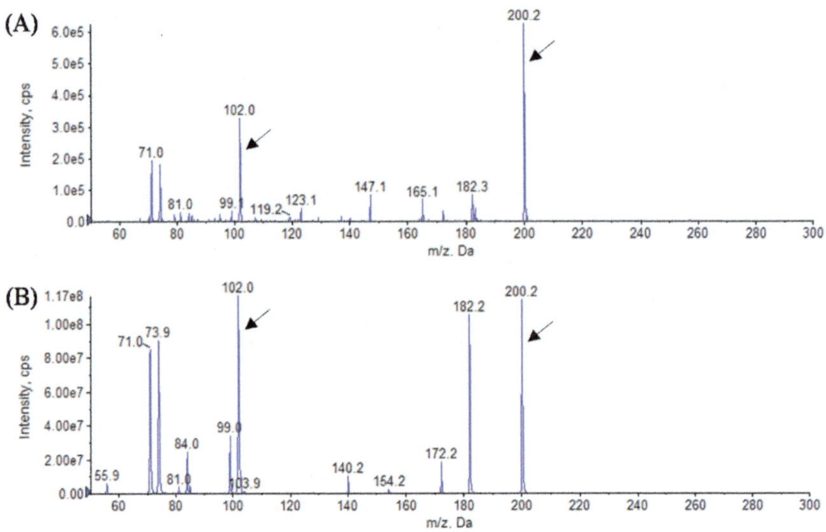

Figure 3. Mass spectra of AHLs extracted from the cell-free supernatant of *B. massiliensis* E9 (**A**) and C6-HSL standard (**B**). All corresponding peaks for respective C6-HSL (m/z 200) along with the product ion peaks (m/z 102) are marked by arrows.

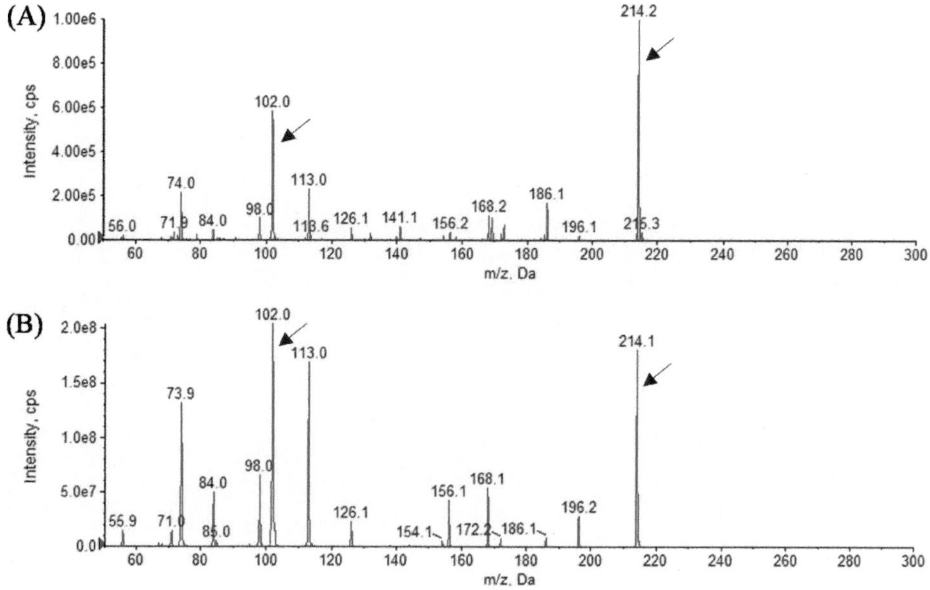

Figure 4. Mass spectra of AHLs extracted from the cell-free supernatant of *B. massiliensis* E9 (**A**) and 3-oxo-C6-HSL standard (**B**). All corresponding peaks for respective 3-oxo-C6-HSL (*m/z* 214) along with the product ion peaks (*m/z* 102) are marked by arrows.

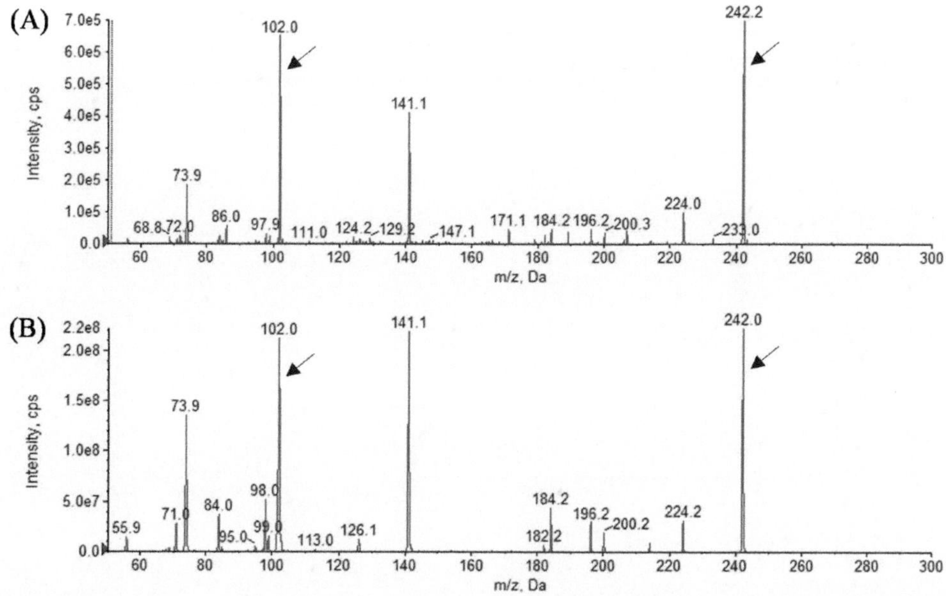

Figure 5. Mass spectra of AHLs extracted from the cell-free supernatant of *B. massiliensis* E9 (**A**) and 3-oxo-C8-HSL standard (**B**). All corresponding peaks for respective 3-oxo-C8-HSL (*m/z* 242) along with the product ion peaks (*m/z* 102) are marked by arrows.

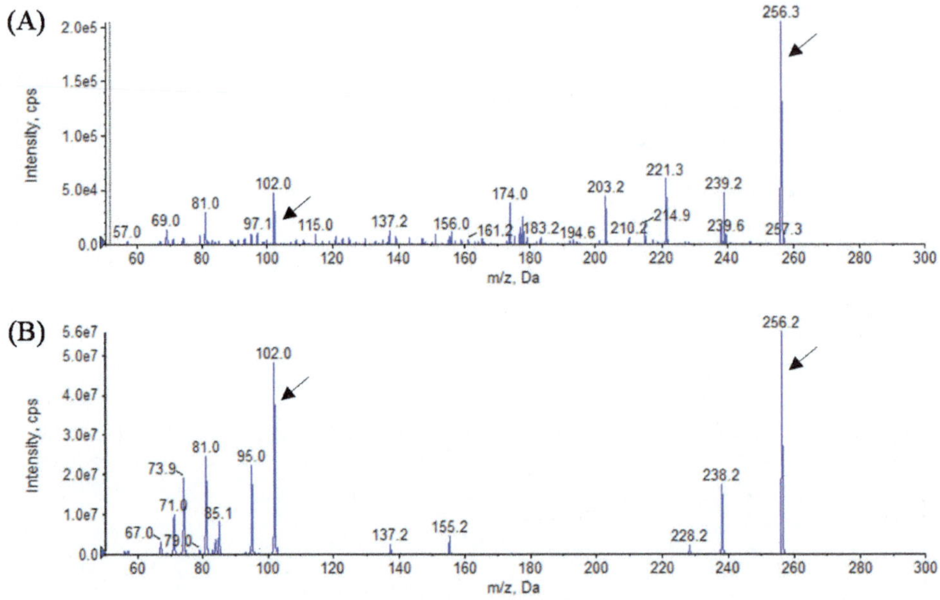

Figure 6. Mass spectra of AHLs extracted from the cell-free supernatant of *Lysobacter* sp. F13 (**A**) and C10-HSL standard (**B**). All corresponding peaks for respective C10-HSL (*m/z* 256) along with the product ion peaks (*m/z* 102) are marked by arrows.

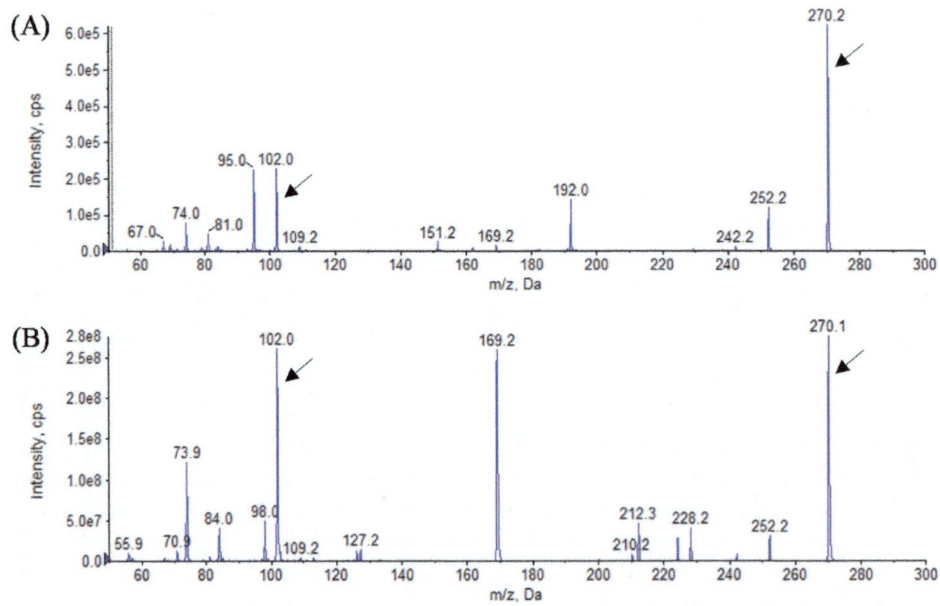

Figure 7. Mass spectra of AHLs extracted from the cell-free supernatant of *Lysobacter* sp. F13 (**A**) and 3-oxo-C10-HSL standard (**B**). All corresponding peaks for respective 3-oxo-C10-HSL (*m/z* 270) along with the product ion peaks (*m/z* 102) are marked by arrows.

4. Conclusions/Outlook

In this study, it was found that there were various AHL-producing bacteria in the industrial cooling water system. *Bosea* have been reported as opine degraders in rhizosphere and an amoeba-resisting bacteria [19]. Interestingly, *Bosea thiooxidans* isolated from the tobacco rhizosphere exhibited the ability to inactivate AHL [20]. *Lysobacter* have been reported as ubiquitous beneficial bacteria emerging as novel biocontrol agents and a new potential source of anti-infectives [21]. Another quorum-sensing system, which was mediated by the diffusible signal factor (DSF), has been identified from *Lysobacter enzymogenes* [21]. However, there is no example of AHL-producing bacteria belonging to genus *Bosea* and *Lysobacter*. Many gram-negative bacteria use AHL-mediated quorum sensing to regulate the secretion of sticky extracellular polysaccharide (EPS), which is cause for biofouling [22]. AHL-degrading techniques have been investigated as an approach to controlling biofouling in membrane bioreactor (MBR) wastewater treatment systems [7,23]. Future studies of the relationship between EPS formation and AHL-mediated quorum sensing in cooling water isolates may provide useful information for the control of biofouling.

Acknowledgments: This work was supported by Core Research for Evolutional Science and Technology (CREST) from the Japan Science and Technology Agency (JST).

Author Contributions: Noriya Okutsu, Tomohiro Morohoshi, Norihiro Kato and Tsukasa Ikeda conceived and designed the experiments; Noriya Okutsu, Tomohiro Morohoshi and Xiaonan Xie performed the experiments, analyzed the data, and wrote the paper.

Conflicts of Interest: The authors declare no conflict of interest.

References

1. Waters, C.M.; Bassler, B.L. Quorum sensing: Cell-to-cell communication in bacteria. *Annu. Rev. Cell. Dev. Biol.* **2005**, *21*, 319–346. [CrossRef] [PubMed]
2. Parsek, M.R.; Greenberg, E.P. Acyl-homoserine lactone quorum sensing in gram-negative bacteria: A signaling mechanism involved in associations with higher organisms. *Proc. Natl. Acad. Sci. USA* **2000**, *97*, 8789–8793. [CrossRef] [PubMed]
3. Schuster, M.; Sexton, D.J.; Diggle, S.P.; Greenberg, E.P. Acyl-homoserine lactone quorum sensing: From evolution to application. *Annu. Rev. Microbiol.* **2013**, *67*, 43–63. [CrossRef] [PubMed]
4. Uroz, S.; Dessaux, Y.; Oger, P. Quorum sensing and quorum quenching: the yin and yang of bacterial communication. *ChemBioChem* **2009**, *10*, 205–216. [CrossRef] [PubMed]
5. Wang, J.; Liu, M.; Xiao, H.; Wu, W.; Xie, M.; Sun, M.; Zhu, C.; Li, P. Bacterial community structure in cooling water and biofilm in an industrial recirculating cooling water system. *Water Sci Technol.* **2013**, *68*, 940–947. [CrossRef] [PubMed]
6. Lade, H.; Paul, D.; Kweon, J.H. Quorum quenching mediated approaches for control of membrane biofouling. *Int. J. Biol. Sci.* **2014**, *10*, 547–562. [CrossRef] [PubMed]
7. Yeon, K.M.; Cheong, W.S.; Oh, H.S.; Lee, W.N.; Hwang, B.K.; Lee, C.H.; Beyenal, H.; Lewandowski, Z. Quorum sensing: A new biofouling control paradigm in a membrane bioreactor for advanced wastewater treatment. *Environ. Sci. Technol.* **2009**, *43*, 380–385. [CrossRef] [PubMed]
8. Case, R.J.; Labbate, M.; Kjelleberg, S. AHL-driven quorum-sensing circuits: their frequency and function among the Proteobacteria. *ISME J.* **2008**, *2*, 345–349. [CrossRef] [PubMed]
9. McClean, K.H.; Winson, M.K.; Fish, L.; Taylor, A.; Chhabra, S.R.; Cámara, M.; Daykin, M.; Lamb, J.H.; Swift, S.; Brcroft, B.W.; *et al.* Quorum sensing and *Chromobacterium violaceum*: Exploitation of violacein production and inhibition for the detection of *N*-acylhomoserine lactones. *Microbiology* **1997**, *143*, 3703–3711. [CrossRef] [PubMed]
10. Morohoshi, T.; Kato, M.; Fukamachi, K.; Kato, N.; Ikeda, T. *N*-acylhomoserine lactone regulates violacein production in *Chromobacterium violaceum* type strain ATCC 12472. *FEMS Microbiol. Lett.* **2008**, *279*, 124–130. [CrossRef] [PubMed]
11. Ochiai, S.; Morohoshi, T.; Kurabeishi, A.; Shinozaki, M.; Fujita, H.; Sawada, I.; Ikeda, T. Production and degradation of *N*-acylhomoserine lactone quorum sensing signal molecules in bacteria isolated from activated sludge. *Biosci. Biotechnol. Biochem.* **2013**, *77*, 2436–2440. [CrossRef] [PubMed]
12. Cole, J.R.; Chai, B.; Farris, R.J.; Wang, Q.; Kulam, S.A.; McGarrell, D.M.; Garrity, G.M.; Tiedje, J.M. The Ribosomal Database Project (RDP-II): Sequences and tools for high-throughput rRNA analysis. *Nucleic Acids Res.* **2005**, *33*, D294–D296. [CrossRef] [PubMed]

13. Tamura, K.; Stecher, G.; Peterson, D.; Filipski, A.; Kumar, S. MEGA6: Molecular Evolutionary Genetics Analysis version 6.0. *Mol. Biol. Evol.* **2013**, *30*, 2725–2729. [CrossRef] [PubMed]
14. Chhabra, S.R.; Harty, C.; Hooi, D.S.; Daykin, M.; Williams, P.; Telford, G.; Pritchard, D.I.; Bycroft, B.W. Synthetic analogues of the bacterial signal (quorum sensing) molecule *N*-(3-oxododecanoyl)-L-homoserine lactone as immune modulators. *J. Med. Chem.* **2003**, *46*, 97–104. [CrossRef] [PubMed]
15. Valle, A.; Bailey, M.J.; Whiteley, A.S.; Manefield, M. *N*-acyl-L-homoserine lactones (AHLs) affect microbial community composition and function in activated sludge. *Environ. Microbiol.* **2004**, *6*, 424–433. [CrossRef] [PubMed]
16. Furuhata, K.; Kato, Y.; Goto, K.; Hara, M.; Yoshida, S.; Fukuyama, M. Isolation and identification of methylobacterium species from the tap water in hospitals in Japan and their antibiotic susceptibility. *Microbiol. Immunol.* **2006**, *50*, 11–17. [CrossRef] [PubMed]
17. Poonguzhali, S.; Madhaiyan, M.; Sa, T. Production of acyl-homoserine lactone quorum-sensing signals is widespread in gram-negative *Methylobacterium*. *J. Microbiol. Biotechnol.* **2007**, *17*, 226–233. [PubMed]
18. Yano, T.; Kubota, H.; Hanai, J.; Hitomi, J.; Tokuda, H. Stress tolerance of *Methylobacterium* biofilms in bathrooms. *Microbes Environ.* **2013**, *28*, 87–95. [CrossRef] [PubMed]
19. Oger, P.M.; Mansouri, H.; Nesme, X.; Dessaux, Y. Engineering root exudation of Lotus toward the production of two novel carbon compounds leads to the selection of distinct microbial populations in the rhizosphere. *Microb. Ecol.* **2014**, *47*, 96–103. [CrossRef] [PubMed]
20. D'Angelo-Picard, C.; Faure, D.; Penot, I.; Dessaux, Y. Diversity of *N*-acyl homoserine lactone-producing and -degrading bacteria in soil and tobacco rhizosphere. *Environ. Microbiol.* **2005**, *7*, 1796–1808. [CrossRef] [PubMed]
21. Qian, G.; Wang, Y.; Liu, Y.; Xu, F.; He, Y.W.; Du, L.; Venturi, V.; Fan, J.; Hu, B.; Liu, F. *Lysobacter enzymogenes* uses two distinct cell-cell signaling systems for differential regulation of secondary-metabolite biosynthesis and colony morphology. *Appl. Environ. Microbiol.* **2013**, *79*, 6604–6616. [CrossRef] [PubMed]
22. Nadell, C.D.; Xavier, J.B.; Levin, S.A.; Foster, K.R. The evolution of quorum sensing in bacterial biofilms. *PLoS Biol.* **2008**, *6*. [CrossRef] [PubMed]
23. Lee, B.; Yeon, K.M.; Shim, J.; Kim, S.R.; Lee, C.H.; Lee, J.; Kim, J. Effective antifouling using quorum-quenching acylase stabilized in magnetically-separable mesoporous silica. *Biomacromolecules* **2014**, *15*, 1153–1159. [CrossRef] [PubMed]

15

Estimation of Gaze Detection Accuracy Using the Calibration Information-Based Fuzzy System

Su Yeong Gwon, Dongwook Jung, Weiyuan Pan and Kang Ryoung Park *

Academic Editor: Vittorio M. N. Passaro

Division of Electronics and Electrical Engineering, Dongguk University, 30, Pildong-ro 1-gil, Jung-gu, Seoul 100-715, Korea; gwonsuyeong@dgu.edu (S.Y.G.); jung4759@gmail.com (D.J.); westlaopan90@gmail.com (W.P.)

* Correspondence: parkgr@dongguk.edu

Abstract: Gaze tracking is a camera-vision based technology for identifying the location where a user is looking. In general, a calibration process is applied at the initial stage of most gaze tracking systems. This process is necessary to calibrate for the differences in the eyeballs and cornea size of the user, as well as the angle kappa, and to find the relationship between the user's eye and screen coordinates. It is applied on the basis of the information of the user's pupil and corneal specular reflection obtained while the user is looking at several predetermined positions on a screen. In previous studies, user calibration was performed using various types of markers and marker display methods. However, studies on estimating the accuracy of gaze detection through the results obtained during the calibration process have yet to be carried out. Therefore, we propose the method for estimating the accuracy of a final gaze tracking system with a near-infrared (NIR) camera by using a fuzzy system based on the user calibration information. Here, the accuracy of the final gaze tracking system ensures the gaze detection accuracy during the testing stage of the gaze tracking system. Experiments were performed using a total of four types of markers and three types of marker display methods. From them, it was found that the proposed method correctly estimated the accuracy of the gaze tracking regardless of the various marker and marker display types applied.

Keywords: gaze tracking; NIR camera; calibration; calibration marker and marker display; fuzzy system

1. Introduction

Studies have been actively carried out to accurately calculate a user's gaze location based on the movement of their face or eyes [1,2]. In addition, studies are being carried out to apply a user's gaze location as a more natural and convenient input device replacing a keyboard and mouse, which are commonly used input devices [3–7]. Because it is intuitive, a user's gaze location can be easily used as an input information without a separate training process. In particular, input devices such as a keyboard, mouse, or remote control, which is commonly used by the disabled who cannot use their hands freely, are difficult to apply properly. Also, even when they can be used, they cannot be conveniently manipulated or precisely controlled.

In general, gaze tracking is a technology for identifying the position on a screen that the user is looking at. In most gaze tracking systems, a user's gaze position is calculated based on the relative positions of the center of the user's pupil and the corneal specular reflection. Here, for an accurate gaze tracking system, it is very important to accurately find the pupil center and corneal specular reflection. Along with this, a user calibration process is necessary for the accurate gaze tracking of the user. In general, a calibration process is applied at the initial stage of most gaze tracking systems. This process is necessary to calibrate for the differences in the eyeballs and cornea size of user, as well as the angle kappa, and to find the relationship between the user's eye and screen coordinates. It is applied on the basis of the information of user's pupil and corneal specular reflection obtained while the user is looking at several predetermined positions on a screen. Here, if the user's calibration process is inaccurate, it will be difficult to accurately calculate the gaze position when a gaze tracking system is executed. Therefore, to achieve a gaze tracking system with high accuracy, an accurate user calibration process is as important as the accurate detection of the user's pupil center and reflection. In other words, if the calibration process is accurate, an accurate gaze tracking can be achieved.

Past studies in this area can be mainly divided into two types: utilizing the estimation model of a user-calibration based gaze tracking result and not utilizing such a model. The latter can be divided into cases applying and not applying a calibration process. For cases using a calibration process, the calibration methods apply static markers, dynamic markers, or a dynamic marker display. For a method not applying a calibration [8], two stereo cameras and five illuminators are applied. In this case, because no calibration process is used, user convenience increases; however, because several cameras and illuminators have to be applied, the system complexity increases, as do the price and system size.

Next, for calibration using static markers, calibration methods applying five or nine points [9], nine calibration points [10–14], a single position calibration [15], and four point calibration [16,17] have been utilized. In general, as the number of reference points for the calibration increases, the gaze tracking error decreases. However, as the calibration points increase in number, the user convenience decreases. In addition, because several calibration points have to be gaze focused, if a single point among several points is not focused, a decrease in the gaze tracking accuracy will occur.

For calibration using dynamic markers and a dynamic display, one such method calibrates while the user's gaze follows a dynamically moving point [18]. In this method, a moving target appears on the screen and the calibration is conducted as the user's gaze follows it. This method can prevent an inaccurate calibration by unexpected external disturbances. That is because the user's concentration is high compared to the use of static markers. However, when the moving path of a calibration marker is so complex that the user cannot predict the movement, the calibration accuracy may decrease because the user's perceptual response velocity slows down. In addition, when the movements of the marker are complex, the user must spend more time training in advance, thereby leading to a decline in user convenience. In these previous studies, a user calibration was conducted by applying various types of markers and marker display methods. However, there have been no studies conducted on estimating the accuracy of a final gaze tracking system through the results obtained from the above calibration processes. Here, the accuracy of final gaze tracking system means the gaze tracking accuracy during the testing stage of the gaze tracking system. Therefore, in this study, using a fuzzy-based estimation model based on the results obtained during the user calibration, a method for estimating the accuracy of the final gaze tracking system is proposed. Because the accuracy estimation of a gaze tracking system is made possible based on the information obtained during the calibration, a quantitative performance can be determined for various calibration markers and display methods. Compared to conventional studies, the present study is novel in the following three ways.

- The Euclidean distance is calculated between the user's calculated gaze and the reference positions obtained during the user calibration. Then, the mean and standard deviations of the Euclidean distances are extracted as the first and second features, respectively. In addition, the change in gaze position for each frame, which is obtained during the user calibration, is extracted as the third feature.
- Developing a fuzzy system that takes these three features as inputs and the accuracy of the final gaze tracking system as an output, the accuracy of the final gaze tracking system is estimated based on the results obtained during the user calibration.
- The validity of the proposed fuzzy-based estimation system is verified experimentally using various types of markers such as static and dynamic markers, and various types of marker displays such as sequential, random, and guiding displays.

The advantages and disadvantages of this study compared to conventional studies are shown in Table 1.

Table 1. Comparison of the advantages/disadvantages between previous studies and the proposed method.

Categories		Method	Advantage	Disadvantage
Not using the estimation model of gaze tracking accuracy	Not conducting a user calibration	Requiring two stereo cameras and five light sources [8]	User convenience increases because of no calibration process	Due to two cameras and multiple illuminators, the system complexity increases, as do the price and system size
	Calibration using static markers	Calibration by selecting five or nine points [9]	As the number of calibration points increases, the gaze tracking accuracy becomes more accurate	Decrease in user convenience
		Nine calibration points [10–14]		Gaze tracking accuracy is affected by the accurate calibration procedure
		One point calibration [15]		
		Four point calibration [16,17]		
	Calibration using dynamic markers and display	Calibration by following a moving target with the eyes [18]	Prevents an inaccurate calibration because the user's concentration is high compared to the use of static markers	When the movement line of the calibration marker is so complex that it cannot be predicted by the user, the calibration accuracy decreases
				When the movements are complex, the user convenience decreases
Using the estimation model of gaze tracking accuracy **(proposed method)**		Gaze tracking accuracy can be estimated using fuzzy system based on the user calibration information from various calibration methods using static, dynamic markers and display	The performance of gaze tracking system can be quantitatively predicted for various calibration markers and display methods	The design of an additional fuzzy rule table and the membership function is necessary

The remainder of this paper is organized as follows. In Section 2, the calibration-based estimation method of gaze tracking is described. In Section 3, the experimental results with various calibration methods and an analysis of these methods are provided. In Section 4, some concluding remarks and the direction for a follow-up study are given.

2. Proposed Method for Estimating the Accuracy of Gaze Tracking System

2.1. Detection of User's Pupil Center and Corneal Specular Reflection, and Calculation of Gaze Position

Figure 1 shows a flowchart of the user calibration-information based gaze tracking estimation system proposed in this study. We refer to the previous research [19] for detecting the centers of pupils and corneal specular reflection.

2.1.1. Pupil Center Region Detection Process

The rough corneal specular reflection of the user is detected based on the bright pixel values in the image. Based on this reflection, the eye region of the user is detected, as shown in Figure 1 step (1) and Figure 2b. The image in Figure 2c is a result of applying a histogram stretching calculation to the image of Figure 2b to obtain a binarization result for a clearer distinction. The image in Figure 2d is a result of applying a binarization calculation to the image in Figure 2c. The image in Figure 2e is a result of removing small noises by applying a morphology calculation to the image in Figure 2d. The image in Figure 2f is a result of leaving only the image of the pupil shape by applying a component labeling to the image in Figure 2e. As shown in the image of Figure 2g, the edge of the pupil shape was detected by applying canny edge detection to the image in Figure 2f. In addition, applying a convex hull method [20–22] to the image in Figure 2g, the image in Figure 2h shows the result of compensating the pupil part damaged from the reflection occurring on the pupil. The image in Figure 2i is a binarized version of the image in Figure 2c centering on the reflection. Figure 2j shows an image, in which the overlapped part of Figure 2i,h images is removed from the image of Figure 2h. In the image in Figure 2j, because the pupil boundary and the corneal specular reflection boundary are touching each other, a concave portion is detected during the pupil boundary detection. Then, a shape different from the real pupil boundary is detected. To resolve this, we remove the overlapped parts in Figure 2h,i from the image in Figure 2h. Then, the pupil boundary that has removed the touching part can be detected, as shown in the image in Figure 2j. Figure 2k shows the detection of the final pupil boundary by applying an ellipse-fitting algorithm using the image in Figure 2j. Figure 2l shows the final resulting image.

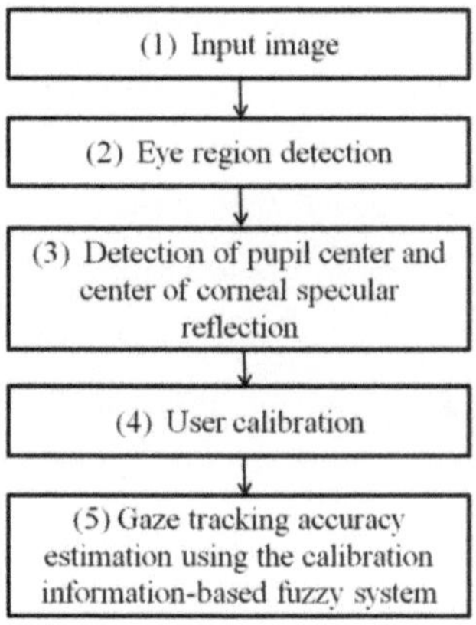

Figure 1. Flowchart of user calibration-information based gaze tracking estimation system.

Based on the detected position of the pupil center, the searching area for detecting corneal specular reflection is defined. Then, the center of corneal specular reflection is located based on binarization and calculating the geometric center within this area.

2.1.2. Calculating the User's Gaze Position

The gaze position of the user is calculated using the detected pupil center and reflection center coordinates. To compensate the differences in each person's eyeball and cornea size, as well as the angle kappa, and to find the relationship between the user's eye and screen coordinates, the calibration process is conducted before the gaze position is calculated. The gaze position is calculated using several geometric transform matrices obtained through the calibration process [19]. When nine reference points are gazed at during the calibration, by using each of the nine pupil center positions obtained as shown in Figure 3, the relationship between the screen and pupil center can be expressed as in Figure 4. As Figure 4 illustrates, when the user is looking at the nine calibration points, the pupil image plane can be mainly divided into four subregions by using the center position of the pupil. Each of the four pupil subregions can be mapped to four monitor subregions, which are divided into the nine calibration points of the monitor image plane. In addition, a geometric transform matrix corresponding to each subregion is obtained and used [19,23]. For example, pupil subregion 3 can be mapped into monitor subregion 3, and here, matrix 3 is used as shown in Figure 4.

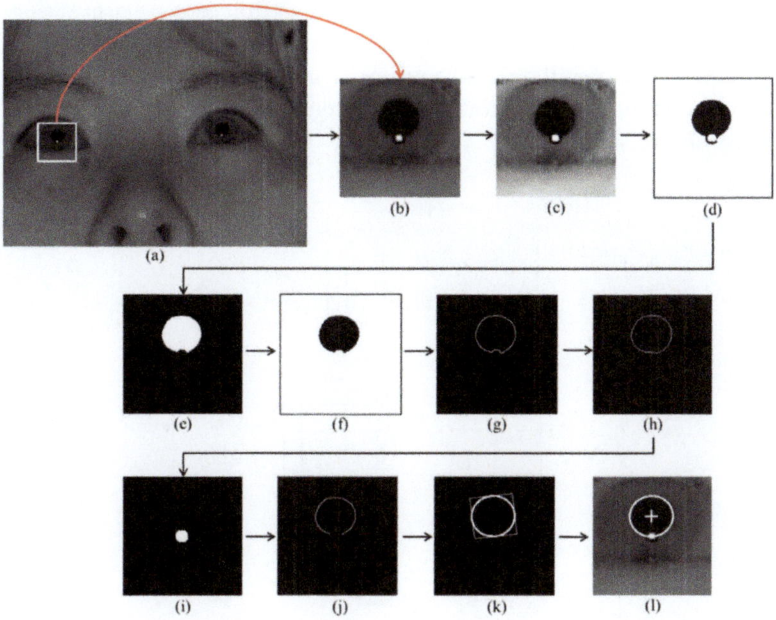

Figure 2. Pupil center region detection process: (**a**) eye detection result based on corneal specular reflection from the original image, (**b**) eye region detected from the original image; (**c**) application of histogram stretching to the image of (b); (**d**) application of binarization to the image of (c); (**e**) application of morphology calculation to the image of (d); (**f**) application of component labeling to the image of (e); (**g**) application of canny edge detection to the image of (f); (**h**) application of a convex hull to the image of (g); (**i**) image that has binarized the reflection with the image of (c); (**j**) image that has removed the overlapped parts of (h,i) from the image of (h); (**k**) application of ellipse-fitting algorithm to the image of (j); and (**l**) the final detection result of the pupil center region.

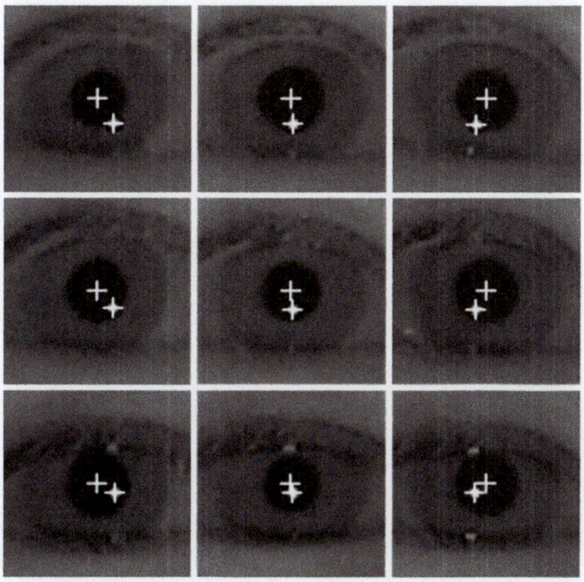

Figure 3. Images of user's eye region when gazing at nine calibration points on the screen.

When calculating the user's gaze position, if the user's pupil center exists at pupil subregion 2 ($P_2P_3P_6P_5$), the user's gaze position is calculated using matrix 2 among the four geometric transforms. Here, the gaze position is calculated through the processes in Equations (1) and (2) [19]. Figure 5 shows the example of correspondence between four pupil centers and four monitor corners when a user is gazing at the four corners of a monitor.

$$\begin{bmatrix} M_{x0} & M_{x1} & M_{x2} & M_{x3} \\ M_{y0} & M_{y1} & M_{y2} & M_{y3} \end{bmatrix} = \begin{bmatrix} a & b & c & d \\ e & f & g & h \end{bmatrix} \begin{bmatrix} P_{x0} & P_{x1} & P_{x2} & P_{x3} \\ P_{y0} & P_{y1} & P_{y2} & P_{y3} \\ P_{x0}P_{yo} & P_{x1}P_{y1} & P_{x2}P_{y2} & P_{x3}P_{y3} \\ 1 & 1 & 1 & 1 \end{bmatrix} \quad (1)$$

$$\begin{bmatrix} M'_x \\ M'_y \end{bmatrix} = \begin{bmatrix} a & b & c & d \\ e & f & g & h \end{bmatrix} \begin{bmatrix} P'_x \\ P'_y \\ P'_xP'_y \\ 1 \end{bmatrix} \quad (2)$$

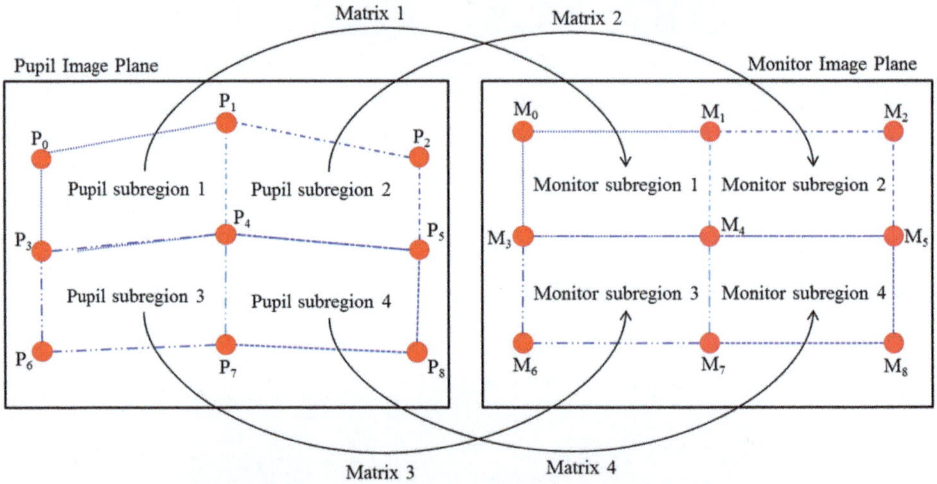

Figure 4. Corresponding relations of four subregions defined by nine pupil centers and four subregions in a monitor defined using calibration points.

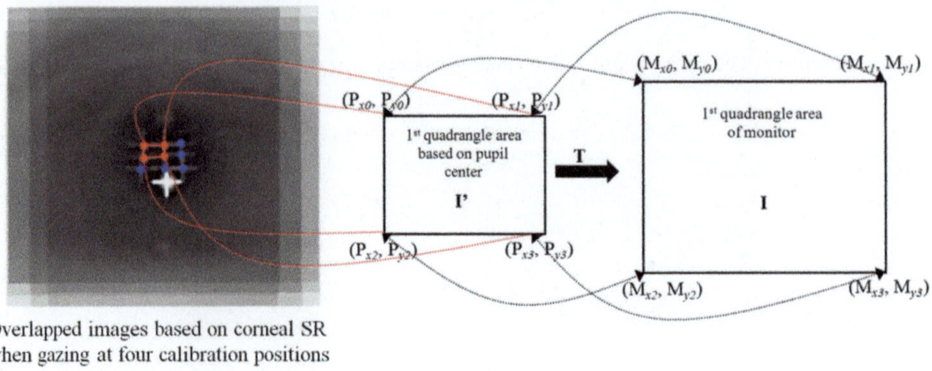

Figure 5. Example of correspondence between four pupil centers and four monitor corners when a user is gazing at the four corners of a monitor.

2.2. User's Gaze Tracking Estimation Using a Fuzzy System

2.2.1. Three Features for the Inputs of Fuzzy System, and Fuzzy Membership Function with Rule Table

In this study, a method for estimating a user's gaze tracking accuracy is proposed using a fuzzy system based on information obtained during the calibration process. As shown in Figure 6, the gaze tracking accuracy of the user is estimated based on the fuzzy algorithm by using three feature values of the user, which can be obtained during the calibration process.

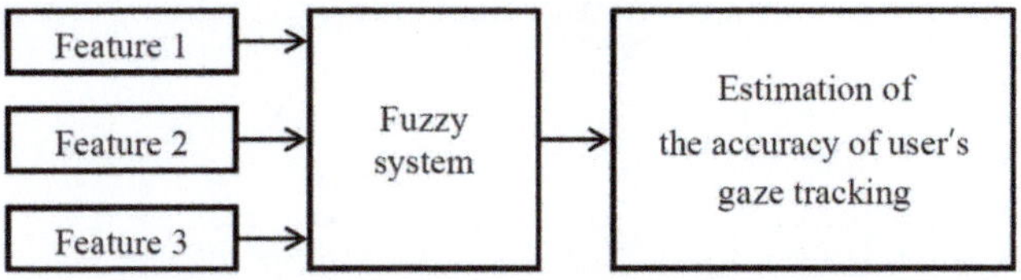

Figure 6. Process of estimating the gaze tracking accuracy of the user by applying a fuzzy algorithm.

Among the eye features of the user that can be obtained during the calibration process, the following three feature values were utilized. In Figure 7, the Euclidean distances are calculated between the reference points that have to be looked at during the calibration, and the positions that the user is actually gazing at. Then, Feature 1 (F_1) indicates the mean value of the Euclidean distances. Feature 2 (F_2) indicates the standard deviation of the Euclidean distances. Feature 3 (F_3) indicates the amount of gaze position movement of the user between the previous and current image frames.

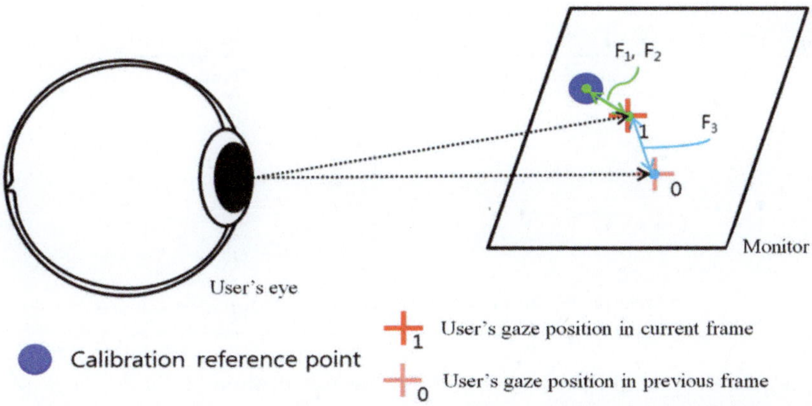

Figure 7. The feature values that can be obtained during the calibration process.

With respect to the use of the fuzzy algorithm, the design of the membership function is a very important factor. Figure 8 shows the input/output membership function configuration of the feature values. As shown in Figure 8, a linear member function or triangular membership function is most commonly used because the system's calculation speed is fast and the complexity of the codes can be relatively reduced [24–26]. Features 1–3 described above each show different values, and the range of each feature value is different. Furthermore, depending on the user's calibration result, the ranges of Features 1–3 change irregularly. Therefore, each feature value was normalized with a value of zero to 1, as shown in Figure 8. The fuzzy membership functions can be mainly divided into two types. As shown in Figure 8a, an input membership function having low (L) and high (H) values was designed. In addition, as shown in Figure 8b, an output membership function having low (L), middle (M), and high (H) values was designed.

Furthermore, a fuzzy rule table (Table 2) was designed. For example, if the mean and standard deviation values are small for the Euclidean distance values between the reference and user's gaze positions, and if the amount of gaze movement of the user is small between image frames, it can be regarded that the error of the user's gaze position is also small. In other words, if the values of Features 1–3 are L, then the error value of the gaze tracking can be estimated to be L as well. If the values of Features 1–3 are high (H), then the error value of the gaze tracking can be also estimated to be high (H).

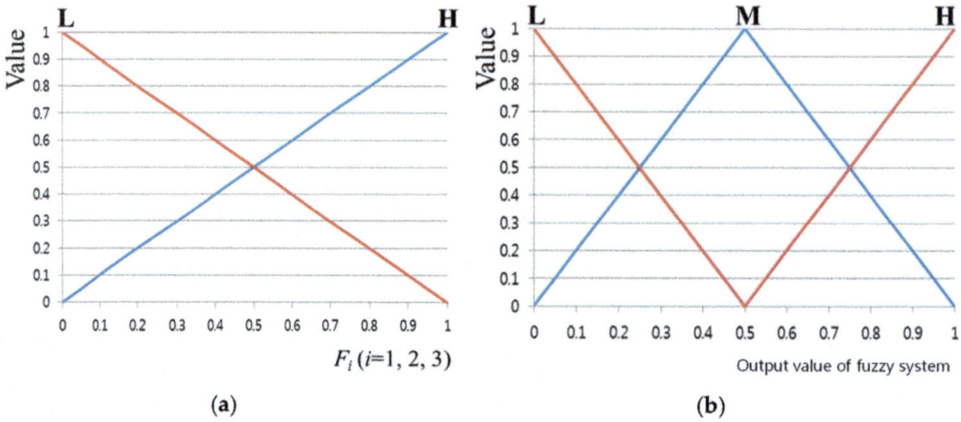

Figure 8. Membership functions of calibration-information based gaze accuracy estimation system: (**a**) input and (**b**) output membership functions.

Table 2. Fuzzy rule table.

Feature 1	Feature 2	Feature 3	Output
L	L	L	L
		H	L
	H	L	L
		H	M
H	L	L	M
		H	H
	H	L	H
		H	H

2.2.2. Obtaining the Output of Fuzzy System Using Defuzzification Methods

In general, an output value of the fuzzy system can be obtained through the output fuzzy membership function and the fuzzy rule. The input function in Figure 8a has two output values and one input value (F_i). In other words, because there are three input feature values, there are a total of eight output values.

For example, as can be seen in Figure 9, the input value of 0.8 has output values of 0.2(L) and 0.8(H). These two output values are because of Feature 1, but also apply to Features 2 and 3. This means that there are three pairs of output values, {0.2 (L), 0.8 (H)}. Based on these three pairs of output values, eight combinations can be obtained: {0.2(L), 0.2(L), 0.2(L)}, {0.2(L), 0.2(L), 0.8(H)}, {0.2(L), 0.8(H), 0.2(L)} ... {0.8(H), 0.8(H), 0.8(H)}. Whenever a single pair of combination values is elicited, one of the values can be determined from (0.2, 0.8) and (L, M, H) through the Min rule, Max rule, and fuzzy rule table [27]. For example, when there is a combination of {0.2(L), 0.8(H), 0.8(H)}, a value of 0.2 can be determined by the Min rule. In addition, in Table 2, because the output value is M when Feature 1 is L and Features 2 and 3 are H, the value of M can also be determined by the fuzzy rule. That is, from the values of {0.2(L), 0.8(H), 0.8(H)}, an inference value of 0.2(M) can be obtained by the Min and fuzzy rules. If the Max rule is used, an inference value of 0.8(M) can be obtained. As mentioned above, because there are a total of eight combinations, the number of inference values is also eight in total.

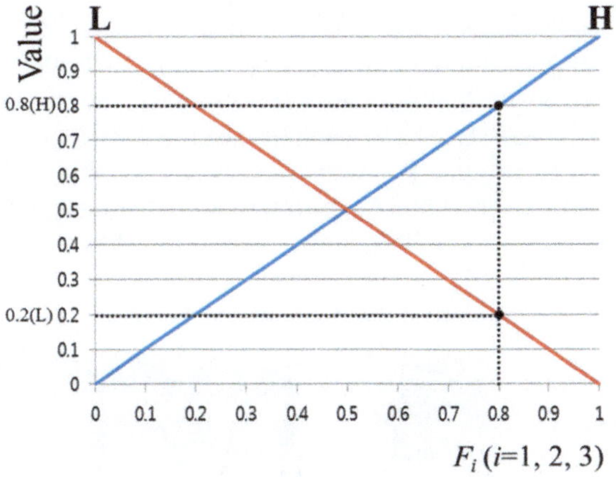

Figure 9. Example of obtaining an output value when one input value is input into the input membership function.

In Figure 10a, several output values can be obtained using one inference value. For example, when the inference value is 0.71(M), the output values O_2, O_3, and O_4 can be obtained. From these several output values, using a defuzzification method, one final output value can be determined [28,29]. This output value is used as the accuracy estimation score of the gaze tracking system.

Conventionally, various defuzzification methods exist, and the performances of five methods were compared: First of Maxima (FOM), Last of Maxima (LOM), Middle of Maxima (MOM), Mean of Maxima (MeOM), and Center of Gravity (COG) [28,29]. FOM selects the first value in the output values calculated with the largest value among the inference values obtained. LOM selects the last value in the output values calculated with the largest value among the inference values obtained. MOM selects the median of the first and last values from the output values calculated with the largest value among the inference values obtained [28]. MeOM selects the mean of the output values calculated with the largest value among the inference values obtained. The final output values obtained according to each method in Figure 10a,b are O_2 for FOM, O_4 for LOM, $((O_2 + O_4)/2)$ for MOM, $((O_2 + O_3 + O_4)/3)$ for MeOM, and O_5 for COG. COG is determined through the mass center point of the region expressed with the slashes shown in Figure 10b.

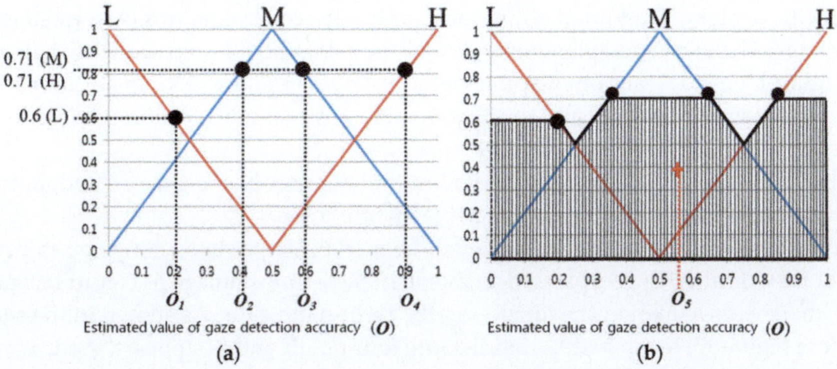

Figure 10. Examples of output value results through various defuzzification methods when an inference value is input: (**a**) FOM, LOM, MOM, MeOM, and (**b**) COG.

3. Experiment Results and Discussion

Figure 11 shows the experimental environment used in this study. For the experiment, a desktop computer containing a 3.5-GHz Intel® core™ i7-3770K CPU and 8 GB of RAM was used. The monitor resolution was 1280 × 1024. For the camera, a 700-nm long pass filter (Kodak Wratten Filter No. 89B, Rochester, NY, USA [30]) and a zoom lens were mounted on a webcam (Logitech C600, Lausanne, Switzerland [31]) having a universal serial bus (USB) interface. In addition, an 850-nm reference illuminator was used in an 8 × 8 array. The camera was positioned right below the center of the monitor, and the illuminator was also positioned below the camera. The distance between the monitor and user was roughly 75 cm. Our program was implemented using a Microsoft foundation class based C++ program, DirectX 9.0 SDK, and OpenCV in a Microsoft Visual Studio C++ 2010 development environment.

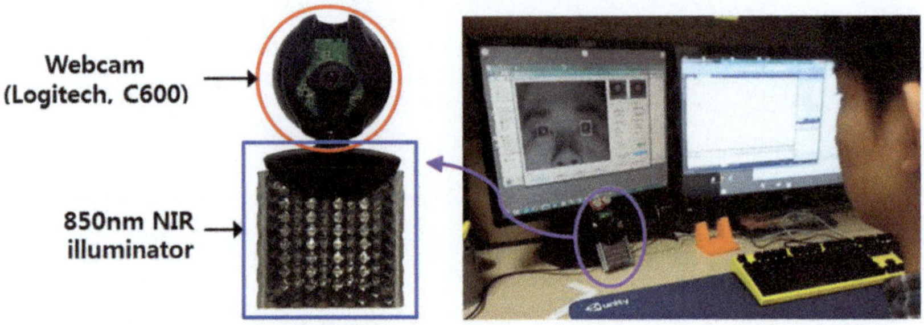

Figure 11. The gaze tracking system and experimental environment used in this study.

The calibration was carried out according to each marker type and display method described in Section 3.1. The gaze position accuracy of the user was measured for each calibration. The experiment was repeated four times for each test subject and a total of seven calibration methods. A total of 15 test subjects were used.

3.1. Various Calibration Marker Types and Display Methods

In this paper, various calibration marker types and display methods are proposed to check the validity of the user-calibration based gaze tracking estimation model. As shown in Figure 1, the calibration marker types are classified into two types: static marker and dynamic marker types.

Among the commonly used calibration markers, Figure 12a shows the static markers of a circle shape and a cross. Figure 12b shows a dynamic marker with a dynamically changing shape, *i.e.*, the color is becoming darker gradually as the shape of the marker decreases from a large circle to a small circle in the actual system. In addition, using each marker, the calibration was performed with different display methods. The marker display methods include a commonly used sequential display method shown in Figure 13a, and a random display method shown in Figure 13b. In addition, they include the method of dynamically guiding to the next display position of the calibration marker, named as guiding display method, shown in Figure 14.

In Figure 13, the red dot is a calibration point that the user has to gaze at, and, in the next step, the calibration finishes with point represented by a white color.

In Figure 14, a dynamic marker is used, and the marker shape changes step by step, as shown in Figure 15. At first, the marker is expressed as shown in the leftmost image of Figure 15, and afterward, the marker changes as shown in the images on the right-hand side. As shown in the second image from the left in Figure 15, the marker is divided into four points and disappears. And, as shown in the third image from the left in Figure 15, at the moment of calibration it blinks as a white dot. Afterward, as shown in the rightmost image of Figure 15, it is changed to an orange dot, and while moving to the next calibration point, it is increased to the marker size shown in the leftmost image of Figure 15. As the marker shape changes in this manner, the calibration is performed at a total of nine positions, as shown in the sequence in Figure 14.

(a) (b)

Figure 12. Various calibration marker types: (**a**) static and (**b**) dynamic markers.

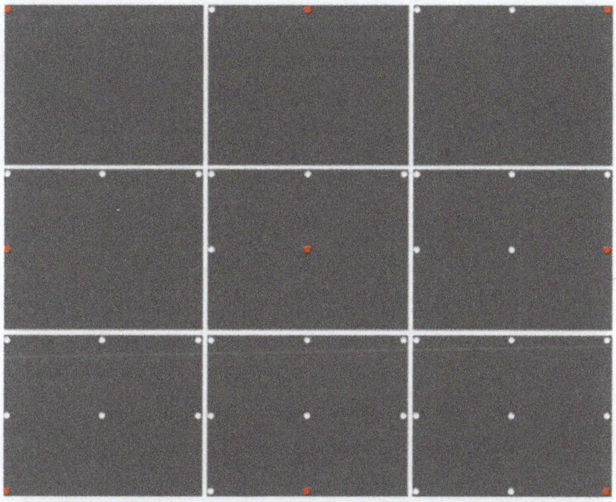

(a)

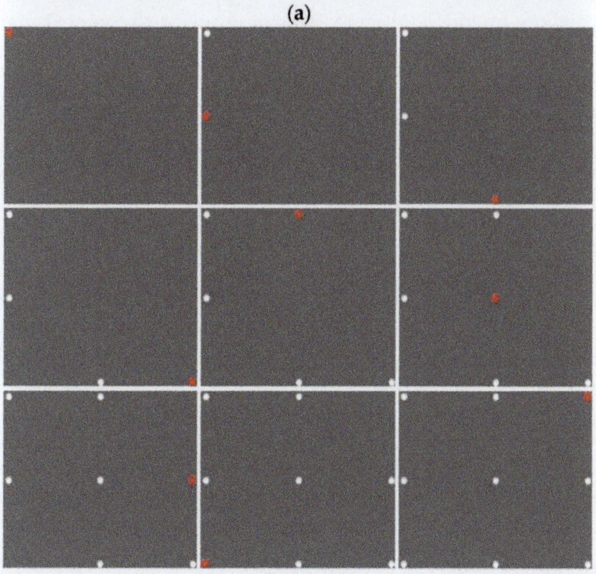

(b)

Figure 13. Marker display methods: (**a**) sequential and (**b**) random marker displays.

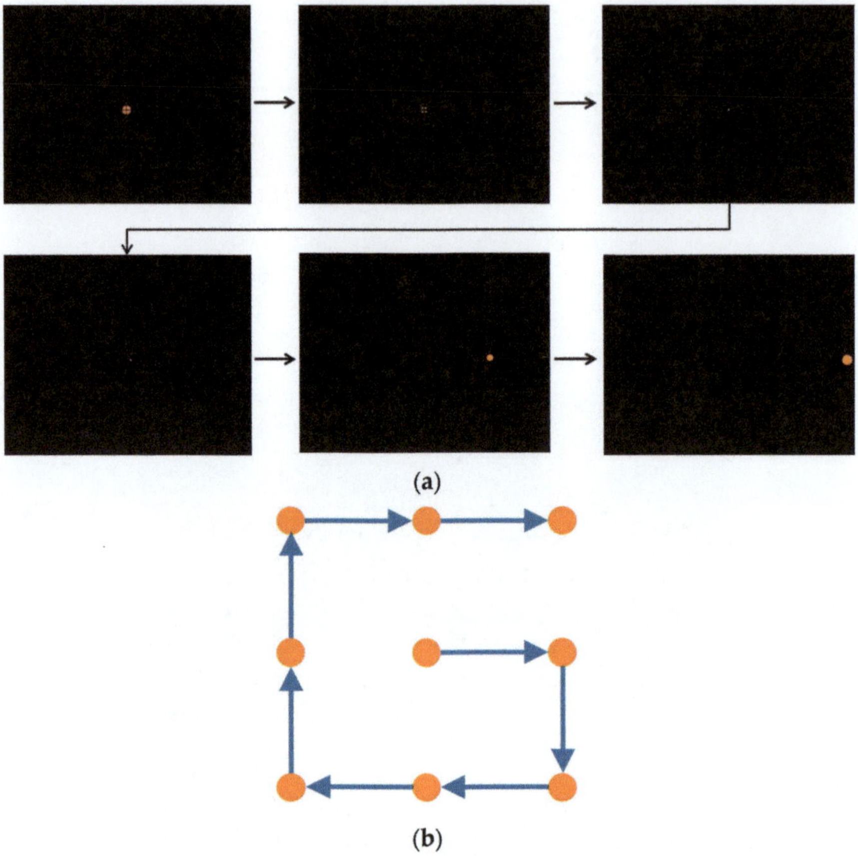

Figure 14. Guiding display method: (a) an expression guiding to the next marker position of the calibration, and (b) an example of a marker's movement path.

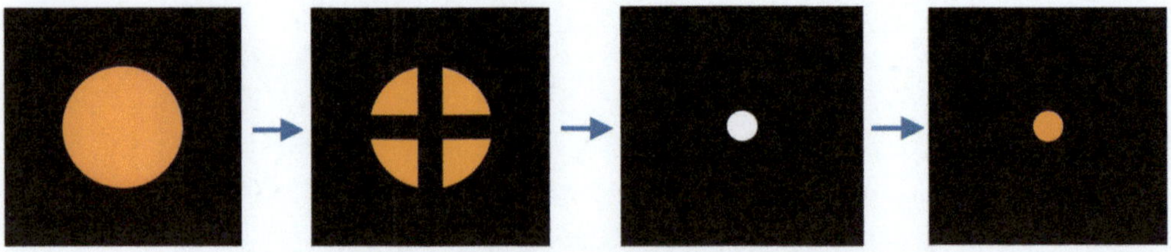

Figure 15. Changes to dynamic marker in the guiding display.

3.2. Calibration Results and Gaze Detection Accuracy

In this section, we show the user calibration results and gaze tracking accuracy. In Figures 16–18 the blue and green dots of (a,c) mark the gaze coordinates of both eyes during the user calibration. The red circles represent the regions that have to be looked at. As shown in (a,c) of Figures 16–18 when the calibration gaze coordinates are concentrated in regions that have to be looked at during the calibration, the error in the gaze tracking accuracy of the users is small, as shown in (b,d). Based on this, it can be seen that the accuracy levels during the user calibration and gaze tracking are related. At the same time, it is confirmed that the gaze tracking estimation model can be obtained using the proposed three features. The first and second features are respectively the mean and standard deviations of the Euclidean distances between the user's calculated gaze and the reference positions obtained during the user calibration. The third one is the amount of gaze movement between the previous and current image frames.

(a)

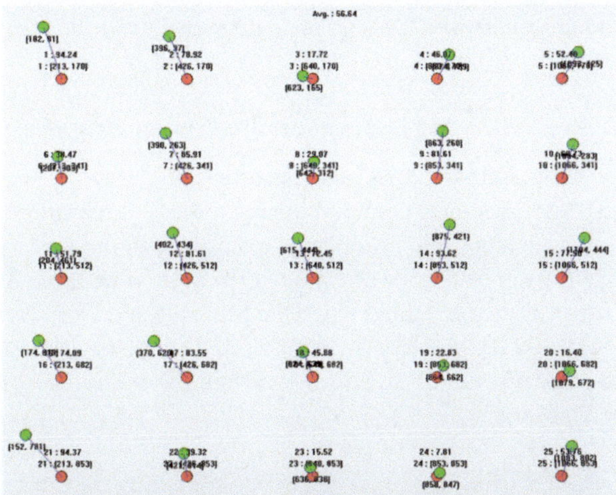

(b)

(c)

Figure 16. *Cont.*

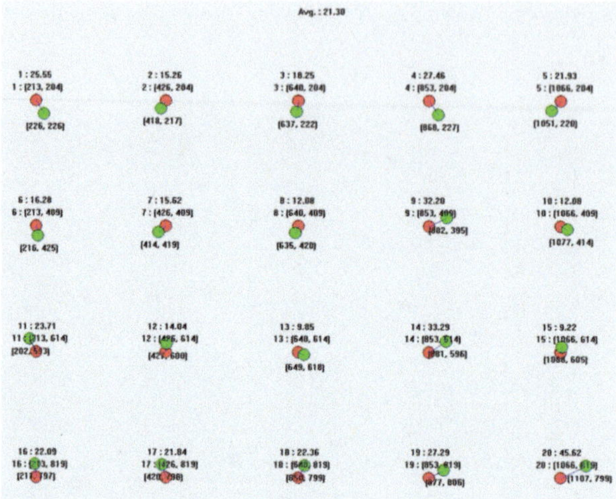

(d)

Figure 16. Results of a circular marker type expressed sequentially: (**a**) an inaccurate calibration, (**b**) an error in the gaze tracking accuracy of 1.223°; (**c**) an accurate calibration; and (**d**) an error in the gaze tracking accuracy of 0.407°.

3.3. *Estimation Analysis of Fuzzy Algorithm for User Gaze Tracking Accuracy Using Calibration Information*

In this study, a method for estimating the user's gaze tracking accuracy was proposed by applying a fuzzy algorithm to three feature values obtained during the user calibration. From Table 3, the gaze tracing accuracy estimation results can be confirmed according to the fuzzy system's Min and Max rules and various defuzzification methods. In the results shown in Table 3, it was found that the estimation value of the fuzzy system obtained when the Min rule and COG were used had the highest correlation with the size of the real gaze error. In other words, when the estimation of the fuzzy system was small, the real gaze error was small. In addition, when the estimation of the fuzzy system was large, the real gaze error was large.

(a)

Figure 17. *Cont.*

(b)

(c)

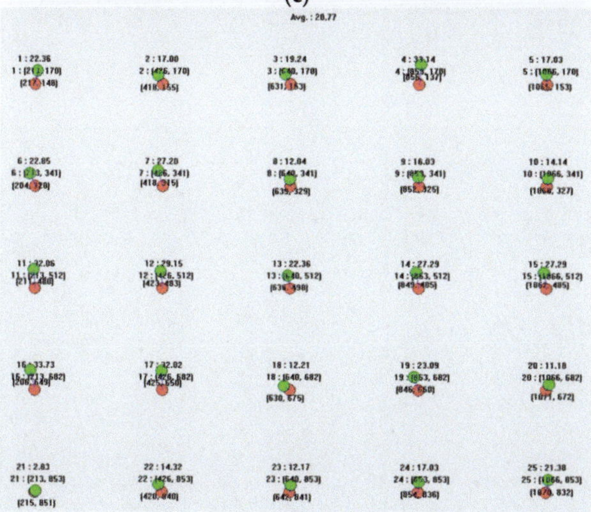

(d)

Figure 17. Results of a cross-shaped marker type expressed sequentially: (**a**) an inaccurate calibration, (**b**) an error in the gaze tracking accuracy of 1.19°; (**c**) an accurate calibration; and (**d**) an error in the gaze tracking accuracy of 0.449°.

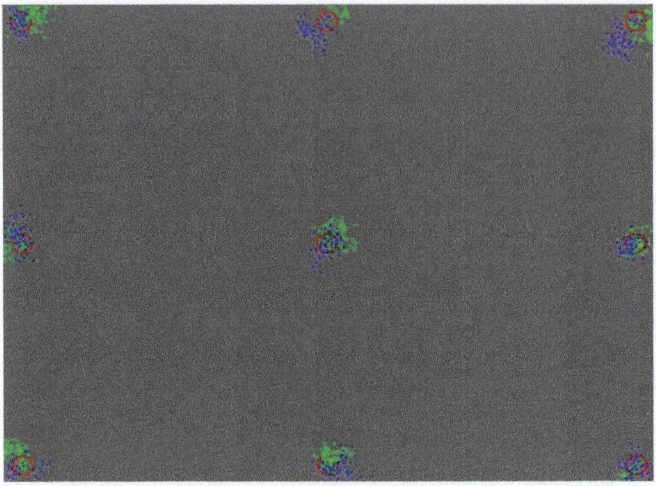

(a)

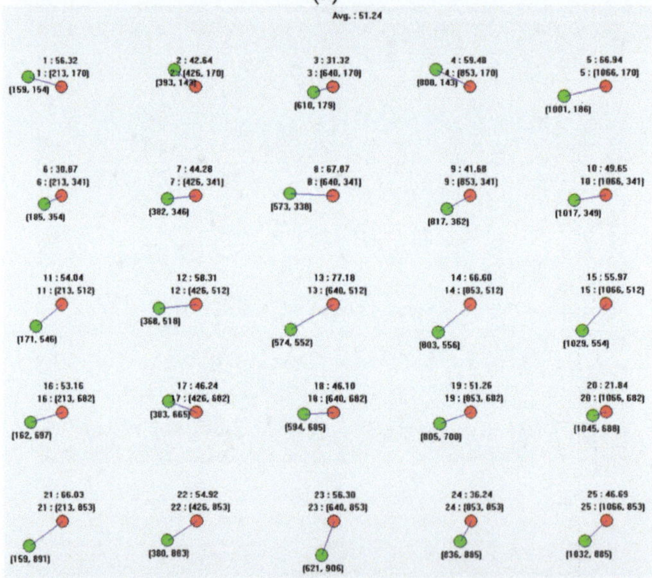

(b)

(c)

Figure 18. *Cont.*

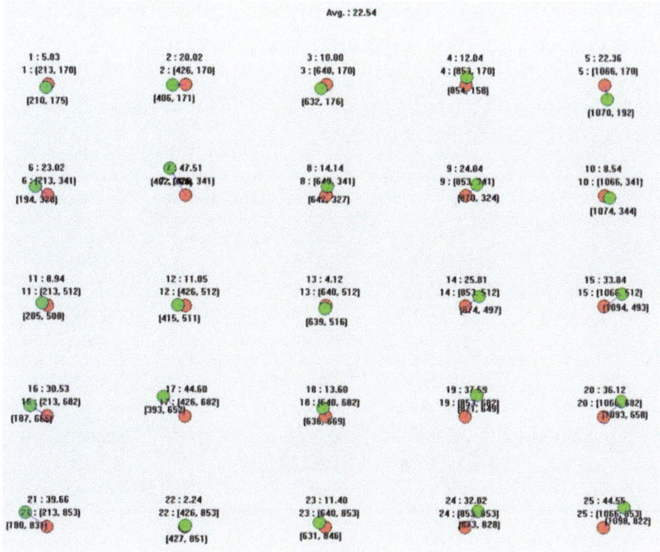

(d)

Figure 18. Results of a gradation marker type expressed sequentially: (**a**) an inaccurate calibration, (**b**) an error in the gaze tracking accuracy of 1.107°; (**c**) an accurate calibration; and (**d**) an error in the gaze tracking accuracy of 0.487°.

Table 3. Estimation results of gaze tracking accuracy according to the marker type and display method (Seq.: Sequential display, Rand.: Random display).

Display Method			**Seq.**	**Rand.**	**Moving**	**Seq.**	**Rand.**	**Seq.**	**Rand.**
Marker Type			**Cross**	**Cross**	**Circle**	**Gradation**		**Gradation Circle**	
Real Gaze Error (°)			**0.6363**	**0.6742**	**0.684**	**0.7527**	**0.7697**	**0.8371**	**0.8498**
Fuzzy system	Min rule	FOM	0.255	0.2299	0.3918	0.3708	0.4461	0.4408	0.1399
		MOM	0.255	0.2768	0.4086	0.4332	0.2763	0.3271	0.1488
		LOM	0.255	0.3237	0.4254	0.4957	0.3422	0.3143	0.1576
		MeOM	0.1948	0.1991	0.2622	0.2671	0.5737	0.5796	0.1259
		COG	**0.3764**	**0.4177**	**0.4688**	**0.4859**	**0.3018**	**0.7411**	**0.3198**
	Max rule	FOM	0.1107	0.1231	0.1451	0.1106	0.4438	0.2446	0.0931
		MOM	0.5357	0.5278	0.5032	0.5224	0.4165	0.4518	0.4136
		LOM	0.9607	0.9324	0.8613	0.9341	0.2692	0.2854	0.7341
		MeOM	0.3122	0.3032	0.3124	0.3201	0.464	0.6109	0.2228
		COG	0.5002	0.5002	0.4991	0.5002	0.5006	0.3433	0.4988

To find the correlation between the real gaze error and each method of defuzzification with the Min or Max rule, the correlation, gradient, and R^2 values were calculated, as shown in Table 4. The range in correlation values was −1 to 1, indicating a negative or positive correlation among the real data. A correlation value of zero indicates that there is no correlation between two sets of data. As shown in Table 4 and Figure 19, when the Min rule and COG were used, the correlation value was the highest between the real gaze error data and the fuzzy system's output estimation value.

Most of the two-dimensional data can be expressed through a linear fitted line, which can express the distribution, and can be calculated using the gradient and R^2. Here, R^2 shows reliability between the data distribution and a linear line. Its value increases as the data are distributed closer to the linear fitted line [32]. As shown in Table 4, when the COG of the Min rule was used, the gradient was closest to 1 and the value of R^2 was the largest. Based on these results, it was found that using the COG method of the Min rule is best when a user is estimating a gaze error.

Table 4. The correlations between the real gaze tracking errors and the estimation values for each method of defuzzification using the Min or Max rule.

Method			Correlation Value	Gradient	R^2
Fuzzy system	Min rule	FOM	0.1093	0.5177	0.068
		MOM	0.1765	0.8036	0.1307
		LOM	0.158	0.7269	0.1464
		MeOM	0.1756	0.5775	0.0776
		COG	0.5184	1.5771	0.2593
	Max rule	FOM	−0.1841	−0.3851	0.1827
		MOM	0.1481	0.3375	0.0784
		LOM	0.0393	0.2843	0.0078
		MeOM	0.1872	0.8416	0.1558
		COG	−0.0134	0.0848	0.0963

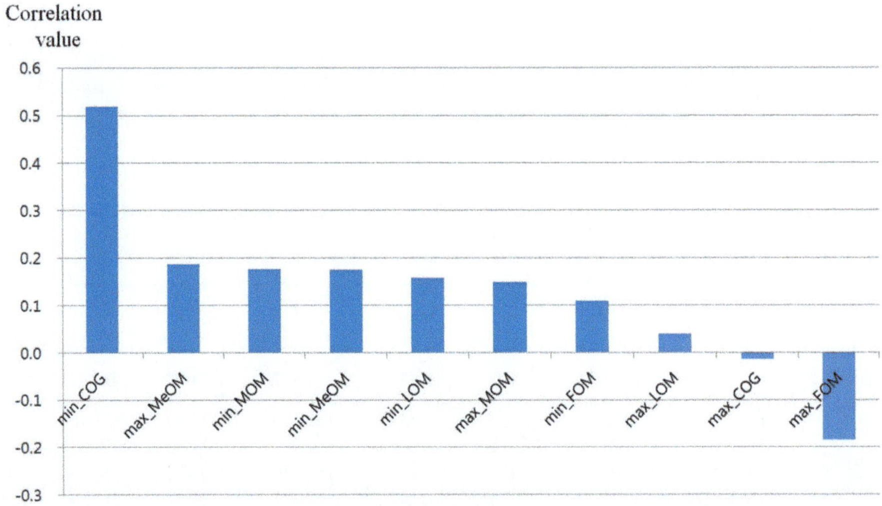

Figure 19. The correlation values between the real gaze errors and each method of defuzzification using the Min or Max rule.

The following is a statistical analysis of the differences between correlation values according to the correlation value ranking and each method of defuzzification using the Min or Max rule.

As shown in Table 4 and Figure 19, the difference between the COG of the Min rule and MeOM of the Max rule, which ranked 1 and 2 on the basis of the correlation values, respectively, was checked through a *t*-test [33]. This is a statistical hypothesis test commonly used in statistical analysis. When the *t*-test was performed using two independent samples, the COG of the Min rule and MeOM of the max rule, the calculated *p*-value was 0.0347, which was smaller than the 95% (0.05) significance level. That is, because the null-hypothesis that there is no difference between two correlation values is rejected, it indicates that there is a significant difference in the correlation values between the two groups, *i.e.*, the COG of the Min rule and the MeOM of the Max rule. This shows that, to estimate the gaze error of the user, it is appropriate to use the COG of the Min rule, which has the top ranking correlation value among the defuzzification methods. As shown in Figure 20b–i, in all other cases of the ranked correlation values, *i.e.*, the No. 2 and 3 through No. 9 and 10 rankings, the *p*-value was larger than 0.05, which does not show a statistically significant difference.

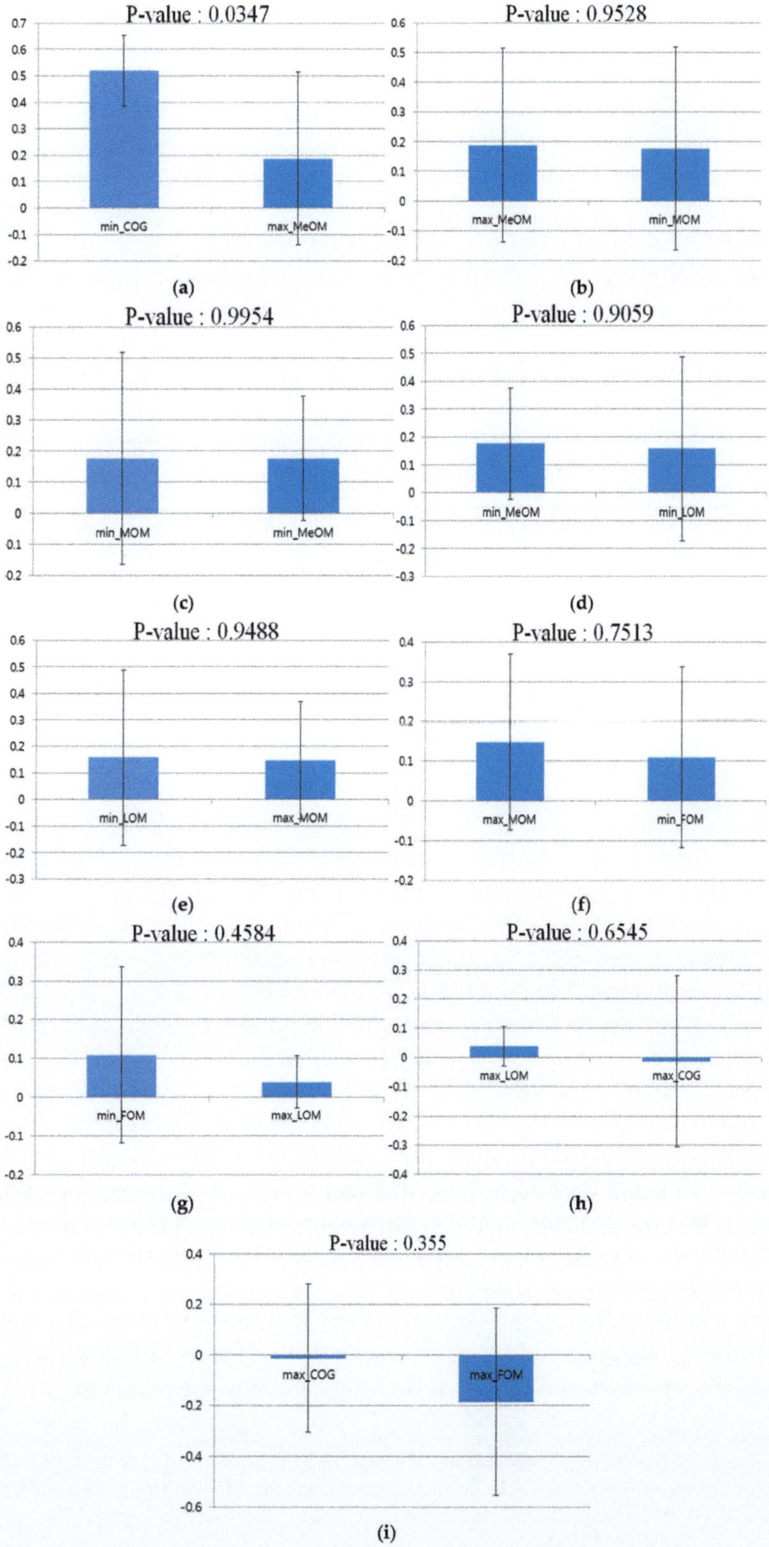

Figure 20. A t-test between defuzzification methods based on the correlation value ranking of Figure 19: t-test between ranks (**a**) 1 and 2, (**b**) 2 and 3; (**c**) 3 and 4; (**d**) 4 and 5; (**e**) 5 and 6; (**f**) 6 and 7; (**g**) 7 and 8; (**h**) 8 and 9; and (**i**) 9 and 10.

If the null-hypothesis between the samples is determined using the p-value through a t-test, the size of the difference between the two groups can be shown using the size of the effect. This can be defined as Cohen's d [34]. Cohen's d is considered as small at around 0.2–0.3, medium at around 0.5, and large at 0.8 or higher. For example, in Table 5, in the case of the COG of the Min rule and MeOM of the Max rule, Cohen's d is 1.2425. Through this, it can be stated that the difference in the correlation values is very large between the COG of the Min rule and MeOM of the Max rule, to the point of having a large effect size. It can be seen from the p-value and Cohen's d that the difference is large between the COG of the Min rule and MeOM of the Max rule, which have No. 1 and 2 ranked correlation values, respectively. Through this, when estimating the gaze position error of the user, it was found to be most appropriate to use the COG method of the Min rule.

Table 5. Cohen's d and effect size calculated between each fuzzy method sequentially according to the correlation value rankings.

Defuzzification Methods with Min or Max Rule	p-Value	Cohen's d	Effect Size
min_COG *vs.* max_MeOM	0.0347	1.2425	Large
max_MeOM *vs.* min_MOM	0.9528	0.0301	Small
min_MOM *vs.* min_MeOM	0.9954	0.003	Small
min_MeOM *vs.* min_LOM	0.9059	0.0604	Small
min_LOM *vs.* max_MOM	0.9488	0.0328	Small
max_MOM *vs.* min_FOM	0.7513	0.1616	Small
min_FOM *vs.* max_LOM	0.4584	0.3895	Small
max_LOM *vs.* max_COG	0.6545	0.2324	Small
max_COG *vs.* max_FOM	0.355	0.4796	Medium

4. Conclusions

In this study, a fuzzy-based estimation model was proposed to measure the accuracy of a gaze tracking system based on the user's calibration information. The accuracy of the gaze tracking system was estimated based on three types of feature values. The first and second features are respectively the mean and standard deviations of the Euclidean distances between the reference and user's gaze positions during the calibration. The third one is the amount of movement in the user's gaze position between the previous and current images. Experiments on the accuracy of various gaze tracking methods were conducted using four types of markers and three types of marker display methods during calibration. Through them, the validity of a fuzzy-based estimation system was verified using the feature values and results according to each calibration method. Through the experimental results of this study, the correlation, gradient, and R^2 were calculated between the real gaze errors and each defuzzification method using the Min or Max rule. From them, the validity of the fuzzy-based gaze error estimation system was proven. In addition, by calculating a t-test and Cohen's d between the defuzzification methods using the Min or Max rule, it was shown that the COG method of the Min rule was the most appropriate.

In the future, a follow-up study will be carried out on the estimation of gaze tracking accuracy based on a neuro-fuzzy system. In addition, the convenience of various calibration methods will be measured through user brainwaves, as well as through a subjective evaluation.

Acknowledgments: This research was supported by Basic Science Research Program through the National Research Foundation of Korea (NRF) funded by the Ministry of Education (NRF-2015R1D1A1A01056761), and in part by Business for Developing the Technologies for Foundation and Growth funded Korea Small and Medium Business Administration in 2015 (Grants No. S2320066).

Author Contributions: S.Y. Gwon and K.R. Park designed the overall system for estimating the accuracy of gaze tracking system based on fuzzy method. In addition, they wrote and revised the paper. D. Jung and W. Pan implemented the calibration algorithm, and helped the experiments.

Conflicts of Interest: The authors declare no conflicts of interest.

References

1. Murphy-Chutorian, E.; Trivedi, M.M. Head pose estimation in computer vision: A survey. *IEEE Trans. Pattern Anal. Mach. Intell.* **2009**, *31*, 607–626. [CrossRef] [PubMed]
2. Morimoto, C.H.; Mimica, M.R.M. Eye gaze tracking techniques for interactive applications. *Comput. Vis. Image Underst.* **2005**, *98*, 4–24. [CrossRef]
3. Amarnag, S.; Kumaran, R.S.; Gowdy, J.N. Real time eye tracking for human computer interfaces. In Proceedings of IEEE International Conference on Multimedia and Exposition, Baltimore, MO, USA, 6–9 July 2003; pp. 557–560.
4. Jacob, R.J.K. Eye movement-based human-computer interaction techniques: Toward non-command interfaces. In *Advances Human-Computer Interaction*; Ablex Publishing Co.: Norwood, NJ, USA, 1993; pp. 151–190.
5. Sibert, L.E.; Jacob, R.J.K. Evaluation of eye gaze interaction. In Proceedings of the SIGCHI Conference on Human Factors in Computing Systems, The Hague, The Netherlands, 1–6 April 2000; pp. 281–288.
6. Ohno, T. Quick menu selection task with eye mark. *Trans. Inf. Process. Soc. Jpn.* **1999**, *40*, 602–612.
7. Cho, D.C.; Kim, W.Y. Long-range gaze tracking system for large movements. *IEEE Trans. Biomed. Eng.* **2013**, *60*, 3432–3440. [CrossRef] [PubMed]
8. Model, D.; Eizenman, M. User-calibration-free remote eye-gaze tracking system with extended tracking range. In Proceedings of the 24th Canadian Conference on Electrical and Computer Engineering, Niagara Falls, ON, Canada, 8–11 May 2011; pp. 1268–1271.
9. Tobii, X2. Available online: http://www.tobii.com (accessed on 2 July 2015).
10. Zhu, Z.; Ji, Q. Novel eye gaze tracking techniques under natural head movement. *IEEE Trans. Biomed. Eng.* **2007**, *54*, 2246–2260. [PubMed]
11. Sigut, J.; Sidha, S.A. Iris center corneal reflection method for gaze tracking using visible light. *IEEE Trans. Biomed. Eng.* **2011**, *58*, 411–419. [CrossRef] [PubMed]
12. Colombo, C.; Bimbo, A.D. Interacting through eyes. *Robot. Auton. Syst.* **1997**, *19*, 359–368. [CrossRef]
13. Morimoto, C.H.; Koons, D.; Amir, A.; Flickner, M.; Zhai, S. Keeping an eye for HCI. In Proceedings of the XII Brazilian Symposium on Computer Graphics and Image Processing, Campinas, Brazil, 17–20 October 1999; pp. 171–176.
14. Mimica, M.R.M.; Morimoto, C.H. A computer vision framework for eye gaze tracking. In Proceedings of the XVI Brazilian Symposium on Computer Graphics and Image Processing, São Carlos, Brazil, 12–15 October 2003; pp. 406–412.
15. Guestrin, E.D.; Eizenman, M. General theory of remote gaze estimation using the pupil center and corneal reflections. *IEEE Trans. Biomed. Eng.* **2006**, *53*, 1124–1133. [CrossRef] [PubMed]
16. Zhu, J.; Yang, J. Subpixel eye gaze tracking. In Proceedings of IEEE International Conference on Automatic Face and Gesture Recognition, Washington, DC, USA, 20–21 May 2002; pp. 124–129.
17. Yoo, D.H.; Chung, M.J. A novel non-intrusive eye gaze estimation using cross-ratio under large head motion. *Comput. Vis. Image Underst.* **2005**, *98*, 25–51. [CrossRef]
18. Pfeuffer, K.; Vidal, M.; Turner, J.; Bulling, A.; Gellersen, H. Pursuit calibration: Making gaze calibration less tedious and more flexible. In Proceedings of the 26th Annual ACM Symposium on User Interface Software and Technology, Andrews, UK, 8–11 October 2013; pp. 261–269.
19. Cho, C.W.; Lee, H.C.; Gwon, S.Y.; Lee, J.M.; Jung, D.; Park, K.R.; Kim, H.C.; Cha, J. Binocular gaze detection method using a fuzzy algorithm based on quality measurements. *Opt. Eng.* **2014**, *53*. [CrossRef]
20. Gonzalez, R.C.; Woods, R.E. *Digital Image Processing*, 2nd ed.; Prentice-Hall: Upper Saddle River, NJ, USA, 2002.
21. Ding, L.; Goshtasby, A. On the canny edge detector. *Pattern Recognit.* **2001**, *34*, 721–725. [CrossRef]
22. Kirkpatrick, D.G.; Seidel, R. The ultimate planar convex hull algorithm. *SIAM J. Comput.* **1986**, *15*, 287–299. [CrossRef]
23. Lee, H.C.; Lee, W.O.; Cho, C.W.; Gwon, S.Y.; Park, K.R.; Lee, H.; Cha, J. Remote gaze tracking system on a large display. *Sensors* **2013**, *13*, 13439–13463. [CrossRef] [PubMed]
24. Bayu, B.S.; Miura, J. Fuzzy-based illumination normalization for face recognition. In Proceedings of IEEE Workshop on Advanced Robotics and Its Social Impacts, Tokyo, Japan, 7–9 November 2013; pp. 131–136.

25. Barua, A.; Mudunuri, L.S.; Kosheleva, O. Why trapezoidal and triangular membership functions work so well: Towards a theoretical explanation. *J. Uncertain. Syst.* **2014**, *8*, 164–168.
26. Zhao, J.; Bose, B.K. Evaluation of membership functions for fuzzy logic controlled induction motor drive. In Proceedings of IEEE 28th Annual Conference of the Industrial Electronics Society, Sevilla, Spain, 5–8 November 2002; pp. 229–234.
27. Klir, G.J.; Yuan, B. *Fuzzy Sets and Fuzzy Logic-Theory and Applications*; Prentice-Hall: Upper Saddle River, NJ, USA, 1995.
28. Leekwijck, W.V.; Kerre, E.E. Defuzzification: Criteria and classification. *Fuzzy Sets Syst.* **1999**, *108*, 159–178. [CrossRef]
29. Broekhoven, E.V.; Baets, B.D. Fast and accurate center of gravity defuzzification of fuzzy system outputs defined on trapezoidal fuzzy partitions. *Fuzzy Sets Syst.* **2006**, *157*, 904–918. [CrossRef]
30. Kodak Professional High-Speed Infrared Film. Available online: http://www.kodak.com/global/en/professional/support/techPubs/f13/f13.pdf (accessed on 2 July 2015).
31. Webcam C600. Available online: https://support.logitech.com/en_us/product/5869 (accessed on 2 July 2015).
32. Heo, H.; Lee, W.O.; Shin, K.Y.; Park, K.R. Quantitative Measurement of Eyestrain on 3D Stereoscopic Display Considering the Eye Foveation Model and Edge Information. *Sensors* **2014**, *14*, 8577–8604. [CrossRef] [PubMed]
33. Student's *T*-test. Available online: http://en.wikipedia.org/wiki/Student's_t-test (accessed on 2 July 2015).
34. Cohen, J. A power primer. *Psychol. Bull.* **1992**, *112*, 155–159. [CrossRef] [PubMed]

Permissions

All chapters in this book were first published in Sensors, by MDPI; hereby published with permission under the Creative Commons Attribution License or equivalent. Every chapter published in this book has been scrutinized by our experts. Their significance has been extensively debated. The topics covered herein carry significant findings which will fuel the growth of the discipline. They may even be implemented as practical applications or may be referred to as a beginning point for another development.

The contributors of this book come from diverse backgrounds, making this book a truly international effort. This book will bring forth new frontiers with its revolutionizing research information and detailed analysis of the nascent developments around the world. We would like to thank all the contributing authors for lending their expertise to make the book truly unique.

We would like to thank all the contributing authors for lending their expertise to make the book truly unique. They have played a crucial role in the development of this book. Without their invaluable contributions this book wouldn't have been possible. They have made vital efforts to compile up to date information on the varied aspects of this subject to make this book a valuable addition to the collection of many professionals and students.

This book was conceptualized with the vision of imparting up-to-date information and advanced data in this field. To ensure the same, a matchless editorial board was set up. Every individual on the board went through rigorous rounds of assessment to prove their worth. After which they invested a large part of their time researching and compiling the most relevant data for our readers.

The editorial board has been involved in producing this book since its inception. They have spent rigorous hours researching and exploring the diverse topics which have resulted in the successful publishing of this book. They have passed on their knowledge of decades through this book. To expedite this challenging task, the publisher supported the team at every step. A small team of assistant editors was also appointed to further simplify the editing procedure and attain best results for the readers.

Apart from the editorial board, the designing team has also invested a significant amount of their time in understanding the subject and creating the most relevant covers. They scrutinized every image to scout for the most suitable representation of the subject and create an appropriate cover for the book.

The publishing team has been an ardent support to the editorial, designing and production team. Their endless efforts to recruit the best for this project, has resulted in the accomplishment of this book. They are a veteran in the field of academics and their pool of knowledge is as vast as their experience in printing. Their expertise and guidance has proved useful at every step. Their uncompromising quality standards have made this book an exceptional effort. Their encouragement from time to time has been an inspiration for everyone.

The publisher and the editorial board hope that this book will prove to be a valuable piece of knowledge for researchers, students, practitioners and scholars across the globe.

List of Contributors

Xiyang He, Xiaohong Zhang, Long Tang and Wanke Liu
School of Geodesy and Geomatics,Wuhan University, 129 Luoyu Road, Wuhan 430079, China
Collaborative Innovation Center for Geospatial Technology, 129 Luoyu Road, Wuhan 430079, China

Botao He
School of Optical and Electronic Information, Huazhong University of Science and Technology, Wuhan 430074, China

Shaohua Yu
Wuhan Research Institute of Posts and Telecommunications, Wuhan 430074, China

Mourad Benoussaad, Benoît Sijobert and Christine Azevedo Coste
INRIA, LIRMM, Rue Saint Priest, Montpellier 34000, France

Katja Mombaur
ORB, IWR, University of Heidelberg, Heidelberg 69117, Germany

Wen-Cheng Wang, Li-Jun Yan, Fan Shi, Xue-Liang Niu, Guo-Lei Huang, Cai-Juan Zheng and Wei Sun
Key Laboratory of Tropical Medicinal Plant Chemistry of Ministry of Education, College of Chemistry and Chemical Engineering, Hainan Normal University, Haikou 571158, China

Kai-Wei Chiang, Jhen-Kai Liao, Guang-Je Tsai and Hsiu-Wen Chang
Department of Geomatics, National Cheng-Kung University, 1 University Road, Tainan 701, Taiwan

Wenhua Cui, Bin Li, Jie Chen and Xinwei Li
School of Mechatronic Engineering and Automation, Shanghai University, Shanghai 200072, China

Dong-Sun Kim and Jinsan Kwon
Embedded Software Convergence Research Center, Korea Electronics Technology Institute, Saenari-ro 25, Bundang-gu, Seongnam-si, Gyeonggi-do 13509, Korea

Christian Bräuer-Burchardt, Matthias Heinze, Ingo Schmidt and Peter Kühmstedt
Fraunhofer Institute Applied Optics and Precision Engineering, Albert-Einstein-Str. 7, D-07745 Jena, Germany

Gunther Notni
Fraunhofer Institute Applied Optics and Precision Engineering, Albert-Einstein-Str. 7, D-07745 Jena, Germany
Technical University Ilmenau, Ehrenbergstraße 29, D-98693 Ilmenau, Germany

Tao Zhang, Liping Chen and Yao Li
Key Laboratory of Micro-Inertial Instrument and Advanced Navigation Technology, Ministry of Education, Southeast University, Nanjing 210096, China
School of Instrument Science and Engineering, Southeast University, Nanjing 210096, China

An Braeken
Department of Industrial Sciences and Technology (INDI), Vrije Universiteit Brussel, Brussels 1000, Belgium

Pawani Porambage and Mika Ylianttila
Centre forWireless Communication (CWC), University of Oulu, Oulu 90014, Finland

Andrei Gurtov
Helsinki Institute of Information Technology (HIIT), Aalto University, Espoo 00076, Finland
Department of "Secure Information Technologies" (SIT), ITMO University, Kronverkskiy prospekt 49, St. Petersburg 197101, Russia

Darius Zizys and Rimvydas Gaidys
Faculty of Mechanical Engineering and Design, Kaunas University of Technology, Studentu 56, Kaunas LT-51368, Lithuania

Rolanas Dauksevicius and Vytautas Ostasevicius
Institute of Mechatronics, Kaunas University of Technology, Studentu 56-123, Kaunas LT-51368, Lithuania

Vytautas Daniulaitis
Faculty of Informatics, Kaunas University of Technology, Studentu 50, Kaunas LT-51368, Lithuania

Giovanni Capellari and Stefano Mariani
Politecnico di Milano, Dipartimento di Ingegneria Civile e Ambientale, Piazza L. da Vinci 32, 20133 Milano, Italy

Saeed Eftekhar Azam
University of Thessaly, Department of Mechanical Engineering, Leoforos Athinon, Pedion Areos, 38334 Volos, Greece

Muhammad Taha Jilani, Wong Peng Wen and Muhammad Zaka ur Rehman
Department of Electrical and Electronics Engineering, University Technology PETRONAS, Bandar Seri Iskandar 32610, Malaysia

Lee Yen Cheong
Department of Fundamental and Applied Science, University Technology PETRONAS, Bandar Seri Iskandar 32610, Malaysia

Noriya Okutsu, Tomohiro Morohoshi, Norihiro Kato and Tsukasa Ikeda
Department of Material and Environmental Chemistry, Graduate School of Engineering, Utsunomiya University, 7-1-2 Yoto, Utsunomiya, Tochigi 321-8585, Japan

Xiaonan Xie
Center for Bioscience Research and Education, Utsunomiya University, 350 Mine-machi, Utsunomiya, Tochigi 321-8505, Japan

Su Yeong Gwon, Dongwook Jung, Weiyuan Pan and Kang Ryoung Park
Division of Electronics and Electrical Engineering, Dongguk University, 30, Pildong-ro 1-gil, Jung-gu, Seoul 100-715, Korea

Index